Toxic Futures

Toxic Futures

South Africa in the Crises of Energy, Environment and Capital

David Hallowes

UNIVERSITY OF KwaZulu-Natal Press

Published in 2011 by University of KwaZulu-Natal Press
Private Bag X01
Scottsville, 3209
South Africa
Email: books@ukzn.ac.za
Website: www.uknpress.co.za

ISBN: 978-1-86914-211-7

Managing editor: Sally Hines
Editor: Jane Argall
Typesetter: Patricia Comrie
Proofreader: Alison Lockhart
Indexer: Christopher Merrett
Cover design: publicide
Cover photograph: South Africa's biggest oil refinery operated by Shell and BP amongst
                  the residential neighbourhoods in south Durban during a plant upset in
                  October 2002 (by groundWork)

groundWork is a non-profit environmental justice service and development organisation
working primarily in South Africa but increasingly in Southern Africa.

groundWork seeks to improve the quality of life of people through assisting civil society
to have a greater impact on environmental governance. groundWork places particular
emphasis on assisting vulnerable and previously disadvantaged people who are most
affected by environmental injustices. groundWork is Friends of the Earth, South Africa.

Printed and bound by Interpak Books, Pietermaritzburg

Contents

Foreword

From its first publication in 2002, the annual groundWork reports have become vital contributions to the unravelling and understanding of the swirl of predatory forces striving to pull the disposed and marginalised into the societal black hole. It is significant to note that the first issue came out in the year when the world marked a decade after the Earth Summit in Rio in 1992. The anniversary was held in Johannesburg and marked the receding of memories of a dream for a sustainable future. The World Summit on Sustainable Development (WSSD) could best have been described as a 'corporate summit'. And so it was that as we stamped the path from Alexandra to Sandton, official participants to the summit were circling a mounted car, the sleek symbol of corporate power, perched conspicuously in the courtyard of their conference venue.

It is fitting that *Toxic Futures* should be released on the eve of another pivotal event dragging the world to South Africa – the 17th Conference of Parties (COP 17) to the United Nations Framework Convention on Climate Change. Many suspect that with a thematic focus on 'green economy' and 'poverty reduction', Rio+20 or the twentieth anniversary of the Earth Summit forthcoming in Rio in 2012 will lean heavily on the planks erected by COP 17. The significance of the piggybacking will be that the world would have set the course of the coming decades on a fabric of false solutions, of negating sustainability, securing a tighter grip of the corporate world on the political and economic spheres of life and placing the struggling people of the world on a dicey battle for survival.

The groundWork reports are life jackets thrown into shark-infested murky waters from which the disposed and the powerless yell for solidarity. As I read through the chapters of this book, I get drawn into the stories and feel as though I was standing on the fencelines with the fighters for environmental justice, be they in south Durban, in the Vaal or in the Karoo. And I have met some of them in person and continue to meet with them as the assaults on their rights persist and the resistance to the contrary waves require eternal vigilance.

I have been to south Durban communities where kids pack inhalers to survive asthmatic attacks as though they were packing lunch boxes for school. Memories of the South African police arresting people and confiscating posters of my poems (including 'We thought it was oil but it was blood') during protests at the WSSD in 2002 remain as the most poignant reminder of the gravity of the challenges faced by dissenters in a period when peoples thought freedom bells were ringing over the tailings pits, the crammed townships and the acid drainages. This book may be based on the South African context, but you will see the direct application in many ways to the situation all over Africa.

Peak oil has not weaned the world from crude oil. Not yet. The decrease in cheap oil has meant the clawing deeper into dirty energy sources. While sweet crude has never been sweet, heavy crude is never too heavy for an industry whose profits continue to soar because they continually externalise the costs to poor people and their environments, while excluding the same marginalised people from the decisions that promote how their territories are accessed and how their resources are extracted and used.

Toxic economies suck the blood of the people – their labour, resources, their well-being as well as their socio-political spaces. This is why one can see this book as a handbook for direct struggles by environmental justice advocates, not just across Africa but across the world. We are united by the attacks on our environments and livelihoods. The struggles against environmental pollution, of land grabs, sea grabs and sky grabs are all one. There are no legislated boundaries to ecological problems. We are all on one planet.

We applaud groundWork and UKZN Press on the achievements accomplished by this publication. David Hallowes's keen insight into the interlinkages of issues shine through and we all benefit from these. The holistic approach adopted in the analyses of the environmental justice deficit in South Africa as presented here gives users a quick guide to figuring out what is happening elsewhere on the continent, from the Eastern to the Western, and from Northern to Southern regions. *Toxic Futures* helps us to connect the dots between the many crises confronting the world today. It also helps us to see why wars and violence are not inevitable occurrences but are necessarily engineered to keep peoples dislocated, disunited and open to further exploitation.

As an imperial force on the continent, South Africa provides examples of what happens at both ends of the pipe of resource extraction and processes. The gold, coal, platinum and other mines show the damage and the levels of toxicity of the sector and that the concept of sustainable mining is nothing but an oxymoron. It also highlights the corruption (in every sense) that sustains the sector. The refining and manufacturing prowess of South Africa has translated into serious health breaches through pollution and at places this has saddled impacted communities with the duties of regulation enforcements. This happens because official regulators appear to have their hands 'tied' and are often unwilling to see or do what needs to be done.

Toxic Futures urges us to understand that we are bound together by our humanity. Competition, dispossession and accumulation will continue to fatten the pockets of the polluters and emasculate the regulators while exposing our people to grave dangers and deeper impoverishment. The fight for environmental justice is not 'the other fight' – it is our fight.

Nnimmo Bassey
Executive Director, Environmental Rights Action, Nigeria
Chair, Friends of the Earth

Acknowledgements

This book is based on the groundWork reports, produced annually since 2002. The first four of those reports were co-authored by Mark Butler and the next three by Victor Munnik. This text uses their work as much as mine although they are not responsible for what I have done with it here.

The groundWork reports had many helpers. Each carries detailed acknowledgements. They were informed and inspired by people from fenceline organisations and the wider environmental justice movement. Particular thanks are due to members of the South Durban Community Environmental Alliance and the Vaal Environmental Justice Alliance who showed us around their neighbourhoods and shared their thinking with us.

It has been a pleasure to work with Bobby Peek, Gill Addison and the groundWork team over the years. They have enabled a sustained documentation of environmental injustice and people's struggles for justice over the first decade of the twenty-first century. They presented us with challenging topics and supported the research process both intellectually and practically with the organisation of workshops, meetings and visits. Best of all, they put these texts to work in their campaigns and their engagement with allied organisations.

Thanks to Sally Hines and the team at UKZN Press for seeing this book through to publication and giving me good advice along the way. Thanks also to the reviewers whose comments and criticisms on the very long first draft were invaluable and whose affirmation of the value of the work helped sustain me through a substantial revision.

Finally, groundWork would like to thank the donors who have supported the groundWork reports, particularly Hivos, Evangelischer Entwicklungsdienst and Sigrid Rausing Trust.

Abbreviations

AbM	Abahlali baseMjondolo
AfDB	African Development Bank
ANC	African National Congress
AR4	Fourth Assessment Report
ASGISA	Accelerated and Shared Growth Initiative for South Africa
ASPO	Association for the Study of Peak Oil
b/d	barrels a day
BEE	Black Economic Empowerment
BOF	basic oxygen furnace
CAER	Community Awareness and Emergency Response
CBD	central business district
CCPA	Canadian Centre for Policy Alternatives
CCS	carbon capture and storage
CDM	Clean Development Mechanism
CEF	Central Energy Fund
CER	Certified Emissions Reduction
CFL	compact fluorescent light
CONNEPP	Consultative National Environmental Policy Process
COSATU	Congress of South African Trade Unions
CSIR	Council for Scientific and Industrial Research
CTL	coal-to-liquid
DEAT	Department of Environmental Affairs and Tourism
DEPP	Developmental Electricity Pricing Programme
DME	Department of Minerals and Energy
DOE	Department of Energy
DRC	Democratic Republic of Congo

DSM	demand-side management
DTI	Department of Trade and Industry
DWA	Department of Water Affairs
DWAF	Department of Water Affairs and Forestry
EAF	electric arc furnace
EIA	Environmental Impact Assessment
EIR	Extractive Industries Review
EJNF	Environmental Justice Networking Forum
EMCA	environmental management co-operation agreement
EPWP	Expanded Public Works Programme
ERMC	Energy Risk Management Committee
EROEI	energy return on energy invested
EU	European Union
FAO	Food and Agriculture Organisation
FTFA	Food and Trees for Africa
GDP	Gross Domestic Product
GEAR	Growth, Employment and Redistribution
Gt	billion tonnes
GTL	gas-to-liquid
GWC	Growth without Constraints
H:H	high hazard
H:h	low hazard
IDC	Industrial Development Corporation
IDP	Integrated Development Plan
IEA	International Energy Agency
IGCC	integrated gasification combined cycle
IMF	International Monetary Fund
IPAP	Industrial Policy Action Plan
IPC	Integrated Pollution Control
IPCC	Intergovernmental Panel on Climate Change
IPP	independent power producer
IRP	Integrated Resource Plan
kWh	kilowatt-hour
LTMS	Long-Term Mitigation Scenarios
mb/d	million barrels a day
MEC	Member of the Executive Committee
MEND	Movement for the Emancipation of the Niger Delta

MERG	Macro-Economic Research Group
mm³	million cubic metres
MOSOP	Movement for the Survival of the Ogoni People
mt	million tonnes
MTRM	Medium-Term Risk Mitigation Plan
MW	megawatt
MWh	megawatt-hour
MYPD	multi-year price determination
NEDLAC	National Economic Development and Labour Council
NEMA	National Environmental Management Act
NEPAD	New Partnership for Africa's Development
NERSA	National Energy Regulator of South Africa
NERT	National Electricity Response Team
NGO	non-governmental organisation
NUMSA	National Union of Metalworkers of South Africa
NWMS	National Waste Management Strategy
OCGT	Open Cycle Gas Turbine
OECD	Organisation for Economic Cooperation and Development
OPEC	Organisation of Petroleum Exporting Countries
PBC	polychlorinated biphenyl
PFC	perfluorocarbon
PFSA	Plastics Federation of South Africa
PJ	petajoules
PMT	Provincial Monitoring Team
ppb	parts per billion
PPC	Pretoria Portland Cement
ppm	parts per million
PV	photovoltaic
PVC	polyvinyl chloride
PWR	pressurised water reactor
RBS	Required by Science
RDP	Reconstruction and Development Programme
RED	Regional Electricity Distributor
REDD	Reducing Emissions from Deforestation and Forest Degradation
SAPIA	South African Petroleum Industry Association
SAQMC	Sasolburg Air Quality Monitoring Committee
SCI	Sasol Chemical Industries

SD	Scenarios Document
SDCEA	South Durban Community Environmental Alliance
SDI	Spatial Development Initiative
SDR	Sustainable Development Report
SRES	Special Report on Emissions Scenarios
Stats SA	Statistics South Africa
SUV	sport utility vehicle
TEEB	The Economics of Ecosystems and Biodiversity
toe	tonnes of oil equivalent
tpa	tonnes per annum
TR	Technical Report
TWh	terawatt hour
UNEP	United Nations Environment Programme
UNFCCC	United Nations Framework Convention on Climate Change
VEJA	Vaal Environmental Justice Alliance
VOC	volatile organic compound
VRESAP	Vaal River Eastern Sub-System Augmentation Project
WBCSD	World Business Council for Sustainable Development
WEO	*World Energy Outlook*
WSSD	World Summit on Sustainable Development
WTO	World Trade Organisation

Introduction

IN THE BROADEST TERMS, environmental injustice in South Africa is evident in that the rich receive the major benefits of development while the poor bear the brunt of environmental degradation caused by development.

The groundWork reports (2002–2008) have identified three ways in which environmental injustice is imposed on people. In the first place, people are polluted, their environments are degraded and they are coerced into working for less than it costs them to live. This is called *externalisation* because corporations get a free ride by offloading costs on to communities, workers, the public purse and the environment. Costs incurred in modern processes of production but not accounted for within the market price are imposed on third parties who are not involved in, and have no benefit from, the transaction.

In the second place, people are dispossessed and common resources or public goods are privatised. This is called *enclosure* because it eliminates or subordinates non-capitalist systems of production whether by direct force, by technological superiority as when modern trawlers compete against traditional fishing techniques, or by commodifying goods that were previously free.

Thirdly, people are *excluded* from the political and economic decisions that lead to their being polluted or dispossessed. Given the weight of economic forces in shaping broader social institutions and relations, political and economic exclusion frequently reinforce each other. The institutions of the market are specifically designed to remove decision making from the public sphere and so exclude all who do not have an interest in profit. Thus, those who are dispossessed or who carry the externalised costs of production are prevented from contesting the theft or contamination of their resources.

These processes are central to the larger process of accumulation that defines capitalist development. Social geographer David Harvey (2005) uses the term

1

'accumulation by dispossession', which effectively includes all three of these mechanisms. This is a highly unequal process as is evident from the growing inequality of people globally and in South Africa. Those who control development do very well out of it and argue that it is for the benefit of all. The evidence does not support this. Rather, as the rich are made richer, the poor are made poorer.

This book looks at how that happens. It is based on the groundWork reports. groundWork is an environmental justice organisation that supports activist groups in communities affected by industrial pollution. It was established in 1999 and started working with people active on the fencelines of the major oil refineries and of waste dumps and incinerators. The groundWork report has come out annually since 2002 and documents the state of environmental injustice in relation to a particular theme. The themes are chosen for their relevance to groundWork's mandate and to explore the context that shapes local struggles.

Thus, the first report came out in 2002, the year of the World Summit on Sustainable Development (WSSD) hosted by the South African government in Johannesburg. Big corporations wield inordinate power in all the fenceline areas, and groundWork director Bobby Peek observed that 'a single thread running through all our community campaigns was the abuse by corporations dished out with impunity from prosecution or penalties'.[1] In the week ahead of the WSSD, groundWork launched a campaign of resistance against corporate power at its Corporate Accountability Week. *The groundWork Report 2002* took up the theme in relation to the petrochemical industry and air pollution. It focused on Durban where the South Durban Community Environmental Alliance (SDCEA) was actively challenging the regime of 'negotiated non-compliance' that characterised the relationship of government regulators to industry.

Many environmental struggles take place around the end of the pipe. *The groundWork Report 2003: Forging the Future,* examined the engine producing environmental injustice with a critique of government's newly minted industrial manufacturing strategy. Alongside the environmental and social devastation of industrial modernisation, the report showed that the engine of global growth was running on empty. The appearance of economic solidity was testimony to the power of an illusion. In South Africa, meanwhile, both government and corporates were closing down the space for participation and dissent. People were denied access to information and gagged by hostile court actions. 'Where is our Constitution?' asked Peek. In response to the urgency of this question, the 2004 report, *The Balance of Rights,* looked at what the Constitution promised and why the promise was not realised by and for the people.

The World Petroleum Congress held its 2005 meeting in Johannesburg on the agenda of 'shaping the energy future'. This was the first World Petroleum Congress meeting in Africa and the venue reflected the increasingly aggressive scramble for African oil. *The groundWork Report 2005: Whose Energy Future?* was launched in opposition to the oil elite's agenda at a gathering of people from the fencelines of the upstream oilfields and the downstream refineries. In the same year, groundWork organised the first exchange visit of people from the South African refinery fencelines to the Niger Delta. They were shown around by Environmental Rights Action and they witnessed the unofficial war on people. The village of Odioma had recently been razed to the ground by the Nigerian army while everywhere the gas flares roared, and spilt oil saturated the ground and slicked over the waters of the delta.

The focus was once more a local one in 2006. The Vaal Environmental Justice Alliance (VEJA) was newly formed and the groundWork report looked at the production of the *Poisoned Spaces* of the Vaal Triangle, using this as a lens on the national and global production of environmental injustice. Energy was again in question in 2007 following the Western Cape power blackouts and the growing evidence that conventional oil production was reaching its global peak. *Peak Poison* found that production was getting dirtier – politically and environmentally – as energy resources were harder to come by. With the Waste Bill going through parliament, the groundWork report returned to the end of the pipe in 2008. *Wasting the Nation* looked at how capitalist production was making trash of people and places.

People living on the fencelines are intensely aware that the industries that directly pollute them are also major carbon sources. The groundWork reports approach climate change in the context created by the themes. This book draws on some shorter papers that address it directly. *A Critique of the LTMS*, written for Earthlife Africa, Johannesburg, responded to the *Long-Term Mitigation Scenarios*, a study commissioned by the Department of Environmental Affairs and Tourism (DEAT) and published in 2007 to inform government climate policy. Eskom, the state-owned power utility, was meanwhile building new coal-fired plants as fast as it could. As the economy turned sour in 2009, it ran into funding difficulties and the news was leaked that the World Bank would rescue it with a very large loan. A groundWork paper, *The World Bank and Eskom*, fed into a campaign that opposed both Eskom's new build and the World Bank loan.

This is a moment of major and rapid historical change. Over the course of the first decade of the twenty-first century, the groundWork reports have documented the unfolding crisis and tried to draw the links to people's experience of environmental injustice. Going back to them for this book has, in some ways, felt a bit like being Cassandra after the fall of Troy. The reports were written as the economy boomed, energy demand pumped, carbon emissions intensified and global inequality gaped ever wider. They anticipated the bust even as the managers of global capital celebrated themselves. In one sense, the boom and the bust are really part of the same moment but the bust also marks a tipping point, both an ending and a beginning. Momentous as this is, the tearing of the planet's ecology is even more so, yet happens on a different time-scale. The 'moment' here is the three or so centuries of imperial capitalism. Nevertheless, some major ecological tipping points are fast approaching and, if not averted, the earth will become uninhabitable.

I wrote the first four reports with Mark Butler and the next three with Victor Munnik. The original content of many of the passages in this book was written by one or other of them. In particular, sections on mining and water come from Victor's pen and he also has the last word in this book with a meditation on 'enough'. Furthermore, the process of writing the reports involved intense conversations and collaborative thinking that shaped the whole of each work and the series as a whole. Thus, Mark and I developed the mechanisms of environmental injustice as a tool for analysis and it has proved to be robust. We did not then have the benefit of David Harvey's concept of accumulation by dispossession but the mechanisms are useful in separating out different moments in the process of dispossession.

We have also wanted the reports to be part of a dialogue with people. Albeit somewhat unevenly, some more than others, they have been deeply informed by conversations with people on the frontlines of the struggle for environmental justice. It was our hope that the reports would ring true to them and contribute to their discussions and debates. But words fall short of experience. As we put it in *Poisoned Spaces*, the 2006 report on the Vaal Triangle:

> This is not an easy story. It is filled with violence that is sometimes direct and brutal but always also insidious – a slow atrocity that periodically produces flashes that glare into publicity. We hope we have done some justice to the history but believe that it is more cruel and more destructive than we can describe (2006: 15).

UNEQUAL SOUTH AFRICA
South Africa remains one of the most unequal countries in the world although it is a little less unequal than the world as a whole. Income inequality has intensified since the first democratic elections in 1994. On the Gini measure – where 0.0

means absolute equality and 1.0 means 1% of households take all income – it rose from 0.68 to 0.73 between 1995 and 2000.[2] Statistics South Africa (Stats SA) puts the 2006 Gini at 0.72 and says, 'If social grants and taxes are excluded, the Gini coefficient . . . would be 0.80 . . .' (2008: 3). Levels of poverty are extreme and poverty is still defined by race, class, gender and geographical location. Thus the poorest people are rural women living in the former Bantustans. Table 1 shows that the richest 20% of South Africans took just short of 75% of household income in 2008, up from 73.5% in 2000 and 71.6% in 1993, the last year of apartheid rule. The poorest 20% increased their share of income between 1993 and 2000, mainly as a result of the equalisation of pensions and other welfare grants. Between 2000 and 2008, the top 10% increased their share at the expense of everyone else (Leibbrandt et al. 2010: 26). The bottom 60% received only 11.4% of all household income in 2008 while the poorest 20% received a mere 1.4%.

Table 1 Household inequality: share of income (percentage).

	1993	2000	2008
Top 20%	71.6	73.5	74.6
Second top 20%	15.8	14.8	13.9
Middle 20%	7.5	6.9	6.4
Second bottom 20%	3.9	3.7	3.6
Bottom 20%	1.3	1.5	1.4

Source: Leibbrandt et al. 2010.

These figures refer only to household inequality, to what the Constitution calls 'natural persons'. It does not refer to 'juristic persons' – that is, to corporations. Since 1994, South Africa's biggest corporations have listed on the London and New York stock exchanges, taking very large sums of capital with them, while more foreign investors and speculators are taking home profits and royalties from money made in South Africa. So part of the difference between global and South African inequality is made up by South Africa's contribution to the global rich.

In 2003 and 2008, the government published its own reviews of the first ten and fifteen years of democracy. The ten-year review claimed a marked decrease in inequality as a result of government's social spending (Presidency 2003: 90). This included increased welfare grants, such as pensions that are now targeted only at the poor, and spending on housing, water, electricity, education and health care. Taking this spending into account, it claimed that the Gini coefficient for 2000 was 0.35.[3] Welfare grants have undoubtedly contributed to alleviating poverty but are already included in the Gini as income. This figure therefore suggested massive benefits to the poor from government spending on housing and service delivery. The figure was met with academic derision while the extraordinary intensity of protest indicates what poor people think of the value of what they have received.

Government's fifteen-year review acknowledged increased inequality but claimed reduced poverty: while the rich benefited most from higher economic growth, 'individuals across the whole spectrum experience[d] positive income growth between 1995 and 2005' (Presidency 2008: 20). Sociologist Jeremy Seekings believes that, although it 'is premature to reach any precise conclusion on poverty trends in the early 2000s', it is 'very likely that weak employment growth and a sharp increase in . . . social assistance programmes did lead to a reduction in income poverty' (2007: 10). Be that as it may, by 2007 escalating food and fuel prices had ripped into any benefit from 'positive income growth' and, in 2008, economic depression evaporated jobs.

Marketing environmental injustice

The gap between what government spends and its value to the poor, and particularly poor women, is amplified by the neglect of environmental justice – or rather, by compounding environmental injustice. Underlying this neglect is a consistent resort to the logic of the market. In respect of its housing programme, government's ten-year review claimed that the value of a house to the occupant was equal to 'the replacement cost' (Presidency 2003: 25). For the most part, however, the poor remain crowded together far from public amenities or job opportunities on land with little market value and many of the new houses are badly built. This is merely reproducing slums.

Government's water and electricity roll-out figures are particularly impressive. Between 1994 and 2008, according to the *South Africa Year Book*, over eighteen million people gained access to clean water, bringing the total to 88% of the population (SAG 2009). The poor, however, are frequently cut off for want of money to pay for the service, as described in Chapter 2. In many places, the water supply has been erratic, whether or not it is paid for, as delivery systems break down. The unpaid 'ecological debts' of past water use are also threatening the supply. In 2003 to 2004 in Limpopo Province, for example, dams and wells ran dry and this was attributed exclusively to drought. Yet it has as much to do with the appropriation of water for irrigated agriculture, which, over several decades, has dramatically lowered the water table.[4] Many community water projects have simply tapped into this diminishing resource. Despite 'integrated' water management policies, 'market opportunity' continues to drive development thinking in Limpopo. The provincial government has supported water-intensive sugar projects in the drought-prone Blyde River area while several dams, or dam extensions, are being built 'to cope with the increasing water demand generated by platinum mining developments in Limpopo and Mpumalanga' (SAG 2004: 660). Such developments are aggravating the ecological debt.

Waste management and sanitation, by contrast, are scarcely registered as priorities. The ten-year review did not mention waste but did promise to eradicate the bucket toilet system by 2007. In 2008, the South African Municipal Workers' Union commented that the 'goalposts were shifted to say that this money was for eradicating the bucket system in "formal" informal settlements only'.[5] By 2009, government claimed to have removed 90% of buckets from formal settlements (SAG 2009: 558).

Meanwhile, poorly maintained sewage systems are breaking down across the country and contaminating water that people use for drinking. In April 2008, 78 children died from diarrhoea in the Ukhahlamba district of the Eastern Cape. The municipality did nothing until the deaths were made public. The provincial government then noted other factors 'including poverty, poor service delivery, environmental health and human resource "challenges"'.[6] It is indeed poor people who die. Yet, when government cites poverty as a cause, the sub-text seems to be that poor people's lives are less valuable.

The ten-year review did include a section on 'preserving the environment' under the social theme. It focused exclusively on nature conservation. Parks and tourism were major themes but the document also recognised the contribution of 'biological resources' to local livelihood strategies and as a buffer against poverty. It claimed that natural resource management has moved 'squarely into an arena concerned with human rights, equity and environmental sustainability' (Presidency 2003: 30).

This is not always evident on the ground where people are increasingly subject to market forces. In many black rural areas, cash crops have displaced diverse food crops as sugar, cotton and forestry corporations have promoted outgrower schemes. Land redistribution remains underfunded and focused on 'fitting emerging black farmers into the existing agricultural sector, without fundamentally restructuring that sector' (Lahiff 2003: 37). Most did not make it and, in 2003, government introduced an agricultural support programme but within the same logic of 'access' to 'a market dominated by established white producers and agribusinesses' (Greenberg 2010: vii). Under the sign of the market, poor people will certainly be excluded from land while a significant proportion of emerging farmers are likely to be bankrupted. Production will be industrialised: it will be capital- and chemical-intensive and will favour mono-cropping.

Neither review mentions the word 'pollution' although the fifteen-year review does hint that environmental degradation may begin undermining economic growth and poverty eradication. Climate change has moved up government's agenda since 2003. It gets a couple of mentions in the fifteen-year review and has belatedly been tagged on to the list of development indicators – curiously under the heading of 'good governance' right after 'ease of doing business' – that government uses to measure its performance. It is the only avowedly environmental indicator out of 76.

Government rhetoric, in South Africa and elsewhere, habitually associates economic growth, development and poverty eradication. Poverty is then represented as the result of an absence of development and, as we will see in Chapter 3, environmental concerns are constructed as getting in the way of development. The groundWork reports have argued to the contrary that poverty and environmental degradation are precisely the products of development as it has been shaped in reality by the powers of state and capital. The next section looks at the defining features of South Africa's polluting economy.

CARBON ECONOMY

South Africa's economy is dominated by the minerals-energy complex (Fine and Rustomjee 1996). This has made for a highly concentrated economy – one in which wealth and the power to direct development is held by a very few large corporations. The concentration of economic power in South Africa has led to one of the most energy- and carbon-intensive economies in the world and it has the dubious distinction of hosting the single largest carbon dioxide emitter in the world, Sasol's coal-to-liquid (CTL) plant at Secunda. Its carbon intensity and high emissions result from two fundamental and related reasons – its reliance on coal as its primary energy source and its policy of supplying cheap electricity to industry.

Table 2 is based on the 2006 *Digest of South African Energy Statistics*[7] and shows where the energy comes from. Primary energy is the original source of energy. Final energy is the form in which energy is actually used. The table shows both the absolute amount of energy in petajoules (PJ)[8] and the proportion of energy (percentage) supplied from each source.

In 2004, South Africa's total primary energy supply came to 5 241 PJ. Seventy-three per cent of this energy came from coal, up from 64% in 2002 but down from nearly 80% in 2000 according to the *Digest*. Coal is the dirtiest possible source of energy. It is used in three ways: it is converted into electricity by Eskom; it is converted into liquid fuels and chemicals by Sasol; or it is used directly as 'final energy' in industrial processes. The best quality coal is exported. Imported crude oil is the next largest source of primary energy and South Africa's largest import item. Its share of the energy supply increased from 9.7% in 2000 to 22% in 2002 but then decreased to 14% in 2004 as oil prices surged. It will have lost more ground to coal through to 2008. Oil is mostly converted into liquid fuels by the oil refineries.

The final energy available for use comes to 2 718 PJ. This means that nearly half the primary energy is lost in the process of converting it into electricity and liquid fuels. A large proportion of the lost energy literally goes up in smoke through the chimney stacks at the power stations and refineries.

Table 2 Primary and final energy in South Africa in 2004.

	Primary energy		Final energy	
	Petajoules	%	Petajoules	%
Coal	3 573	73.0	788	29.0
Crude oil	1 017	14.0	n/a	n/a
Renewables	418	9.0	190	7.0
Natural gas	84	1.0	54	2.0
Nuclear	145	3.0	n/a	n/a
Hydro	3	0	n/a	n/a
Electricity	n/a	n/a	815	30.0
Liquid Fuels	n/a	n/a	870	32.0
Total	5 241	100.0	2 718	100.0

Compiled from DME 2006.

Box 1   Greenwashing renewables

The figure given for renewable energy in Table 2 is deceptive. It is almost entirely accounted for by biomass while the supply from wind and solar energy is minute. Over half the biomass supply is from sugar and wood-pulp wastes used to generate energy for sugar and pulp mills. Biomass is properly renewable only if its production is sustainable. High-energy mono-crop sugar and plantation forestry do not meet this criterion.

The rest of the biomass supply is from firewood used for domestic consumption. Information on this is very unreliable and the figures may be exaggerated. The use of firewood is sustainable only if harvesting is balanced by new growth. In many areas of rural South Africa, where people are starved of energy, this is not so. The burden of collecting wood falls mainly on women who have to walk further and further as supplies are depleted. This results from the unequal distribution of energy resources and the long history of repeated dispossession.

Cheap electricity has been central to South Africa's industrial expansion strategies throughout its history and was written into the 1928 law that established Eskom as a state-owned power utility. Cheap electricity relies on the abundance of coal in South Africa, cheap labour, extensive externalities and huge additional historical and current subsidies. Industry uses the largest part of South Africa's available energy, as shown in Table 3, and this share will have increased with the commodities boom through to 2008. Consistent with the concentration of economic power, the 36 members of the energy-intensive users group consume 40% of electricity. All but six of the group are in mining and mineral processing or fuels and chemicals.

Within the industrial sector, the iron and steel (29%) and petrochemicals plants (22%) are the two biggest users. Over 45% of the energy used in steelmaking comes directly from coal and coke with a further 23% coming from electricity (DME 2002: v). ArcelorMittal's four South African plants consumed about 169 PJ and the Vanderbijlpark plant alone consumed a massive 76 PJ in 2005.[9] Other metal smelters are also very intensive users. Aluminium is notable for the high proportion of electricity in the energy mix. Bauxite is not mined in southern Africa and BHP Billiton's three smelters were located in the region specifically for the low-priced electricity. In 2006, they consumed a total of 98 PJ of energy including 74 PJ of electric energy or about 11% of Eskom's total production.[10]

Sasol's coal-based processes are largely responsible for the extraordinary intensity of energy use in the petrochemicals sector. Over 80% of the energy used to make liquid fuels and chemicals is directly supplied by coal and Sasol is the only producer that uses coal to drive its plants. Sasol's global energy use is 443 PJ excluding coal, oil or gas converted into liquid fuels and chemicals. Most of this energy is consumed in its South African plants at Sasolburg and Secunda but Sasol's Sustainable Development Reports do not give separate figures for South Africa. The crude-oil refineries are also intensive energy users by any measure other than comparison

Table 3   Final energy demand by sector in 2004.

	Total energy		Electricity	
	Petajoules	%	Petajoules	%
Industry	983	36.2	484	59.3
Mining	190	7.0		
Transport	697	25.7	22	2.7
Residential	487	17.9	130	15.9
Agriculture	78	2.9	22	2.7
Commerce	183	6.7	90	11.0
Other	79	2.9	67	8.2
Non-energy*	20	0.7		
Total	2 718	100.0	815	100.0

Compiled from DME 2006.

* 'Non-energy' includes chemicals, plastics and paper made from coal,
oil, gas or wood.

Electricity consumption figures exclude energy producers. Including the
oil refineries, but not Eskom's own use, adds 27 PJ and increases
industry's share to 60.6% in 2004.

with Sasol. The cost of electricity to energy-intensive industries is the lowest in the world. The cost to households is relatively high and higher still for poor people on prepaid meter systems. Access to domestic energy and electricity is highly unequal. Table 3 shows that households use 16% of all electricity but most of this is used by the richest 40% of households. A large proportion of the population are 'energy-poor'; 20% do not have access to electricity and many who do use very little because they can afford electricity only for lights, TV and radio. For many people, access to electricity is intermittent. Millions of South Africans are regularly cut off because they cannot pay their bills and, with the introduction of prepaid meters, uncounted numbers are cut off every month when they run out of money to feed the meters (Dugard 2010).

Emissions

South Africa positions itself as a victim of climate change and this will indeed prove to be the case. It is also the largest emitter of greenhouse gases in Africa and is ranked as the twelfth largest emitter in the world[11] – up from the fifteenth in the mid-1990s. This compares with its global economic ranking in twenty-ninth place. According to the United Nations Environment Programme (UNEP) it was responsible for 42% of Africa's total carbon emissions in 1998 (2002: 218).

The National Climate Response Strategy for South Africa, authored by the DEAT in 2004, acknowledges the reality of climate change and emphasises the dangers to South Africa's economy over the next 50 years. Health experts expect increased water-borne diseases including malaria and bilharzia. South Africa is a semi-arid country and water resources will be increasingly stressed through reduced rainfall and increased evaporation, desertification, droughts and flood events. Rangelands will become drier and produce less food. Maize production, which provides 70% of total grains, is expected to decrease by up to 20% while pests and diseases are also likely to increase. Biodiversity will be dramatically diminished to the detriment of tourism. The fynbos and karoo biomes as well as large parts of the flagship Kruger National Park will have transformed unrecognisably by 2050, according to science writer Leonie Joubert (2006).

The second set of threats that the response strategy identifies, is that South Africa's mining and energy industries are particularly vulnerable to climate change mitigation measures. Exports of fossil fuels, especially coal, and carbon-intensive products could in future be penalised. Table 4 puts the carbon intensity of the South African economy into perspective although the latest figures indicate that it is even worse than this.[12]

Table 4   Energy sector carbon dioxide emission intensity in 2002.

	CO_2/cap	CO_2/GDP	Cumulative energy CO_2 emissions from 1950 to 2000	
	t/capita	kg/1995 US$	Mt CO_2	Proportion of world total %
South Africa	6.65	1.65	10 165	1.29
Africa	0.89	1.16	13 867	1.75
Non-OECD	1.65	1.33	318 117	40.23
OECD	10.96	0.44	472 635	59.77
World	3.89	0.68	790 753	100.00

Source: Winkler 2007.

CO_2 includes emissions from fossil-fuel use and cement manufacture but excludes industrial process emissions.

The per capita carbon intensity is misleading, first because of the unequal access to domestic energy and second because of the intensity of industrial energy use. In effect, South Africa exports energy and carbon embedded in minerals to the benefit of capital but at the cost of the majority of people. The carbon intensity per unit of production signifies South Africa's structural location within the global economic order. It is not about a phase of development through which the country will pass to higher value production and reduced carbon intensity.

The energy sector leads on carbon and also pumps out pollutants that directly affect people's health. Eskom has consistently resisted installing pollution controls and the results show in Table 5. In absolute terms, it stands out even in the company of South Africa's other world-class polluters. Sasol's coal-based processes are largely responsible for the extraordinary intensity of energy use in the petrochemicals sector. In terms of usable energy produced, Sasol is more pollution-intensive than Eskom. By any other standard of comparison, the crude-oil refineries are also pollution-intensive.

Table 5 **Air emissions from main energy and chemical producers (tonnes) in 2006.**

Pollutant	Eskom	Sasol global	Durban refineries*
Carbon dioxide	203 700 000	60 009 000	1 860 774
Sulphur dioxide	1 763 000	223 000	8 683
Nitrogen oxide	877 000	162 000	3 236
Particulates	55 760	7 560	–
VOCs	–	461 000	4 500
Hydrogen sulphide	–	78 000	–

Based on industry reporting.

* The Durban refineries are Sapref and Engen. The disgraceful Chevron Refinery in Cape Town gives no public account of its emissions. Blanks may indicate the absence of data rather than of pollution.

Eskom and Sasol are particularly vulnerable to mitigation measures. Both have committed to reducing their greenhouse gas emissions but neither has done so. Both corporations expanded production through to 2008. At Eskom this was accompanied by increased carbon intensity as will be discussed in Chapter 7. At Sasol, increased production offset efficiency gains and these gains were reversed when production declined on lower demand in 2009. We will take a closer look at this in Chapter 6.

Air pollution is matched by ground and water pollution. No one actually knows how much waste is produced, recycled or dumped. Developing a waste information system has been consistently identified as a priority since the early 1990s and has as consistently been neglected. Figures confidently given in the official *Environment Outlook* (DEAT 2006) are recycled from a report produced by the Council for Scientific and Industrial Research (CSIR) in 1992. There has been no update since but the figures have been turned from cautious estimates to 'facts' even as they have become increasingly meaningless.[13] What can be said is that South Africa's mining and industrial corporations produce mountains of solid waste and rivers of liquid waste, much of it toxic. In addition to the pollution of water used in production, mining turns groundwater into toxic 'acid mine drainage', discussed in Chapter 4. The large-scale destruction and contamination of aquifers, wetlands and rivers now presents the immanent prospect of an environmental catastrophe that will, for South Africa, be of the same order as catastrophic climate change.

THE GLOBAL SCALE OF ECOLOGICAL DEBT

While South Africa is one of the most unequal countries in the world, the world as a whole is even more unequal. The richest 20% of the world's people 'account for 86% of total private consumption expenditure' (UNEP 2002: 35). They consume '68% of all electricity, 84% of all paper, and own 87% of all automobiles' (Sachs et al. 2002: 19). It follows that they produce a similar proportion of polluting waste. This creates an ecological debt owed by the rich to the poor.

Counting carbon emissions alone, Christian Aid (1999) calculated that this debt is growing by $13 trillion per year using 1990 figures. Despite international agreements to reduce emissions, the gap between rich and poor country emissions continues to grow. Since the industrialised countries have been burning fossil fuels for far longer than poor countries, the historical debt is obviously enormous. Emissions from industrialising countries, particularly in East Asia, have grown substantially over the last 50 years. Rich countries remain responsible for most of the increased concentration of carbon in the air but a number of middle-income countries, including South Africa, should now be accounted as debtors to the poor countries. The poor, however, are most vulnerable to the consequences of climate change:

> Poor people in poor countries suffer first and worst from extreme weather conditions linked to climate change. Today, 96% of all deaths from natural disasters occur in developing countries. By 2025, over half of all people living in developing countries will be 'highly vulnerable' to floods and storms (Simms 2001).

The cost of production is thus much greater than the costs paid by producers and consumers. It is, in short, externalised and so produces a form of ecological debt.

Capitalism grew up alongside imperialism. Its development depended on appropriating the resources of other people and other systems of production. In the first place, the imperial powers took people's land and labour. People were forced to work either by being captured and sold as slaves, or because they were dispossessed of any other means of survival. In most cases, those who were not killed defending their resources then had to take work that paid them less than the cost of living. This is the ecological debt from enclosure. The historical debt here cannot be calculated because the process of enclosure involves putting a monetary value on resources that were not previously valued by money.

The ecological debt is growing rapidly. The debtors, however, have no intention of paying. The reason for this is simple. Even if the historical debt is cancelled, capitalist production makes massive losses if it is held responsible for its year-on-year ecological debt. The price of sustaining this form of production is that the creditors, whether as poor countries or as poor people, must be impoverished. Unsustainable development is visible not only in the extinction of species or the melting of glaciers, but also in poverty and inequality. Conversely, sustainable development is not possible except on the foundation of environmental justice.

Towards environmental justice

Environmental justice is both a battle-cry and a way of thinking about people in their relationship to the environment. It contests the dominant discourses of environmental management. The neo-liberal discourse – think ExxonMobil or George Bush – disregards external costs, particularly when those costs are imposed on anyone without the power to make a fuss. Ecological modernisation was effectively endorsed by governmental negotiators at the United Nations Earth Summit in 1992. It is promoted by the World Bank and the World Business Council for Sustainable Development to proclaim that corporate capital is 'part of the solution'. It allows for state regulation to compensate for 'market failures' but promotes the use of voluntary market mechanisms. It also advances the model of stakeholder participation but in a way that obscures unequal relations of power between social actors. Environmental justice is a rights- or values-based discourse which locates environmental degradation within the relations of power that determine development. It marks a point of resistance within the struggle for the control of natural and labour resources that we call development.

Environmental injustice is thus produced through the social and economic relations which constitute development and through the relation of development to the environment. The call for justice is a call to change these relations. This

15

opens the question of what relations would produce environmental justice. The groundWork reports' working definition of environmental justice encompasses the idea of empowered people in relations of solidarity and equity with each other and in non-degrading and positive relationships with their environments.

Central to this working definition, and to the idea of environmental justice, is the understanding that 'environment' is about relationships – it is not just something 'out there'. People are part of the web of life. In April 2010 the People's Conference on Climate Change held in Cochabamba, Bolivia, adopted a draft Universal Declaration of the Rights of Mother Earth claiming that '. . . we are all part of Mother Earth, an indivisible, living community of interrelated and interdependent beings with a common destiny'.

* * *

In putting together this book, I was reluctantly brought to realise that, short of producing a tome at two or three times the length, I would have to leave out much more of the groundWork reports than I put in. Some of the reports are merely referred to in the text and none are fully covered. Each of the originals retains its own value and interested readers can access them from groundWork's website. This book focuses on the contemporary crises. It is written in the conviction that the world is changing willy-nilly and that the character of this change depends on people's actions.

The first chapter outlines three dimensions of the elite crisis: the crisis of imperial capitalism; the crisis of energy resource depletion; and the environmental crisis. It concludes with a section on the crisis in the lives of the poor seen in relation to waste. Chapter 2 enters South Africa through the Vaal Triangle. It explores this space at the polluting heart of South Africa's economy in the company of the people who are struggling on the fronts of environmental injustice. These fronts are created by the powers of state and corporate capital, manifest in the minerals-energy complex, which have defined development over the last 150 years but also by peoples' resistance to those powers.

The political transition to majority rule was full of hope for a fundamental change in the relations of power and indeed the walls of secrecy erected under apartheid did begin to crumble. Chapter 3, however, shows the post-apartheid government managing economic and industrial development, and South Africa's re-entry into the 'international community', largely on the terms dictated by capital. It contrasts this top-down imposition with the initially more open process, driven by a host of local struggles that shaped environmental policy. Nevertheless, the environment has consistently been subordinated to economic development and industrial policy has implicitly left environmental management to the self-regulation of the market. The chapter concludes with an interrogation of the claims made for green capitalism. It argues that the appearance of cleaner production in the North results from a global restructuring of production that has concentrated the lower-value and dirtier end of the production chain in the South. In South Africa, that means the energy-intensive production of primary mineral resources and Chapter 4 focuses on the first two links in the value chain. It opens with mining and then discusses selected industries at the next link in the chain: iron and steel, aluminium and cement. It then relates the threatened environmental ruin to the conspicuous consumption that is symptomatic of the contemporary crisis of capitalism.

The next four chapters are concerned with energy. Oil is the world's premier source of energy and Chapter 5 shows that peak oil marks the beginning of a deepening energy crisis and the intensification of the environmental crisis. It considers the implication for all the major energy sources and concludes by showing why, in a capitalist economy, energy efficiency does not save the day. Chapter 6 then looks at the petrochemicals production chain. It opens with a brief account of the scramble for Africa's oil but focuses on refining and coal-to-liquids production in South Africa and, further along the value/waste chain, plastics.

South Africa's power tripped out first in the Western Cape in 2006 and then nationally in 2009. Chapter 7 shows the root of the crisis in Eskom's history and then looks at how the crisis played out in the Western Cape, first in terms of the politics of energy and second in terms of economic impact, focusing on the Cape's globally integrated agriculture. Finally, it draws some conclusions about the broader implications of an overall decline in the energy system following peak oil. The national crisis, described in Chapter 8, opened some space for public dissent but also confirmed the deep-rooted instincts of state and capital. The chapter looks at the immediate response to the crisis and locates it in relation to the larger crises of the times into which South Africa also walks with eyes wide shut. The second part looks at the future of power, now under construction, and at how it is being contested. It concludes that this is not a viable future.

Chapter 9 is on the politics of climate change. It gives a brief critical review of the history of the international negotiation process that staggered to its knees in Copenhagen. Underlying the conflict between North and South, it finds a common interest in a dysfunctional climate regime that avoids any challenge to economic growth and the never-ending accumulation of capital. At Copenhagen, South Africa made an offer to reduce carbon emissions and the second half of the chapter looks at how that offer stacks up against the research that is said to underpin it. A critique of that research, the Long-Term Mitigation Scenarios, wraps up the chapter.

The final chapter draws the conclusion that the global elites – what used to be called the ruling classes – are incapable of confronting the crises into which they have led the world. Another world is necessary if there is to be a liveable future for the people of the Earth. If it is to be brought into being, new life must spring from people's creativity and their resistance to the economy of death presided over by the lords of capital.

1

Elite crisis

INDUSTRIAL DEVELOPMENT, ENERGY and pollution go together. In the nineteenth century, Britain became the first properly industrial power and was driven by coal. In the twentieth century, the US took the industrial lead and oil was, and still is, the fuel of choice. The growth of industrial and economic power throughout these two centuries has been staggering and the world is now made to work on the assumption that growth is never ending.

The global regime of accumulation presided over by the US is now faltering for both political and economic reasons. This is one of three dimensions of a larger crisis that haunts the world of plenty. The second is 'peak oil', a global energy crisis that has been deferred by the recessionary cut in demand but which waits to blight any 'green shoots' of economic recovery. Third, climate change is gathering momentum and is just one aspect of a broader environmental crisis. The three dimensions of the crisis are profoundly interlinked: the extravagant use of fossil energy has been essential to, and driven by, economic growth and accumulation that is the foundation of capitalist and imperial power. This use of fossil energy is also the primary cause of the increased concentration of greenhouse gases in the earth's atmosphere. The effects of climate change and peak oil will rebound in very powerful ways on the economy. At the same time, each of the three dimensions of crisis has its own logic. The 'internal' crisis of imperial capital is happening irrespective of climate change and peak oil. Similarly, the coincidence of peak oil and accelerating climate change is arbitrary. Even while the use of fossil fuels drives climate change, the logic of peak oil works independently of the effect of carbon emissions on the climate.

The industrial production of abundance has been accompanied by the production of want on an even greater scale. If coal, oil and gas fuel industrial growth, food remains the fundamental source of energy for people. Famine and hunger marked the origin of Britain's imperial capitalism as the market centralised control and used a series of droughts in the nineteenth century to dispossess peasants in the colonial world. Food was then linked to fossil energy through the steam trains and ships. Merchants used the new railways to transport what grain was produced in drought areas to central stores while 'the telegraph ensured that price hikes were coordinated' across the empire. Famine spread even into areas where rain had fallen while large quantities of grain were exported to Britain as the market supplied those best able to pay (Davis 2002: 26). Industrial energy now saturates the food chain, providing fuel or feedstocks for everything from farm to plate: agricultural machinery, fertilizers, pesticides and herbicides, processing and packaging, transport and refrigeration. The successive waves of modernisation in the production of food and everything else have been accompanied by the ever more intense concentration of market power in the hands of transnational corporations. Industrial production ends in an abundance of waste. Following the introduction of the dimensions of the elite crisis, this chapter ends in 'dust and ashes' and the never-ending crisis in the lives of the poor.

IMPERIAL CRISIS

George W. Bush's war on terror headed straight for the oil lands. Following the break-up of the Soviet Union, the major oil corporations had already moved to cut deals with the new republics surrounding the Caspian Sea. The US government had also established a strong diplomatic presence in the area and military links with countries such as Georgia. With the justification of the invasion of Afghanistan, it consolidated its growing influence with military bases in Uzbekistan and Kyrgyzstan.

In the nineteenth century, these were the original oil lands of the Russian Empire. To the south was Persia (now Iran), which Britain marked as within its sphere of influence to keep Russia away from the Gulf Sea ports. That stand-off was about control of trade routes to the east. But the border that it established between 'West' and 'East' remained essentially unchanged throughout the twentieth century, even as the US supplanted Britain as the leading Western power and as the Russian Empire was transformed into the Soviet Union. The US advance across this border seemed to confirm its victory in the Cold War and to shift the longer-term boundaries of international power established by the 'great game' of nineteenth-century imperial rivalries. In 2008, however, the US made no response when Russia invaded Georgia to reassert its primacy in the region.

The war on terror also provided the spurious justification for the invasion of Iraq. There is little doubt that oil was central to the strategic calculations behind the invasion. The US moved quickly to secure the oil wells and preserved Iraq's oil administration while targeting the rest of the civil service – from health and education to water, sewage and energy services – for destruction. The war profiteers were led by oil services corporation Halliburton and were closely linked to Bush's administration.

The war on terror provided the justification for the US military moving in on Africa too. In the process, it is displacing the former colonial powers as the primary military 'partner' for most African countries. Various official reports and statements emphasise the link with oil. For example, 'the report of Vice-President Cheney's Energy Task Force stressed the importance of gaining and maintaining access to African oil resources, which US Intelligence assessments expect to increase to as much as 25% of US oil imports by the year 2020'.[1]

In the east, a major base in Djibouti was established in 2001 and overlooks the Middle East. Other bases, such as in Uganda, Senegal and Botswana, are designed to service a 'rapid response' strategy. The permanent US troop contingent is light but maintains an infrastructure to enable a rapid build-up of troops when required. At the same time, joint military exercises and training programmes in 43 African nations provide for a regular US military presence across the continent. This is backed by military aid funding to a more select group of countries. In 2003, the top two recipients of this aid in sub-Saharan Africa were Nigeria and South Africa. In 2007, a separate US military command – Africom – was established. African countries have refused to host it and it remains headquartered at the European Command in Germany. Nevertheless, taking the example of Somalia and noting the scale of overt and covert co-operation, Ba Karang argues that African forces are now sub-contracted to fight America's wars on the continent.[2]

While spending in Africa has risen sharply, it is dwarfed by US military spending in the rest of the world. Including the Iraq war budget of $82 billion, the Pentagon spent $500 billion in 2005 – rising to $600 billion in 2008 – about the same as the rest of the world put together.[3] This spending has brought the US massive supremacy in military technologies as well as a global military presence with troops and military facilities located in foreign and supposedly sovereign countries around the world. With the Soviet Union out of the way, US military power cannot be challenged and successive US administrations have said they will keep things that way.

The war on terror was the legitimizing label given to the neo-conservatives who came to power with George W. Bush after the spectacular 9/11 attacks on New York and Washington in 2001. As sociologist Giovanni Arrighi notes, the attacks 'scared hell out of the American people' and so served the same purpose as the Cold War: justifying the US's global role and, more immediately, providing a reason for war 'that made sense to the American public'(2005a: 54). The 'neo-cons' had already published their agenda before Bush's election under the title of the 'Project for a New American Century'.[4] This project was rebranded as the war on terror, a war without end and with no defined enemy, a declaration that any political group or organisation or any country may be defined as outlaws at any time convenient to the US.

Far from protecting liberty, this looks like a protection racket and the legitimacy of US global leadership has eroded in the face of the naked self-interest of its actions. Thus, the invasion of Iraq is widely and rightly seen as an oil grab. But it is much more than this. As the anti-war Retort group argues (2005), big oil is articulated with other 'centres of capital' with interests in war, most immediately the 'military-industrial' complex, the giants of construction given corrupt contracts – largely paid for with Iraqi money – for 'reconstruction' and, 'not least, financial services and banking capital' looking for a flood of petrodollars from high-priced oil. War provided an '"extra-economic" restructuring of the conditions necessary for ex-panded profitability – paving the way for new rounds of American-led dispossession and capital accumulation . . . It was intended as the prototype for a new form of military neo-liberalism' (Retort 2005: 71, 72).

Disaster capitalism

Activist academic Naomi Klein calls it 'disaster capitalism'. Iraq is not alone, nor was it the first to be 'reconstructed' as a neo-liberal economy with a client government tricked out in the rags of democracy. The same prescriptions are applied both to 'post-conflict' societies and to countries hit by natural disasters: 'disaster capitalism really hit its stride with Hurricane Mitch', which devastated Central America in 1998. The International Monetary Fund (IMF) and World Bank aggressively pushed the radical 'opening' of the domestic economies to foreign capital and, according to *The Wall Street Journal*, made privatisation 'a condition for release of roughly $47 million in aid annually over three years and linking it to about $4.4 billion in foreign debt relief for Nicaragua'.[5] Reconstruction following the Asian tsunami of 2004 was similarly used to appropriate local people's beachfront sites and fisheries and turn them over to transnational corporations. Shalmali Guttal of Focus on the Global South argues that 'failed states' are now a structural requirement of capitalism. 'Poor governance' is used to justify privatisation and the contracting out of 'reconstruction' to transnational corporations. The structural and historical causes of failure – the collusion of the imperial powers and their agencies with dictatorships and the 'draining of national wealth through colonial structures of production, debilitating debt repayment burdens and the structural adjustment programmes' – are ignored.[6]

Indeed, the conflict within failed states is frequently manufactured by the imperial powers. Haiti's elected president, Jean-Bertrand Aristide, was deposed following US-sponsored agitation. A similar coup, plotted by the same US groups, against Venezuela's anti-imperialist president, Hugo Chávez, failed in 2002 when the poor flooded on to the streets in support of him. Haiti provides a kind of history of what might have been on a much larger scale in Venezuela. Guttal relates that a client government, 'hand-picked by an eight person "Council of Eminent Persons" backed by the US', was installed and adopted a social and economic reconstruction plan drawn up 'behind closed doors' under direction from the World Bank and US. According to the World Bank, '[t]he Transitional Government provide[s] a window of opportunity for implementing economic governance reforms . . . that may be hard for a future government to undo'.[7] UN troops then occupied Haiti to provide a multilateral cover for US interests and they systematically attacked the poor in the slums of Cité Soleil and Bel Air, centres of support for Aristide and of opposition to the occupation and the client government.

Capitalism is famously flexible. It is not merely that it has the capacity to adapt to crisis but that it both creates and feeds off crisis. Disaster capitalism appears as one of the ways that capital is able to respond to climate change, feeding from a crisis it cannot address.

The great consumer

Apart from its sheer military power, the US retains immense power by virtue of its economic dominance. It is by far the world's biggest economy and its premier market – the great consumer. Producers everywhere, most notably China, still rely on it to buy their goods. Until 2008, the US managed a series of 'bubbles' by passing them off on to foreigners and on to domestic consumer debt. In 2007, according to Arrighi, its economy required $2.5 billion per day from the rest of the world to keep afloat, up from an already unsustainable $1 billion a day in 2003.[8] Much of it came from China, anxious to keep the consumption pump going. As Walden Bello put it, the US and China are chained together in an unsustainable relationship:

China's breakneck growth has increasingly depended on the ability of American consumers to continue their consumption of much of the output of China's production brought about by excessive investment. On the other hand, America's high consumption rate depends on Beijing's lending the US private and public sectors a significant portion of the trillion-plus dollars it has accumulated from its yawning trade surplus with Washington.[9]

China's production is subsidised by cheap labour supplied from an enormous pool of dispossessed peasants and by large-scale trashing of environments. Yet, while China tries to create the jobs that will soak up those it has dispossessed, in 2006 it was estimated that '75% of China's industries are currently plagued by overcapacity' – they were producing more than they could sell even as the bull markets roared. Investments in over-producing industries accounted for '40–50% of China's GDP growth' and much of it came from US and other transnational corporations searching for higher profits, says Bello. America's consumers, on the other hand, have paid for the goods by mortgaging their mortgages.[10] This was sustained by rising house prices and hard-sell tactics by moneylenders who were themselves encouraged by the US central bank. When the housing market crashed, people were stranded in houses worth less than their debt. The poor in the US were the first to feel the heat but they were joined by the 'refugees of the middle class, drowning in debt, and frequently wondering how they fell so far so fast'.[11]

The subsidy to America was and is supplemented by the windfall of petrodollars created by the escalation of crude prices from 2004 to 2008. Yet this merely compounds the problem for the root of the crisis lies in the logic of an over-accumulation of capital – there is more money than there are safe and profitable locations to invest in – resulting in declining profits. Since the 1970s this has resulted in 'financialisation': a shift of power within global capital from production to finance capital accompanied by a growing volatility of global markets. In this context, production capital itself turned increasingly to financial instruments to show profit.

The collapse of US energy giant Enron was a symptom of this shift. It could not make enough profit from producing energy to attract the investment from finance capital necessary to keep it in the top rank of corporations. Instead, with the collusion of the world's top finance houses, it conjured profits – mostly illusory – from financial dealing and trading. It received real money from California where it engineered a series of blackouts and so created an energy crisis that boosted profits from trading energy. It then blamed the California state regulators for the blackouts and called for total 'deregulation' – meaning total power to regulate the market in its own interest, provided it could maintain its position as the dominant energy corporation. Controlling information was critical to its dominance. For the most part, Enron's spin was what the financial press wanted to believe – Enron embodied the virtues of privatised 'wealth creation'. Once it lost control of information, it collapsed in a matter of weeks.[12]

At the end of the American century

The US proclaimed its global leadership during the twentieth century under the banner of The American Century. At the dawn of the twenty-first century, it remains without question the leading power in the world. Indeed, the 1980s and 1990s saw the defeat of its global rival for power, the collapse of Third World resistance to its economic policies and the retreat of labour unions. At the same time, the political and economic elite – the capitalist class – never had it so good. Everywhere the rich got richer and nowhere more so than in America. The dramatic failure of the Project for a New American Century seems to run against this endless flow of accumulating wealth and power. But a longer historical view suggests that this is not simply the aberration of a strategic error.

The US regime is the latest in a line of four global regimes of accumulation that link territorial dominance with the economic power of capital. Arrighi (1994) shows that, thus far, these regimes have followed a similar pattern of growth and decline. In each case, a 'golden period' of growth is interrupted by a 'signal crisis', which is the first symptom of over-accumulation. The economic power of the centre is then revived through financialisation during what he calls the 'belle epoch', a period of extravagant concentration of wealth in the hands of the rich and growing inequality. Financialisation, however, merely masks the underlying problem of over-accumulation and the regime is confronted with growing political and economic instability that leads into a 'terminal crisis'.

The signal crisis of the British regime of accumulation was the economic depression that lasted from 1873 to 1896. From this time on, the growing power of the US became increasingly evident. US power, however, developed within the global order of capital commanded by Britain and was subordinate to it. At the turn of the nineteenth century, the British regime enjoyed a resurgence that seemed to guarantee its continued leadership. This belle epoch was followed by its terminal crisis that extended from 1914 to 1945 and was marked by the two world wars, with the great economic depression of the 1930s in between. Britain won the wars but lost the world to its key ally as the US increasingly assumed leadership of the global capitalist system and finally re-ordered that system in its own image.

The signal crisis of the US regime came with its defeat in Vietnam, the economic stagflation of the 1970s and the 'oil shocks' that ended the golden age of post-war growth. The victory of a poor people over the world's greatest military power gave hope to the dispossessed of the world and encouraged the assertion of Third World nationalism. The US and its First World allies were rudely confronted by Third World states acting as if the legal sovereignty and equality of nations proffered by US leadership was for real. For the first time, the Organisation of Petroleum Exporting Countries (OPEC) states actually acted together to increase their share of oil revenues and Arab producers subsequently went so far as to impose oil sanctions on the West in support of Egypt in the 1973 Yom Kippur War with Israel.

The economic dimension of the crisis came in the form of a recession that resulted from increasing competition between the dominant economies – the US, Europe and Japan. After the Second World War, the economic growth in these economies was mutually reinforcing. From the early 1970s, however, their combined production had overtaken the growth in markets: they were producing more than they could sell at a profit even as profits were squeezed by the successful demands of Northern labour and the rising price of Third World commodities. This marks the origin of over-accumulation and the oil shocks played into this crisis in two ways. First, they stoked inflation because the rising price of petrol fed into all other prices and second, the windfall profits to oil-producing countries created a glut of petrodollars – more capital with nowhere to go. This was partly managed by laying it off on to the Third World. Bankers, led by the World Bank, rushed to sell cheap loans to Third World governments who were only too eager to take them. Oil producers in particular spent on arms and prestige mega-projects which recycled the money back into the profits of Northern corporations.

Imperial power was restored by economic means with the adoption of neoliberal policies – what came to be called the 'Washington Consensus' – in the early 1980s. It did so by engineering a recession on the principle pronounced by banker Andrew Mellon earlier in the twentieth century: 'In a depression, property returns to its rightful owner.' For Mellon, the rightful owner was finance capital. Chris Sanders suggests another version of this business principle: 'Making the other guy pay.'[13] The International Monetary Fund (IMF) and World Bank were turned into enforcers of the new policies by acting as the global arm of the US Treasury Department.[14] Those made to pay were labour and the countries of the global South who found that the easy money of the 1970s had turned into the debt trap of the 1980s. The fabulous concentration of wealth in the hands of finance capital is matched by growing inequality in the world and in all countries, including the US and China. At the dawn of the twenty-first century, the resurgence of opposition at all levels to imperial capitalism is contesting 'rightful ownership' while the crisis of over-accumulation deepens.

The crash

In July 2007, it was the bubble that returned to where it belongs. Several hedge funds dealing in dodgy housing loans were revealed to be empty. These funds were operated by major US finance houses and were at the heart of deals spun across the world of high finance and tangled in such complexity that, even now, no one really knows who owns what or who owes who. By August, banks in Europe and the Far East were draining money. The financiers and dealers who had conjured vast fortunes from the tangle at the expense of others, and who had insisted on the rights of 'the market', then ran to the state central banks to bail them out. The central banks did indeed come to their rescue with billions of dollars, pounds, euros and yen. Nevertheless, the five Wall Street investment banks, the masters of the universe, were either wiped out or forced to change their spots. In May 2008, the US Federal Reserve handed Bear Stearns to JP Morgan which took it only on condition that the Federal Reserve would guarantee its debts. In September, the Federal Reserve let Lehman's go bust. 'The market' was horrified as trillions of dollars were written off the world's stock exchanges. Merrill Lynch then sold itself to Bank of America and only Goldman Sachs and Morgan Stanley were left standing. They were effectively bailed out – along with the major European banks – when the Federal Reserve took over insurance giant AIG the day after Lehman's collapsed and guaranteed massive payouts to AIG's 'counterparties', including $13 billion to Goldmans, from national taxes. Shortly thereafter, the two investment banks redefined themselves as ordinary banks, so submitting to state regulation such as it is but getting access to the Federal Reserve's 'liquidity support'.

Meanwhile, 57 smaller US banks failed between January 2008 and May 2009. More failed in Europe and were effectively nationalised or merged with rivals on the basis of the state guaranteeing the bad debts. All told, the IMF calculated in 2009 that 'total support for the financial system from the governments and central banks of the US, the Eurozone and the UK has amounted to $8,955 billion – $1,950 billion in liquidity support, $2,525 billion in asset purchases and $4,480 billion in guarantees'.[15] Liquidity support is more or less free money from the central bank 'borrowing windows'. According to business columnist Jeremy Thomas, Goldman Sachs used this money to puff up junk rated equities,[16] the implication being that it would dump them as soon as enough naïve investors were suckered by the appearance of a bull run. That is, it is making a Ponzi scheme of the market and, with fewer rivals on the scene, has more power to do so.

This, however, is precisely what the US Treasury and the Federal Reserve themselves have been doing since the 1990s. They used low interest rates to pass on the money lent by the rest of the world at virtually no cost, gnomic statements to assure everyone of the imponderable wisdom of markets, and self-regulation as the first article of faith in those markets. In December 2008, it was revealed that Wall Street luminary Bernie Madoff, a key proponent of self-regulation, had been running a Ponzi scheme, taking in $50 billion, for over two decades. He has since been jailed. But the system that enabled this massive and sustained fraud was itself systemically fraudulent. Sustaining the belle epoch of global finance capital required a constant bull run to keep 'compounding value' and keep the suckers coming into the pyramid base. With finance capital unmoored from production because the latter could not provide the return on capital necessary for growth, this was the other side of the coin of accumulation by dispossession and was necessary for continued economic growth. Consequently central bankers, led by the US Federal Reserve, blew up one bubble after another to absorb surplus capital, pump up Northern (and Southern elite) consumption, and sustain the bullish sentiment on stock markets.

It was to this end that state regulation was suspended in favour of market regulation. Market functions that had been strictly demarcated were merged, even in defiance of remaining laws, to provide one-stop diversified and innovative financial services. As one financier declared: 'What used to be a conflict of interest is now a synergy'.[17] The finance houses that commanded the world's economy competed with ever more innovations to give higher returns to ever more demanding investors. They took to spinning financial assets based on debt through ever-more complex derivatives through which the original debt could be sold off several times over. The global value of 'securities' exceeded world production several times over.[18] Money begat money. Value became the creation of mathematical algorithms, scarcely understood even by finance-house bosses, for calculating tradable risks. The formulae themselves, however, were confined within the rationality of the market, which took ever-expanding global values as axiomatic. Systemic risk – the simultaneous popping of all major bubbles – was placed outside the bounds of rationality. So the maths failed when the markets went down. Suddenly, nothing could be valued. Meaning drained from the language of banking.

Despite appearances, the Wall Street era is not exactly over. The banks deemed too big to fail are now bigger than ever following innumerable state-guaranteed takeovers. They have found the 'other guy' in the world's taxpayers (present and future) and are still calling the shots in the halls of power. The revolving door between the US Treasury, the Federal Reserve and Goldman Sachs is particularly notorious, leading former IMF chief economist Simon Johnson to denounce the regulatory capture of the US state by special interests.[19] And the trillions of stimulus funding have not escaped the depleted logic that gave rise to the crisis. A satirical headline in *The Onion* reads, 'Recession-plagued nation demands new bubble to invest in'.[20] The world's leaders are doing everything possible to provide it.

Meanwhile, China and others have been questioning the value of their holdings in US dollars. Dollar devaluation amounts to defaulting on a large part of the debt. This strategy is available only to the US because the dollar is the world's reserve currency and required for most international trade. Everyone else pays foreign debt in dollars and cannot write off their debt through devaluation of their own currencies. They are thus subject to US monetary policy, thereby expanding the reach of US economic control as well as giving it considerable leverage over global flows of oil and other commodities.

In 2000, Saddam Hussein announced that Iraq would trade its oil in euros. There followed well-informed speculation that the US invasion was intended to prevent the euro from usurping the dollar as the world's premier reserve currency by warning off any oil producer thinking to copy Iraq. It did not, however, have this effect. To the contrary, Janet Bush (no relation) reported in 2004 that Arab disapproval of the war was creating a growing 'consensus for switching out of dollars . . . OPEC has openly discussed the option and even Saudi Arabia, once America's staunchest Middle Eastern ally, is reported to be considering rejecting the dollar'.[21] Venezuela and Iran, both confronted with US hostility, declared that they would no longer trade in dollars.

The run-down of the dollar over the next few years proved more significant than anti-imperialist sentiment for big exporters with major dollar holdings and currencies pegged to the dollar. China and Saudi Arabia both started diversifying their investments, although cautious not to provoke a dollar rout. The massive expansion of US debt in 2009 increased fears that, apart from yielding no return, US Treasury bonds are no longer safe. In October 2009, journalist Robert Fisk reported that secret meetings between the Gulf Arab states and China, Japan, Russia, Brazil and France were aimed at ending the dollar regime for oil trades and moving to a basket of currencies over the next decade.[22] It seems unlikely that the US would intentionally risk its imperial status through a default by stealth. More likely, it no longer has the power to sustain it. It could restore the dollar by ramping up interest rates as it did in the early 1980s but would then deepen the depression and catch itself in a debt crunch. In 2010, it chose rather to launch a second round of 'quantitative easing' – printing dollars that are immediately put into international circulation – in what appears to be an attempt to pass the bill to China by forcing the revaluation of its currency.

ENERGY IN CRISIS

The British regime of accumulation was the first to develop an industrial production base and to do so it depended on a massive supply of cheap energy which it found in coal. As the nineteenth-century economist William Jevons remarked, coal stands 'entirely above all other commodities. It is . . . the universal aid, the factor in everything we do' (quoted in Yergin 1991: 543). For the American regime of accumulation, oil is the universal aid that powers never-ending accumulation. Within the next few years, however, global oil production will be in decline and there is no alternative energy source available to compensate for that loss. This is the meaning of 'peak oil'.

The oil industry is the largest in the world and, for most countries, it is the biggest single import item. Furthermore, power within the industry is highly concentrated. Throughout the twentieth century it was dominated by a handful of 'majors' and, following a series of mergers, there are now just six 'supermajors'. Yet the meaning of 'big oil' is shifting as these corporations lose ground to large state-owned corporations that control the bulk of reserves in producing countries. For all their ideological differences and conflicting interests, however, they are as likely to collude as to instigate war.

At the 2005 World Petroleum Congress in Johannesburg, the world's oil elite promised a future of abundant, cheap and clean energy. To start with 'clean': the stench of blood, oil and corruption affronts the sky all along the production chain. The cheap price of oil was always at the cost of the people who live on the fenceline of production. In September 2007, there was yet another reminder of this when a series of explosions ripped through the Island View chemical storage tanks at Durban docks. Flames fed by a toxic mix of chemicals leapt high in the night sky and melted eight tanks. Residents living across the road were left to evacuate themselves. The very next morning, industry and government officials claimed that there was minimal environmental impact. Three days later, dead fish floated to the surface of the bay and reporters at the scene said the air was still thick with the smell of chemicals.

Oil is no longer cheap. Through the 1990s, it traded at around $18 to the barrel, dropping to $10 in 1999 following the 'Asian crisis'. From 2000, the price started rising and was then stoked by Bush's wars. Even as war receded into background noise for the market, the price remained high and volatile while the big-oil corporations raked in record profits. As the financial markets tumbled in 2008, it spiked to the record price of $145 before crashing precipitately to around $35 and then pushing back up to swing between $60 and $85. In late 2010, the price broke through $90 and soon touched $100. The Arab spring uprisings then pushed it to a high of $124. Oil in particular and fossil energy in general have been extravagantly abundant for the world's rich nations and people. Even the poor in most countries have come to rely on what trickles down from this abundance: paraffin or coal for cooking and often dangerous transport affordable to some. Growing abundance has a limited future as declining oil production will not only overturn the cheap energy regime, but will provoke a crisis of energy in general.

As an energy source, petroleum has unique qualities that are not easily replaced. As Richard Heinberg (2005: 138) summarises: it is easier and cheaper to transport – by pipeline, ship or road tanker – than any other energy source; it has a very high-energy density, meaning that a little does a lot of work; it can be refined into different fuels – gas, petrol, diesel, paraffin, etcetera, and these fuels can be put to a range of uses, providing energy for transport, industrial processes, generating electricity, cooking and heating. Besides energy, oil provides the basis for the massive chemicals and plastics industries – the products of petroleum are all around us.

Peak oil

Oil is a finite resource. Peak oil is the moment when half of what can be pumped from the earth has been used. It is, more importantly, also the point of maximum production. Through most of the twentieth century, the consumption of oil increased by leaps and bounds but potential production from the discovery of new oilfields grew even faster. In other words, the potential supply was mostly far greater than the demand. In this decade, the potential supply has been very little more than demand, and demand rose rapidly through to 2008. After the peak, production will decline so that potential demand on a rising market becomes greater than the supply. Consumption must then be forcibly reduced.

Any individual oilfield goes through a typical pattern of production from discovery to final closure. First, the rate of oil extraction accelerates, it then reaches maximum – or peak – production when half the recoverable oil has been used, after which production starts to decline until no more oil can be extracted and the well is closed. In 1956, M. King Hubbert showed that the same pattern applies to any oil-producing region and, by implication, to the world as a whole.

Hubbert, one of the top US oil geologists working at the Shell laboratory in Texas, developed this model from an intensive study of geological and production data and predicted that oil production in the mainland US would peak in 1971. In fact, he was one year out. It peaked in 1970. His conclusion was not welcome. Shell attempted to silence him and the US Geological Survey, under pressure from the Department of Energy, ran a long campaign to discredit him. At one level, previous predictions of scarcity had proved unfounded but generated panic in the oil markets. The notion of peak oil thus suggested unwelcome instability. It also threatened the industry's power by indicating a limit to its ability to deliver cheap and reliable energy into the future.

At another level, the dispute reflected an argument between geologists and economists.[23] For the former, physical constraints were the bottom-line reality. Oil can only be found in specific geological formations and, once those are exploited, there is no more. For the latter, the only admissible constraint was the level of investment driven by anticipation of profit. Any shortage on the market would raise prices and so drive investment. This would inevitably result in new finds and better recovery from existing fields. However, in a paper for the US Department of Energy, Hirsch, Bezdek and Wendling note that very substantial investments in the US following the US peak yielded very modest returns and did not reverse the overall pattern of declining production. They conclude that, once world oil production peaks, 'higher prices and improved technology are unlikely to yield dramatically higher conventional oil production' (2005: 17).

The theory of peak oil is no longer in dispute. What is now disputed is when it will happen and whether it matters. Economists, along with the industry establishment, argue that investment will secure a plentiful supply for decades and the market will find alternatives when needed. Thus, the World Petroleum Congress dismisses any notion of limits to the supply of energy. This position was well summarised by Euan Baird of the Schlumberger oil services corporation:

> Fossil fuels are the only credible candidate for cheap, clean energy, in the required quantities, over the next 50 years. This will buy valuable time for the world to move cost effectively to alternative energies as they become competitive and as the cost of exploiting depleting reserves of oil and gas increases (Baird 2003: 40).

Oil geologists working in Hubbert's tradition formed the Association for the Study of Peak Oil (ASPO) in 2001 and argue that peak oil is already upon us. There is little time to develop alternatives and, because oil is the world's largest source of energy, the peak will create a more generalised energy crisis. This will be reflected first in further steep increases in the price of oil that will then drag up the price of gas, coal and other energy sources. The argument of the mid-twentieth century is thus repeated in the early twenty-first century. Now, however, the stakes are higher since it concerns the global peak and not just the US peak.

Peaking production

The controversy on the timing of peak oil is fed by unreliable data. Whereas Hubbert worked with reasonably reliable figures, oil corporations and producing countries now tend to lie about how big their reserves are.[24] In 2004, for example, Shell was forced to admit that its oilfields had 25% less oil than it had claimed. It had inflated its reserve figure in order to keep its share price up and it is likely that other big oil corporations have similarly massaged their figures. For their part, the OPEC countries treat their technical production data as state secrets. They have a vested interest in inflating their reserve figures because OPEC production quotas are linked to reserves. Despite pumping millions of barrels of oil every day, and without finding new oilfields, most OPEC countries reported increases in their reserves in the late 1980s. This was really a bidding war for quotas. OPEC was then trying to restrict production to defend the price, but many individual OPEC countries were in financial trouble and desperate to export more to compensate for the low price. Jeremy Leggett (2005) cites evidence that OPEC reserves are over-stated by about 300 billion barrels – ten years' worth of production at current rates.

In 2007, Hirsch listed a growing number of credible oil experts, inside and outside of ASPO, who put peak oil within the next decade.[25] Several of them believed that peak had already occurred or was occurring. Several factors explain why this might be possible without world markets noticing. Firstly, 'experience from oil fields and large oil producing regions demonstrates that maximum oil production is sometimes characterised by a few-year-long gentle rollover' (Hirsch 2007). Secondly, Hirsch, Bezdek and Wendling note that 'geological realities are clearest after the fact' as was evident from the decline in US production after 1970 (2005: 36). Thirdly, there is a great deal of 'noise' in the evidence. For example, peak oil is expected to be heralded by volatile prices, but this volatility cannot be separated out from that caused by the broader context of political and economic instability.[26] Finally, production information may be smoke-screened. Thus, some commentators believe that OPEC announced a cutback in October 2006 to cover for its inability to maintain production. Saudi Arabia had already cut production from 9.5 to 9.1 million barrels a day (mb/d), most of the reduction being in very low-quality crude.[27] The implication is that the Saudis were scraping the bottom of the barrel even at a level of production well below their nominal capacity of 10.5 mb/d. Saudi Aramco has since developed new wells raising nominal capacity to just over 12 mb/d but, with demand having dropped since mid-2008, its capacity to produce at this level is yet to be tested.

ASPO analysts have used various techniques to correct the data but remain open to revising their projections as new information becomes available. Colin

Campbell (2007), the doyen of peak oil studies, puts the peaking of 'regular' oil in 2005. Regular oil excludes very deep sea reserves, extra heavy oils, tar-sands and other sources where production is very expensive and the energy return on energy invested (EROEI) is low (see Box 1.1).

Box 1.1   EROEI

EROEI is the acronym for 'energy return on energy invested'. It is the measure of how much energy is used in the production process as against how much energy is contained in the product. Thus, a very high EROEI of 100 means that 1 unit of energy is used to produce 100 units. An EROEI of 1 would mean that the product contains only as much energy as was used to produce it. An EROEI of 0.5 means that the product contains only half the energy used to produce it. By extension, if production is based on non-renewables, a diminishing EROEI means higher carbon emissions.

In 2007, Campbell predicted the peak of all oil production, including from non-conventional sources, at 2011. And while he projected gas production expanding until about 2045, this would not compensate for the decline in oil production. Thus, he showed the peak of oil and gas combined also at 2011 as shown in Figure 1.1. Andrew McKillop (2006), however, sees peak gas production riding hard on the heels of peak oil. In this case, the impact of peak oil will be even more dramatic as the 'gas bridge' to a post-oil energy future collapses. At the height of the crude-oil market in mid-2008, Campbell revised his projection and put the peak in 2008 at 85.3 mb/d.

In Figure 1.1, the dip in production from the late 1970s reflects temporary declines in consumption resulting from economic recession as well as a modicum of energy conservation in the 1980s. In June 2008, as the financial crisis gathered momentum, the oil price topped out at $147 a barrel and then crashed. On top of tight supply margins, the price was pushed up by speculators who saw commodities and oil in particular as a better bet than equities. This was part opportunism and part desperation as investors searched for safe havens. The crisis, however, was moving beyond the financial sector to the 'real economy' and cutting into demand. Supplies were no longer tight and the oil price crashed. Speculation in oil and other

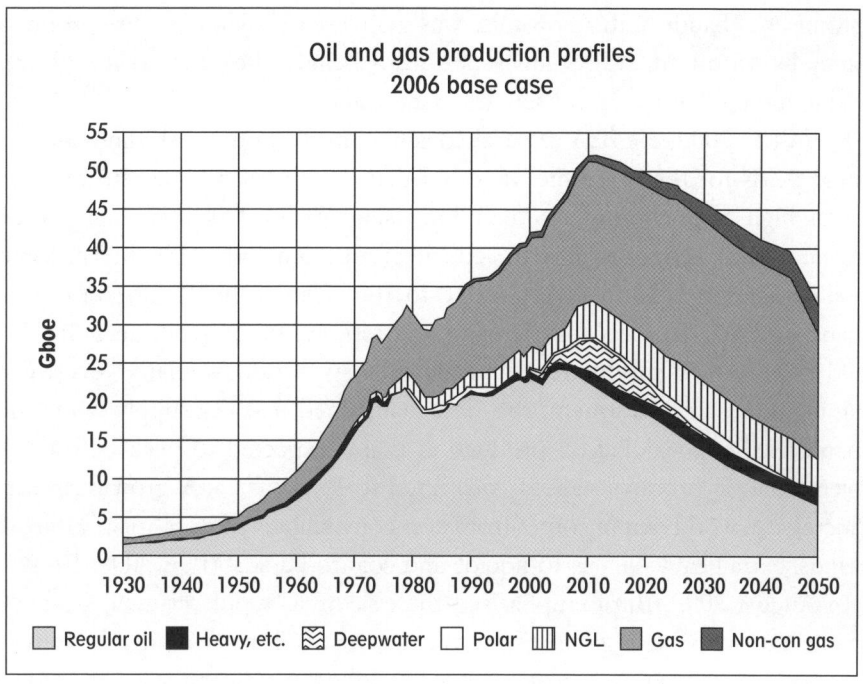

Figure 1.1 ASPO production profiles – history and projections.
Source: Compiled by C.J. Campbell, Staball Hill, Ballydehob, Co. Cork, Ireland.

commodities thus fed off the economic crisis even as escalating prices fed into it. Until recently, the industry establishment has held a common position that there were three or more decades to go before peak oil. This was the official position of the International Energy Agency (IEA), which serves a membership of rich OECD (Organisation for Economic Cooperation and Development) countries. Nevertheless, its *World Energy Outlook* (WEO) 2006 warned that massive energy investments – in oil, gas, coal, nuclear and renewables – would be needed to avert an energy supply crisis. The then IEA boss Claude Mandil opened the report by saying that the 'energy future we are creating is unsustainable. If we continue as before,' he said, 'the energy supply to meet the needs of the next twenty five years is too vulnerable to failure from under-investment, environmental catastrophe or sudden supply interruption' (IEA 2006: 3).

Subsequent IEA pronouncements have been less and less confident. In June 2007, chief economist Fatih Birol said that unless Iraqi production rises 'exponentially by 2015, we have a very big problem, even if Saudi Arabia fulfils all its promises'.[28] Saudi Arabia's promise was to invest $55 billion to raise production capacity by 15 mb/d. ASPO analysts do not believe they can do it and Birol's phrasing hinted that the IEA itself was sceptical.

For WEO 2008, the IEA researched actual production and decline rates from existing fields for the first time. It found that the global decline rate was nearly twice as high as previously assumed and rising. WEO 2008 saw 'higher energy prices and slower economic growth' reducing future demand and slashed its forecast for oil demand in 2030 from 116 mb/d to 106. That is still 20 mb/d more than demand in 2007. To meet this demand and replace the oil from declining fields would require new production equivalent to six Saudi Arabias. Nevertheless, assuming adequate investment, the IEA maintained that the supply would meet demand: 'Although global oil production is not expected to peak before 2030, conventional oil production ... is projected to level off ... A growing share of the increase in world output comes from non-conventional sources, mainly Canadian oil sands, extra heavy oil, gas-to-liquids and coal-to-liquids' (IEA 2008: 103).

In August 2009, Birol emphasised the risk of a 'supply crunch' from 2011 when he anticipated that global economic recovery will revive demand. Many new oil projects had been delayed or cancelled following the price crash and new production was unlikely to compensate for declining production from existing wells. Further, he was reported by the *Independent* as saying 'that the oil ... is running out far faster than previously predicted and that global production is likely to peak in about 10 years – at least a decade earlier than most governments had estimated'.[29] In fact, most governments have not made any estimates whatsoever. The IEA subsequently said the 2020 date referred to conventional oil only and it anticipated total production to peak 'around 2030'.[30]

From 2005, the supermajors started making conflicting statements. Chevron kicked off with an advertising campaign announcing the end of the cheap-oil era. In January 2006, Shell boss Jeroen van der Veer said that '"easy" oil has probably passed its peak'.[31] In June 2006, Total gave 2020 as the likely date for peak oil.[32] A couple of months earlier, Total had said there was enough oil in the ground but that the demand forecast by the IEA to 2030 could not be met because the human and technical resources could not be developed to keep up with the increase in demand.[33] These statements can, of course, be interpreted as attempts to justify high prices and record corporate profits. Thus Retort, who are sceptical of peak oil, note the history of 'organised scarcity' aimed at keeping 'prices low enough for capitalist growth ... but high enough for corporate profitability ...' (2005: 60). On the other hand, BP and Exxon together with the US Energy Information

Administration maintain the position that there is plenty of oil – virtually unlimited in Exxon's view. These statements are also positioned by interest, being calculated to increase political pressure for corporate access to reserves under national management.

Either way, the consensus of the establishment has broken up. Contrary to its earlier statements, WEO 2010 casually remarked that 'crude oil output reaches an undulating plateau of around 68–9 mb/d by 2020 but never regains its all-time peak of 70 mb/d reached in 2006' (IEA 2010: 6). Peak oil was suddenly in the past tense. With demand forecast at 107 mb/d by 2035, natural gas liquids and unconventional oil were made to cover the 37 mb/d difference along with oil yet to be discovered. This last category, according to Campbell, is IEA code for shortage.[34]

ENVIRONMENTAL CRISIS

Climate change is just one dimension of global ecological change forced by the massive scale of fossil-fuelled industrialisation. The scale of change is such that Steffen et al. conclude that 'a new geological era, the *Anthropocene*, has begun'(2004: 6). That is, it is an era in which the basic functioning of earth's ecological systems is decisively influenced by human actions.

Troubled skies

Global warming and climate change are driven by the increasing concentration of carbon dioxide in the atmosphere. Earth's climate has never been stable. Over the last million-odd years, it has fluctuated between cold ice ages and warmer temperate periods that have defined the previous geological eras. The difference in average global temperatures between an ice age and a temperate age has been around 5 °C. These fluctuations in temperature have been accompanied by the fluctuation of carbon dioxide concentrations in the atmosphere, ranging from 180 parts per million (ppm) during the cold periods to about 280 ppm in the warm periods. Concentrations topped 390 ppm in 2010, well outside earth's normal operating range. The rate of increase is around 2 ppm a year and was higher than that in the boom years before the 2008 economic meltdown (Levin and Pershing 2007: 2). Temperature rise lags behind the rise in carbon dioxide concentrations. The earth is now 0.8 °C warmer than in 1900 and the pace of warming is accelerating. It now averages about 0.2 °C every decade. Because of the time lag, this probably reflects CO_2 concentrations in the 1980s or earlier and a further 0.6 °C rise is still to come in response to past industrial carbon emissions.

The effects are already evident. The melting of glaciers and polar ice is beginning to raise sea levels and, once it gets going, ice melt can raise sea levels by 'one metre every twenty years for centuries' (Hansen 2006).[35] More extreme weather events are experienced across the world and some areas, including much of Africa, are becoming drier overall while others are becoming wetter. 'By 2020, between 75 and 250 million people [in Africa] are projected to be exposed to an increase of water stress' while agriculture and food security 'in many African countries and regions is projected to be severely compromised,' according to the Intergovernmental Panel on Climate Change (IPCC 2007b: 10). There is also a strong probability that environmental systems will 'flip': the environment absorbs a variety of pressures until a threshold is reached at which point very abrupt change takes place. In this case, rainfall patterns are likely to change dramatically.

The 'greenhouse gases' are accompanied by a cocktail of industrial emissions in the atmosphere. The impacts of sulphur dioxide (SO_2), nitrogen oxides, hydrogen sulphide, particulates, metals and the exotic mix of volatile organic compounds on local people and their environments have been documented in successive groundWork reports. At the regional scale, acid fallout acidifies seas, rivers and land, and 'soil acidification is a non-reversible change over anything other than very long time scales' (Steffen et al. 2004: 163). Even where the direct effects of pollution remain regional, as in the case of sulphates, Ulrike Lohmann shows that they can precipitate a 'cascade [of effects] through the earth system' (in Steffen et al. 2004: 169). Thus, high sulphur dioxide emissions in Europe and North America during the 1960s and 1970s produced regional cooling[36] sufficient to change atmospheric circulation patterns and is likely to have contributed to drought in the African Sahel during those decades with severe consequences for peasant agriculture.

Ruin on earth

Land change has a long history throughout the world. By the sixteenth century, Europe was largely deforested for naval timber as well as clearance for cultivation. Imperial expansion drove deforestation throughout the colonies. It also displaced indigenous environmental management and production systems that relied on a diversity of biological resources with capitalist production technologies and food crops favoured in European markets. The scale of change increased dramatically in the twentieth century: '[I]n little more than a century the amount of forest that fell was equivalent to the entire previous historical conversion of forests over thousands of years' (Steffen et al. 2004: 96). Grasslands were ploughed up even faster, soils were destructured through mechanisation and massive chemical inputs, and water resources were sucked out for irrigation while being polluted by chemical run-off. During the twentieth century, cities began to sprawl across ever more land, particularly in coastal areas, and the process is now accelerating with the development of mega-cities. The scale of land disturbance by the extractive industries – mining and oil – is locally devastating and increasingly significant globally.

Fresh water hydrology has been modified on an equal scale. Land conversion affects the rate of evaporation sufficiently to affect local climates and rainfall. Groundwater aquifers have been depleted and wetlands, together with the 'eco-service' they provide in filtering and cleaning water, are everywhere threatened. Up to 45 000 large dams interrupt the flow of rivers and of sediments and nutrients formerly deposited in estuaries, deltas and coasts. Two islands formed from the sediment flow of the Ganges have been lost to the rising sea, creating 6 000 refugees.[37] The loss of sediments to the Niger Delta has reduced the fecundity of its fisheries and increased its vulnerability to sea level rise. Niger Delta fish, as well as marine fish that have their nurseries in the Delta, are also poisoned by the appalling pollution of the oil industry. In South Africa, the industrial pollution of rivers, making them unfit even for industrial consumption, is part of the motivation for building more dams upstream to capture clean water and transfer it across watersheds.

Species extinction has accelerated rapidly during the industrial period, to the point that 'the earth is now in the middle of the sixth major extinction event in its history' (Steffen et al. 2004: 118). The previous five extinctions were caused by natural events such as major volcanic eruptions and ice ages. This is the first to be caused by the actions of a living species. Historically, the main cause was loss of habitat as people turned more land over to cultivation. More recently, industrial fishing has driven a number of marine species to the edge of extinction.

Climate change is now the most serious threat to species. On land, species are migrating towards the poles to keep ahead of rising temperatures, but the pace of change is so rapid that plants in particular cannot keep up. Others are running out of space. The Western Cape fynbos, an entire floral kingdom, has nowhere to go. At sea, the warming of the oceans is compounded by the fact that the oceans have absorbed a large proportion of carbon dioxide emissions, making them more acid. Corals that act as marine nurseries are gravely threatened and some populations of plankton species at the bottom of the food chain are in sharp decline. Consequently, whole ocean food chains may collapse, thus wiping out fisheries.

Valuing loss

The loss of eco-services from the degradation of forests alone comes to between $2 trillion and $4.5 trillion a year, according to a study by The Economics of Ecosystems and Biodiversity project (TEEB 2009). This is of the same order of magnitude as the losses from the financial crisis but is not accounted for in GDP figures, is imposed most directly on poor people who depend on forest services and is repeated year after year. It could thus be taken as an indication of the scale of the ecological debt from forest losses but it is a conservative estimate because eco-services are not fully understood and many cannot be monetised.

The project sees economic trade-offs following from ethical choices. For example, downstream subsistence farmers are exposed to flood and drought when forests are destroyed. It would be 'ethically difficult to justify destroying such a forest watershed in order to release economic value which has utility for the agents of destruction (for example profits from minerals and timber, related employment, etcetera)' if the cost to farming communities is 'impossible to bear in human terms' (TEEB 2008: 32). Their broad argument assumes that recognising the value of eco-services will result in markets internalising costs that are presently externalised and, in consideration of the Millennium Development Goals, that state policy will protect both ecosystems and poor people's rights in them.

In the act of costing the loss, however, ecological systems are framed within the market. Eco-services are monetised, so making them available for sale. The project cites the example of a private equity firm that 'recently bought the rights to environmental services generated by a 370 000 hectare rainforest reserve in Guyana recognising that such services – water storage, biodiversity maintenance and rainfall regulation – will eventually be worth something on international markets' (TEEB 2008: 11).

This compounds the problem at the heart of capital's relationship to people and their environments. As analysed by Karl Polanyi, in a critique of conventional economics written in the 1940s, '[a] market economy must comprise all elements of industry, including labour, land, and money . . . But labour and land are no other than the human beings themselves of which every society consists and the natural surroundings in which it exists. To include them in the market mechanism means to subordinate the substance of society itself to the laws of the market' (2001: 75). Labour, land (or 'nature'), and money are not properly commodities in that they are not produced for sale but precede all production. Capital nevertheless requires that they be treated as commodities because they provide critical inputs that must be subordinated to protect investments in increasingly expensive technologies of production. The end result in a self-regulated market driven by profit is the destruction of society and nature. While TEEB's initiative aims to conserve, it ultimately puts ecological knowledge at the service of capital, opening ecological systems to the deeper penetration of market logic.

Local communities get 80% of revenues from the Guyana deal according to TEEB. This provides a justification for the enclosure of eco-services as private property. But international markets will only find worth in that property when it is sold, bought and sold again. The interests of the poor will then be appropriated, for market economics flow not from ethics but from relations of power.

DUST AND ASHES

Waste used to be something of the past, a part of life turned to dust and ash. For much of the nineteenth century, dust and ash was all that went into the domestic dust bin. Everything else was separated and recycled in one way or another. Even shit – politely known as 'night soil' – was taken out along with organic wastes to fertilize fields. Or at least some of it was. The rest was thrown into the streets where waste-pickers competed with dogs, pigs and crows for anything of value.

The business of waste was neither clean nor orderly. In the rapidly growing cities of the industrialising world, the luxurious houses of the elite classes rose above the filth and contrasted with the jerry-built tenements housing the mass of working people. In Manchester, at the centre of imperial Britain's industrial revolution, about one-sixth of the population lived in cellars 'with walls oozing human waste from nearby cesspools' (Pichtel 2005: 26). Such conditions were replicated in the 'old world' of Europe and the 'new world' in America.

Waste-pickers, scavenging for bones, clothes and coal, were amongst the poorest. Most did not have a secure roof over their heads and worked and lived in the filth of the city, vulnerable to diseases and periodic epidemics of cholera and dysentery. Epidemics were not confined to the poor, however, and once the link between disease and dirt was made,[38] middle-class activism demanded sanitary improvements from city authorities. This marked the origins of modern waste management and the construction of what US researcher Heather Rogers describes as 'a border separating the clean and useful from the unclean and dangerous' (2005a: 3). Moreover, cleanliness was found to be good for business. The middle classes no longer deserted the city in the face of epidemics and clean streets enhanced property prices, made for easier transport of goods and workers and for an altogether more pleasurable shopping experience. From the start, cities prioritised the service to business and middle-class areas and 'left the poor, working class and immigrants to live with a disproportionate amount of waste' (2005a: 64).

In the twentieth century, the nature of waste was to change. The mass manu-facture of plastic goods began to expand. Packaging started to displace the practice of measuring out groceries such as sugar, flour and milk at the shop counter. The shops themselves were reorganised as the forerunners of the modern supermarket replaced the counter with check-out tills and channelled customers down aisles to select pre-packaged items from the shelves. These changes took time but, by the 1930s, household bins were filling with rubbish that does not biodegrade. And they positively bulged with plastic and paper when the packaging and marketing industries took hold after the Second World War. Separation and recycling were entirely abandoned as household goods flooded the market, things broken could not be repaired or were not worth the effort, chemical fertilizers displaced organic wastes on the fields and packaging was made for instant dumping.

Surveying England's biggest tip, Andrew O'Hagan observes: 'A dumped bath, a heap of carpet, a thousand empty bottles of orange squash, a hundred thousand legs of lamb, a million bottles of shampoo: it was all the stuff of life and it was all evidence of death' (2007). The business of burying or cremating[39] the wastes of consumer abundance was and is accompanied by the stench of industrial-scale rot and decay. Writing for Greenpeace, industrial economist Robin Murray observes:

> Throughout the twentieth century, waste was the terminus of industrial production. Like night cleaners, the waste industry had the task of removing the debris from the main stage of daily activity . . . The principle was to keep it out of sight. Whereas consumer industries seek publicity, this post-consumer industry prided itself on its invisibility (2002: 5).

The sheer scale of waste is staggering and this is just what we throw away. For every bin of consumer waste, says Annie Leonard (2008), another seventy are dumped by corporations in the process of production – from mining and extraction to manufacture, distribution and marketing. This waste is kept on the other side of the boundary between clean and unclean. It lies behind the bright new goods displayed in bright clean shopping malls and must be concealed from the consumer. Increasingly, the dirty part of the 'value chain' is located in 'developing countries' while the economies of 'post-industrial' nations are said to become cleaner as their economies are 'dematerialised'. The wastes of manufacturing at the lowest possible cost fill the air and poison the water in the rapidly growing mega-cities of the East. And upstream from manufacturing, mining waste is dumped right next to the mines, smothering the land, choking the rivers and laying waste to the people who used them and must be thrust aside.

Meanwhile, what is thrown away and supposed to disappear overflows the dumps. It leaches into the water; it blows on the wind; it contaminates the food chain. Everywhere, countries and municipalities are running out of space for landfills and both landfills and incinerators are meeting with determined opposition from local communities. Ultimately, says O'Hagan (2007), we find that 'there is no such place as "away"'. What we throw away comes back to us, our past catches up with us.

Box 1.2 Sea trash

The sea is one kind of 'away'. The North Pacific sub-tropical gyre is a vast area of the ocean where the wind hardly blows. It is called a gyre because the atmosphere and ocean circulate – very slowly – towards the centre. So things that drift in to the edge – on the wind or in the sea – tend to get stuck in the system. In the days of sail, it was known as the doldrums and terrified sailors who feared being becalmed with never enough wind to sail out again. It remains outside of the main shipping routes so hardly anyone goes there. In 1997, US ocean researcher Charles Moore took his boat through the gyre. He expected to see pristine ocean but 'was confronted, as far as the eye could see, with the sight of plastic . . . In the week it took to cross the subtropical high, no matter what time of day I looked, plastic debris was floating everywhere: bottles, bottle caps, wrappers, fragments' (2003).

The gyre has become the world's unseen dump as ever more rubbish accumulates. Over time, the plastic breaks up into smaller pieces but, even when microscopic in size, it is still plastic. The result is a plastic soup mixed up with the plankton that is the basis of the ocean food chain. There is now more plastic than plankton in the gyre.

The North Pacific gyre is the biggest of six subtropical gyres covering about 40% of the world's oceans. All are accumulating trash. But sea trash is not restricted to the gyres. Greenpeace (2006) reports that plastic can be found floating everywhere in the world's oceans, including the Arctic and Antarctic seas, and litters the world's coasts, even the coasts of remote and uninhabited islands. Much of it does not float on the surface. It is either suspended in the water or sinks into the sediments on the sea bed, particularly in coastal areas. The trash enters the food chain via filter feeders and fish and birds that mistake plastic objects for food. The toxicity is enhanced because plastic absorbs and concentrates other chemicals polluting the seas. Toxicity is then further concentrated up the food chain until it returns to people in the fish on their plates.

Dumping on the poor

For industry and the middle classes, 'away' is mostly where poor people live. Observing and fighting against this gave rise to the idea of environmental injustice and racism in the US. As activist Dana Alston put it, 'We have learned . . . that communities of colour are targets for the siting of toxic waste dumps and most hazardous industries' not wanted in white, middle-class communities (1993: 188). This targeting was accompanied by the promise of jobs in areas with high unemployment. But 'the few jobs that we did get were lower paying and more hazardous jobs' (1993: 189). The US environmental justice movement thus saw 'putting it in black people's back yards' as the other side of the coin to 'not in my back yard'. The principle has now gone global as corporations export waste from North to South in search of cheaper and less protected recycling labour or unregulated dumping. In many cases, recycling is merely a cover for dumping.

South Africa's apartheid planners similarly located poor and black communities next to polluting mines, industries and waste dumps. Waste services were well developed in white areas and the waste dumped in black areas, while basic services for waste and sanitation and water and energy were systematically neglected in black areas. Formal townships received partial and perfunctory services that were not expanded even as population growth was stimulated by the policy of removals from 'white areas'. Removals also led to the creation of large and completely unserviced settlements in rural areas or on the distant peripheries of the cities. Human waste and garbage accumulated, smoke filled the air and water sources were contaminated or difficult to access.

Poor people are still living with the dumps fed by the wastes of the rich and of industry. Indeed, most dumps now have shack settlements alongside them because, like other environmentally hazardous locations, this land has next to no value on the market. It thus appears as open land on which poor people can establish a place to live. Some also find the means of a bare livelihood by picking through the rubbish.

This pattern of injustice is not only a feature of societies with a history of racist exclusion. It is part of the global ordering of power relations necessary for the conduct of business. State investments in infrastructure are designed to defend high-value locations, cleansed of dirt and poverty, in a global competition for private sector investment, and both private and state investments are increasingly concentrated in wealthy areas. South Africa's metropolitan municipalities are now all focused on creating competitive 'world-class cities', producing 'development corridors' linking prestigious industrial clusters, high-value residential enclaves and airports, all wired up for global connection.

In Bénit and Gervais-Lambony's analysis, these spaces are produced as glittering 'shop windows' specifically designed to attract international investments. Thus Johannesburg's Security Strategy focuses on 'areas which are visible to investors and will have an impact on their perceptions' (quoted in Bénit and Gervais-Lambony 2005: 6). As part of 'cleaning up' these visible areas, the poor are driven out to spaces on the periphery where the language of 'participatory democracy' is invoked, with more or less sincerity, to manage poverty in the decay at the 'back of the shop'. The wastes of these investments must also be cleaned away. In Cape Town, taking residential wastes alone, the richest 16% of households[40] produce over half the waste while the dumps are located in poor areas (Swilling 2006). Dumps are expensive but this is an investment that destroys value. The object then, is to invest in removing the waste from wealthy areas and to invest as cheaply as possible in disposing it at the back of the shop.

War on the poor

Yet the relation between poverty and waste goes deeper than this. Development has, since the Second World War, been associated with geopolitical strategies. Thus, the green revolution promised a better life for the rural poor in Third World countries who might otherwise be inclined to revolt under the flag of the red revolution. For the most part, it delivered new markets for corporate agri-business in alliance with local elites while the dispossession of peasants and rural workers was naturalised in the language of development as part of 'the urban transition'. Policies that supported the accumulation of wealth in urban areas would, it was promised, create industrial jobs to absorb the flow of migrants. Nevertheless, permanent urban migration was restricted in many countries, including South Africa, in order to subsidise low wages for migrant workers with the shrinking product of peasant farming. Rural insurgencies resisted dispossession across much of the Third World and were contained by the deployment of counter-insurgency strategies framed in Cold War terms. The defeat of this strategy in Vietnam was central to the crisis of US power in the 1970s. The empire fought back. In the 1980s, the US used the economic instruments of neo-liberalism to reclaim power and reframe development as a function of 'the market'.

There are now more people in the cities than in the country and one third of them live in slums with little hope of secure work as economic growth yields fewer jobs at lower wages. The urban poor are now at the centre of a development discourse that expects them to create their own jobs through entrepreneurial enterprise. This follows the World Bank's prognosis that, throughout the 'developing' world, the informal sector will provide the jobs that the formal sector no longer offers. In South Africa, it has been formalised in the language of the 'two economies' adopted by the Accelerated and Shared Growth Initiative for South Africa (ASGISA). Even dump-picking is now counted as a job in employment statistics and so contributes to government's claims for employment growth. As urban scholar Mike Davis comments, 'it makes more obvious sense to consider most informal workers as the "active" unemployed, who have no choice but to subsist by some means or starve' (2004: 25).

The poor have not gone quietly to the back of the shop. Across the world, local resistance has manifested in protest actions: against removals from homes or from street-trading sites, against restricted and unaffordable essential services, against pollution by industries and waste dumps, against rising prices of energy and food, against exclusion from decisions concerning their own futures. Confronted by armed security deployed by the state, many protests turn into riots. They are not exclusively urban but it is the urban terrain that is now given strategic significance. At global level, the war on terror has replaced the Cold War as the organising principle of violence directed at maintaining the conditions for capital accumulation. The US Pentagon now draws on the theorists of 'fourth generation war' against 'non-state enemies'. These enemies may be international migrants or the urban poor who are held to threaten state order and incubate or shelter terrorists. US battlefield training grounds are therefore being made over, transformed from rural terrains for the tanks to roll across to replicas of Third World slums – with a little help from Hollywood set designers (Graham 2007).

The global sphere is not the sole reserve of the US nor even of the Northern states in general. Raúl Zibechi, a Latin American researcher, notes that the Brazilian army has admitted to using the same techniques in its occupation of Brazilian *favelas* as it uses in its peacekeeping mission in Haiti. Zibechi comments that the admission 'largely explains the interest of Lula da Silva's government in keeping that country's troops on the Caribbean island: to test, in the poor neighbourhoods of Haiti's capital, Port-au-Prince, containment strategies designed for application in the slums of Rio de Janeiro, São Paulo, and other large cities' (2007).

Force is not enough against non-state enemies. US commanders in Iraq see it as one dimension of 'total war' in which traditionally civilian functions of service delivery, political legitimacy and capitalist economic development are deployed. In this context, Zibechi observes: 'Electoral democracy and development are necessary to prevent terrorism, but they are not objectives in and of themselves'. They are rather the obverse of the walls built to contain those who refuse subordination.

Gaza is the final image of the walled-in slum, cut off from all development and made into a free-fire zone for the Israeli Army, which has specifically targeted its capacity to deliver municipal services. Yet in much of the world, the walls are as often symbolic as made of concrete and razor wire.

> Control mechanisms – whether dressed in military garb, or as NGOs for development, or promoting market economy and electoral democracy – are interlaced and, in extreme cases like the suburbs of Baghdad, the slums of Rio de Janeiro, or the shanty towns of Port-au-Prince, they are sub-ordinated to military planning (Zibechi 2007).

In South Africa, Ashwin Desai and Richard Pithouse observe that the urban poor have found themselves 'under armed assault from the state' (2004: 2). In Durban, '[t]he police that do this work are equipped and conduct themselves like soldiers and are popularly known in fear as *amaSosha* . . .' (Pithouse 2006: 8). Elsewhere, the 'red ants' have come to symbolise forced removals but are themselves impoverished casual workers hired by firms contracted by local government. Even removals are privatised. The objectives, observed in all South Africa's cities, are to exclude the poor from the centres where the cities hope to sell themselves to foreign investors and to discipline their consumption of essentials. The scale of confrontation is escalating. In 2007, over 10 000 protests were officially registered.

The people so excluded have been made the waste of the global economic system as shown by the repeated use of the metaphor of cleansing to justify the removals of street traders and poor residents. Robert Mugabe's government in Zimbabwe made the political stakes clear when it named its assault on people's livelihoods and dwelling places Operation Murambatsvina. This was given the English title of 'Operation Restore Order' but was also known as 'drive out the rubbish'. The Zimbabwean government was widely condemned for the action, including by institutions such as the World Bank. Yet this institution itself has been widely associated with similar operations justified in the more moderate language of globally sanctioned development.

People who are seen as waste understand it very well. At Sasolburg, the people who pick waste from the dump told researcher Melanie Samson why the local council did not consult them when it handed out a recycling contract to a private company: 'They say you are just people from the dumpsite. You are just scrap' (Samson 2008: 27). This echoed the view of casualised workers in Johannesburg's waste system: 'You are like the thing, which is inside that dustbin. You are just stupid' (Samson 2004: 1).

The Kennedy Road settlement in Durban is located next to the city's Bisasar Road dump. The people there initiated the formation of Durban's shack-dwellers' movement, Abahlali baseMjondolo (AbM), whose central demand is that they should be addressed as equals, capable of expressing their own will, and should take the central role in deciding their own future. They make the point that they are seen as 'stupid, dirty, lazy, criminal and dangerous' (Pithouse 2006: 21), a stereotyping that associates them with waste – unclean and dangerous – and makes them appear less than human and incapable of thinking and acting for themselves. Hence it is used to exclude them from the city authority's decision-making process which has the intention to remove them from the central city. For several years the residents of Kennedy Road acted within the official process. They moved to protesting both the process and the agenda when the promises made to them were repeatedly broken. Two things followed from this: first, the city authorities effectively branded them as enemies of the state and, second, they started organising for effective resistance within the settlement and other shack settlements across the city.

Since its inception, AbM has insisted on democratic practice and on people thinking and speaking for themselves. In May 2008, people from other African countries were subject to a series of xenophobic attacks by South African citizens. The attacks took place mainly in poor areas because, it was said, foreigners were taking what properly belonged to South Africans. In its response, AbM emphasised that its membership, and indeed its leadership, includes 'people born in other countries'.[41] At meetings called in response to the crisis, it opened up the issue to debate but questioned those who attributed anti-social behaviour only to foreigners. The message was that people should respond to the behaviour, not to the identity of the person. 'An action can be illegal. A person cannot be illegal. A person is a person wherever they may find themselves. If you live in a settlement you are from that settlement and you are a neighbour and a comrade in that settlement.' At the same time, AbM asked, 'why it is that money and rich people can move freely around the world while everywhere the poor must confront razor wire, corrupt and violent police, queues and relocation or deportation?'[42]

In September 2009, some 40 men armed with an assortment of weapons rampaged through Kennedy Road. Two men were killed in uncertain circumstances on that night. In AbM's account, the armed men were shouting, 'The AmaMpondo are taking over Kennedy. Kennedy is for the AmaZulu'.[43] They demolished and looted a number of houses, specifically targeting those of the Abahlali leadership irrespective of their ethnic origins. Those targeted and other witnesses believe that the attacks were instigated by local African National Congress (ANC) politicians and, in the following days, the ANC took control of the settlement – including the community hall and AbM's office. Thousands of people fled the settlement while a local ANC councillor claimed that 'harmony' had been restored. Abahlali president S'bu Zikode, whose own home was destroyed, responded: 'For the ANC harmony means their power and our silence. For us our silence means evictions, shack fires, children dying of diarrhoea and the organised contempt that we face day after day . . . Our crime is a simple one. We are guilty of giving the poor the courage to organise the poor'.[44] This, it seems, challenged the ANC's possession of the poor as a political asset, not just for the votes but for the claim to represent the interests of the poor.

That claim is reflected at all levels in the global discourse of development. The poor are at once excluded from the rights of citizenship and targeted as the objects of development aid. The ultimate image of this is 'the American warplane flying above Afghanistan – one is never sure what it will drop, bombs or food parcels' (Zizek 2002: 94). The war on terror already provides a proxy for great power global rivalry over oil and other resources, shaping the diplomatic as much as the military terrain. It justifies violence at any scale including outright invasion (as in Iraq), counter-insurgency (as in the Niger Delta) or local actions aimed at containing dissent and protest by workers and citizens. Yet it is also the symptom of the failing power of imperial capitalism. That power is rapidly being overwhelmed by the crises of its own making but is unlikely to be any less brutal in chaotic decline than it was in its compulsive expansion.

2

The Vaal in South Africa

F ROM A DISTANCE IT seems that a number of hills rise prominently over the landscape of the Vaal Triangle. Coming closer, the hills turn to black and barren slag or to grey ash with a thin covering of vegetation. They are toxic solid-waste dumps and at the foot of each hill of waste is the industrial plant that made it – Eskom's Lethabo Power Station just south of Vereeniging, ArcelorMittal's Vanderbijlpark steelworks (formerly Iscor), Sasol's coal-based chemicals industries.[1] These plants themselves are impressive for their sheer size and their smokestacks and flares dominate the urban skyline. Enormous volumes of gas flow up these stacks and carry millions more tons of waste into the air. Hidden within the landscape are the lakes and pools of liquid waste and, beneath the ground, poisoned aquifers.

What is turned to waste comes from the ground. The towns of the Vaal Triangle are built on coal and black valleys are cut in the opencast mines while vast caves are dug out underground. Coal is moved by heavy trucks or conveyor belts, some stretching over twenty kilometres across the countryside, to feed a voracious industrial appetite for energy. Remote from the Vaal, but linked to it by the heavy-industry infrastructure of railways, pipelines, power lines and roads, are the iron mines of Sishen and Thabazimbi, the manganese mines of Hotazel in the Northern Cape, the coking coalmines of Witbank, the gasfields of Temane off the Mozambique coast, and the oilfields of the Middle East and West Africa linked through Durban. The infrastructure also carries the product to market. The dominant domestic market of the Johannesburg conurbation is just 50 kilometres to the north and the northern Free State goldfields lie just to the west. Much of the product is exported through Durban, Richards Bay and Saldanha to the wealthy North or the booming economies of China and India.

This industrial space is also linked to the far corners of southern Africa, to Asia and to Europe in the lives and histories of the people. Its construction was a profoundly masculine as well as a racist enterprise and the men who designed, managed and built it came for opportunity or were driven to work there by coercion. Many of those who are settled in the area have families in rural South Africa, in Lesotho and Mozambique and a part of the labour force still migrates for work.

THE CONCENTRATION OF POWER

The Vaal Triangle is a major centre of the minerals-energy complex but, for most of its history, has had a subordinate place within that complex; the poor relation expected to deliver cheap inputs for the greater profits of gold production. It has also been central to state strategies, first for 'inward industrialisation' which, in the 1970s and 1980s, was reinforced by apartheid's security needs, and more recently for export-oriented production. Its economy is still reliant on the primary industries of energy and steel and on cheap labour.

The concentration of power is magnified in the Vaal where the giant state-owned corporations have largely dictated the production of space and built instant towns on the open veld to serve their needs. The relationship between these state corporations and private mining and industrial corporations as well as state and private finance corporations has been close. It was founded on often tense negotiation and deal-making relating to such issues as the price of energy and steel to the mines as well as the cosier co-operation on a variety of joint projects – 'public-private partnerships' as they would now be called. The 'mega-projects' gave physical form to the concentration of power, first in the Vaal Triangle and later at Secunda (energy and chemicals), Richards Bay (aluminium smelting and coal exports), Saldanha (steel and iron ore exports) and Maputo in Mozambique (aluminium smelting). The Coega Industrial Development Zone near Port Elizabeth is the latest initiative in this line but, having devoured much money with little return, looks like a white elephant.

The major towns of the Vaal Triangle are named for the industries that founded them and dominate their economies.

Vereeniging pre-dates the age of the mega-project. It was founded in 1892 on the vast Vereeniging Estate belonging to the partnership of Lewis and Marks.[2] The town only really developed into more than a coalmining village after 1910 with the building of the first Vaal power station and the Union Steel Corporation works. Lewis and Marks sold the steelworks to Iscor in the 1930s while their interest in the Free State goldfields, their collieries and the Vereeniging Estate itself were taken over by Anglo American in 1945. AngloCoal remains South Africa's largest collier and one of the top global producers. It owns the New Vaal Colliery, a vast

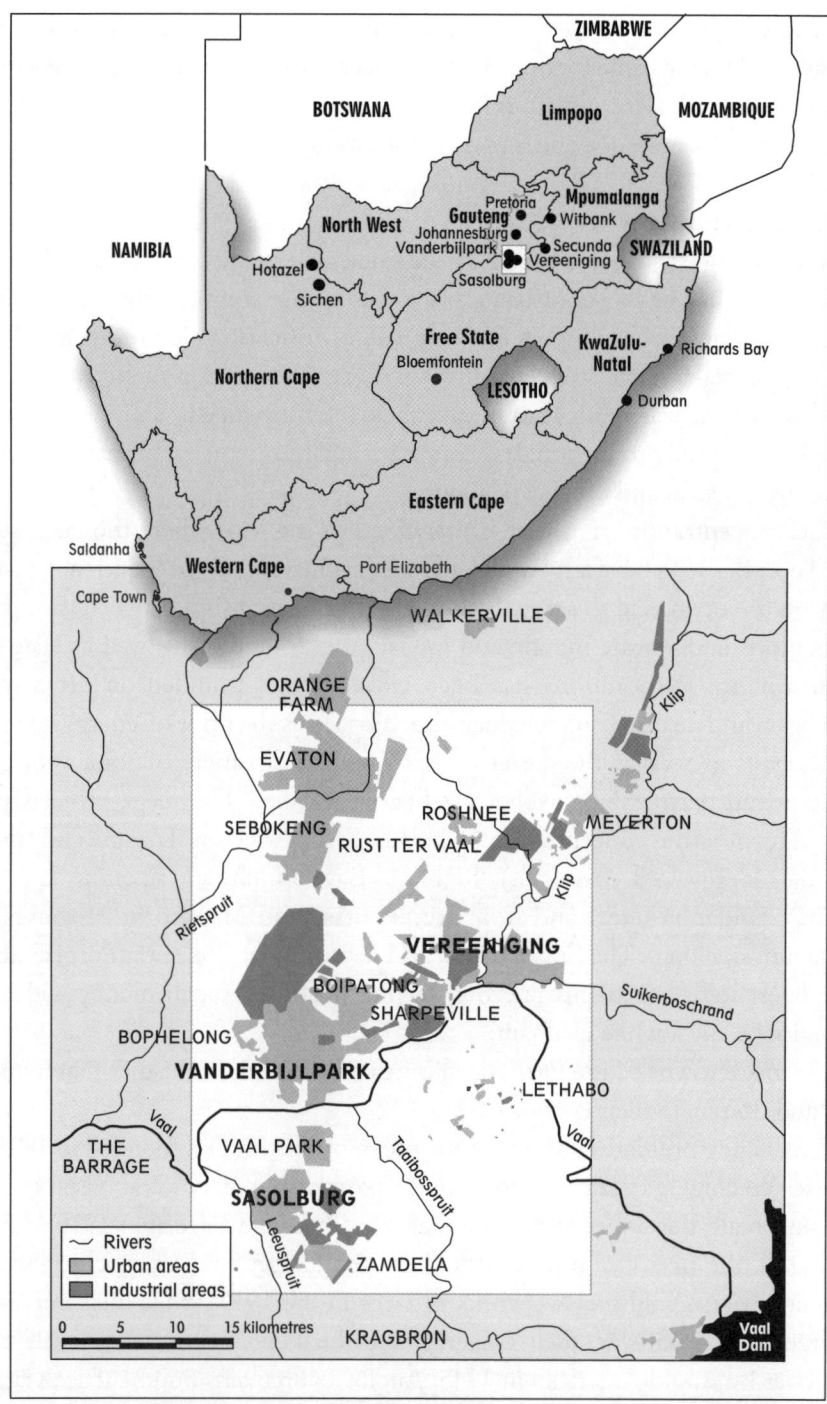

Figure 2.1   Map of South Africa and the Vaal Triangle.

opencast mine covering 2 275 hectares on the bank of the Vaal River opposite Vereeniging. The colliery supplies Eskom's very large Lethabo Power Station, the latest of a succession of power plants in the area. The transnational Mittal corporation took over Iscor in 2004 and now as ArcelorMittal its Vereeniging plant produces 'long steel' products. There are several downstream metal and engineering works in the town as well as refractory, ceramics and brick and tile industries.

'Top Location' in Vereeniging was once the social melting pot of the Vaal area but now stands empty apart from the local museum. People were moved to the 'Sharpe Native Township', later Sharpeville, between Vereeniging and Vanderbijlpark. The specific intention was to move black African workers away from the town. Shortly afterwards, Indian and 'coloured' workers were moved from Top Location to Roshnee and Rust ter Vaal well north of the town. The elite of Vereeniging, meanwhile, have settled on the banks of the Vaal River upstream from the town.

The smaller town of Meyerton, north of Vereeniging, originates in late nineteenth-century land speculation. Samancor Manganese, jointly owned by global mining giants BHP Billiton and Anglo American, is the most significant industry with two plants: Metalloys produces manganese used as an alloy in steel production and DMS Powders makes ferrosilicon powders also for steel producers. Other industries in Meyerton produce bricks, tiles and domestic ceramics.

Vanderbijlpark is a company town planned by Hendrik van der Bijl to house workers for the giant Iscor works constructed in the 1940s. This is South Africa's original mega-project. Van der Bijl set up the state-owned Iron and Steel Corporation in the late 1920s and the Vanderbijlpark plant represented the massive expansion of capacity necessary to establish the corporation as an 'integrated' steel producer controlling production from the iron ore mines, through iron smelting and raw steel production to the manufacture of finished steel for sale to industry and the mines. Iscor was privatised in 1989. Now owned by ArcelorMittal, which controls 10% of global steel production, the plant produces 'flat steel' products. It occupies a massive site astride a ridge above the town. Several lesser, but nevertheless substantial, downstream metal and engineering plants are clustered around it.

The formerly white town stretches south to the river some eight or nine kilometres from the plant. It starts with white working-class housing, separated from the steel plant by a light-industry buffer zone, and gets richer with the distance from the plant. Nearest to the river and furthest from the blast of air pollution, wealth is visibly displayed in opulent houses. The brassy Emerald Casino, occupying a good stretch of river front, fits the neighbourhood. The real wealth of

Vanderbijlpark, however, is overseas in London where corporate boss Lakshmi Mittal splashed R840 million on a house of unrivalled extravagance.

Bophelong and Boipatong, located west and east of the steel plant, are Vanderbijlpark's original townships. They were designed to house black workers close to work in a way that would not take them through the white town south of the plant. Boipatong is next to Sharpeville and both are downwind of the plant in the path of pollution. Sebokeng lies to the north of ArcelorMittal. It started with worker hostels built in the 1960s and 1970s and recently converted to residential units. To the north again, Sebokeng merges with the older settlement of Evaton where a history of black freehold has created a mix of owners and tenants and of middle and working-class residents. Beyond this is Orange Farm, a settlement of iron 'shacks' that was originally a last refuge for people who had nowhere else to go and which remains at the economic periphery of the Vaal to the south and greater Johannesburg to the north.

Sasolburg was established four years after Vanderbijlpark on the other side of the Vaal River. It is also a company town, taking its name from Sasol, the South African Oil and Gas Corporation initiated as a state-owned corporation. The town is now a major hub for the petrochemical and chemical industries. The major plants include the Natref oil refinery, Sasol Gas, Sasol Chemical Industries (SCI) – producing olefins and surfactants, fertilizers and explosives, waxes and a variety of other chemicals – Sasol Polymers, Karbochem and Safripol (formerly Dow Chemicals). Sasol's Sigma Colliery supplies SCI as well as Sasol's own power plants which supply electricity and steam to the chemical works. Sasol was privatised in 1979 and, since 1994, has developed into a substantial transnational corporation. It is now tied into petrochemical global production networks through a web of partnerships that include the oil supermajors ChevronTexaco and Total, state-owned Qatar Petroleum, and chemical giant Mitsubishi.

As at Vanderbijlpark, Sasolburg's white suburbs were designed as a 'garden city' with tree-lined avenues. The white working-class areas are closest to the chemical plants and waste dumps while wealthier Vaalpark is to the north, closest to the river. In the planning of Sasolburg, a careful study was made of wind directions to minimise the impact of pollution on the white town. Zamdela, the black township, is separated from Sasolburg by the dumps. It lies in a triangle of land formed by the chemical works to its north and mines and dumps to the west. To satisfy the criteria of proximity to the plant and separation from the town, it was knowingly placed in the path of the prevailing plume of pollution.

The big industries of the Vaal Triangle are both major producers and consumers of energy. Eskom's 3 700 megawatt (MW) 'six pack' Lethabo Power Station is one of the largest in the country. The Sasol 1 plant in Sasolburg started off making synfuels from coal but now uses the same basic process to make heavy chemicals. Sasol and Total own the Natref crude-oil refinery in Sasolburg, the oil being piped up from Durban. ArcelorMittal's Vanderbijlpark plant's total energy consumption is nearly equivalent to Lethabo's annual output, while the Vereeniging plant, at one-tenth the size, is a significant consumer.

FRONTS OF ENVIRONMENTAL INJUSTICE

The process of producing the space of the Vaal Triangle, of turning what was the open veld landscape of the pre-colonial Tlhaping people into an industrial and urban space, was dominated by the powers of the state and of capital and driven by conflict. It has not been a tidy process as different elements within the state and within capital have come into conflict with each other or made alliances according to the contingencies of the day. More broadly, these powers have sought to control labour and people and have, at every turn, met with resistances that have profoundly influenced the process. This history of development has created many fronts of environmental injustice. The costs have been mostly displaced on to the poor but the economy as a whole will soon start feeling the pinch as development pushes up against ecological limits while making those limits ever tighter.

The Vaal Environmental Justice Alliance (VEJA) was formed in 2005 at a meeting of community-based organisations. The meeting drew groups from across the political spectrum because, as the people say, 'everybody in the Vaal is polluted'. Taking inspiration from the well-established South Durban Community Environmental Alliance (SDCEA), VEJA is inclusive. It is composed of twelve organisations of varying shapes and sizes that have been active on different fronts of environmental justice and so brings both the people and the issues together in a common group for the first time. VEJA's expressed demands are that the pollution must stop, the damage to the environment must be repaired and people must be compensated for the damage to their health and livelihoods. People are well aware that these demands have far-reaching implications and the idea of 'another Vaal', echoing the World Social Forum demand for 'another world', is very much part of their debates.

In 2006, when VEJA participants showed us around the Vaal, the commodity boom was in full swing. It started off at a low point in 1999 and, with some precipitous drops on the way, gathered pace through to 2008 and drove growth in the South African economy to about 5.5%. This chapter describes conditions in the Vaal at a time when the captains of industry were confidently proclaiming a commodity super-cycle and finance minister Trevor Manuel said the economy was hitting the 'sweet spot'. Within the logic of capital, this is about as good as it gets for the majority of the Vaal's people. This chapter is largely based on our conversations with people – not all of them VEJA activists – who shared their formidable analytic understanding as well as their experience of living at the cruel heart of the minerals-energy complex.

Toxic externalities

Bad air on the fencelines

In their annual reports, the Vaal Triangle's big corporations all state their commitment to reducing carbon emissions. None has done so. In the boom years to 2008, their carbon emissions rose with production except where production was interrupted by plant failures. Sasol and ArcelorMittal focus on reducing carbon intensity – emissions per unit of production – but whatever is gained is more than lost to increased production. Real reductions were forced on them when the power tripped out in early 2008 and, more significantly, the economy tripped out later that year.

Along with carbon comes a cocktail of other air pollutants with immediate consequences for people's health and well-being and for the productivity of natural resources. The Vaal Triangle was the first air pollution hot spot to be declared a 'priority area' by the national Department of Environmental Affairs and Tourism (DEAT). Yvonne Scorgie produced a comprehensive report on air quality in the area in 2004 based on available information. She warns that this information is far from complete, mostly not validated and often dated. It should be added that all information on industrial source emissions and most information on ambient air quality comes from industry and that, throughout the world, industry commonly under-reports or conceals emissions. Scorgie lists a total of 58 polluting industrial and mining activities and the top polluters for particulates, sulphur dioxide and carbon dioxide are ranked in Table 2.1 based on information that dates from 2000. All emissions have increased significantly since then. The Vaal Triangle totals at the bottom of the table include emissions from all the industries listed by Scorgie.

Other big-ticket pollutants are nitrogen oxides from all the big plants and hydrogen sulphide from Sasol's coal-based processes. Sasol is South Africa's biggest source of volatile organic compounds (VOCs)[3] while ArcelorMittal also emits significant amounts but does not report them. VOCs include a heady range of chemicals that evaporate easily into the air and most of them are highly toxic.

Table 2.1   Ranking of top industrial polluters in the Vaal Triangle in 2000. Emissions given in tonnes per annum (tpa).

Particulates (PM10) / tpa		Sulphur dioxide / tpa		Carbon dioxide / tpa	
Iscor Vanderbijlpark	8 990	Eskom Lethabo	219 868	Eskom Lethabo	21 920 000
Eskom Lethabo	8 150	SCI	33 061	SCI	7 100 000
Iscor Vereeniging	8 046	Iscor Vanderbijlpark	23 203	Iscor Vanderbijlpark	6 244 000
SCI	6 618	Sasol/ Total Natref	19 144	Sasol/ Total Natref	3 076 950
Vaal Triangle Totals	43 040		298 624		38 565 422

Compiled from Scorgie 2004.

Sasolburg Air Quality Monitoring Committee (SAQMC) activists, using low-tech 'bucket' sampling, revealed some sixteen different VOCs in Zamdela's air in 2000. Several of these compounds had not previously been reported in South Africa. Samples showed dangerously high levels of benzene and high levels of toluene and xylenes at some sites.

Sasol switched from coal to gas piped from Mozambique to provide the feedstock for chemical production in 2005 and said this reduced its sulphur dioxide and nitrogen oxide emissions in South Africa and 'eliminated' hydrogen sulphide odours at Sasolburg.[4] It also promised to reduce emissions of eight VOCs – benzene, butadiene, ethylene oxide, propylene oxide, vinyl chloride monomer, acetaldehyde and formaldehyde – by 50% over the next ten years. Five years later, it had reduced VOC emissions by just 12%. In 2005, ArcelorMittal reported only its greenhouse gas emissions. Following the tighter regulation promised with the declaration of the Vaal as a priority area, its 2008 report identified sulphur dioxide and particulates as its most significant emissions. It outlined three projects to reduce them but two were under review because of cost escalations and the crash in sales revenues. Its recorded emissions intensity was actually higher because it introduced 'more accurate emission inventories'.[5] In other words, it had previously under-reported. Eskom said it reduced particulate emissions per megawatt-hour in 2005 by installing filter bags at Hendrina and Arnot stations and by 'optimisation of the sulphur trioxide flue gas conditioning plant at Lethabo Power Station'.[6] The improvement has, however, been largely offset by increased production. Its sulphur dioxide and nitrogen oxide emissions have increased in line with production.

Scorgie shows that industry emits 90% of total air pollution in the Vaal Triangle. Much of it is emitted from high stacks claimed to reduce the local impact. During winter, however, temperature inversions trap pollutants in the lower atmosphere, creating a visible brown haze, and down-drafting brings the pollution down to earth. Most high-stack emissions in fact come to earth within a ten-kilometre radius. Further, particulates from ArcelorMittal and VOCs from Sasol are emitted close to the ground while dust from coal, slag and ash heaps blows across neighbouring settlements. Spontaneous combustion at New Vaal Colliery results in repeated fires at ground level. Scorgie notes that such fires are estimated to burn as much coal as Eskom and are associated with 'elevated sulphur dioxide concentrations . . . in the Witbank and Vaal Triangle areas' (2004: Section 3: 64). They burn without any pollution abatement whatsoever and under conditions that produce a high percentage of incompletely combusted VOCs, such as the carcinogenic benzene.

Throughout the Vaal Triangle, people complain of itching eyes and burning mucous membranes whenever the wind is in their direction. Zamdela, across the road from the Sasol 1 chemical plant and downwind of it, is particularly hard hit. Even following Sasol's conversion to gas, the air has a sharp chemical smell and people complain of constant headaches.[7] Health impacts, and struggles for relevant information, are reported in more detail in Chapter 3.

Metal pollutants are a growing area of concern. Samancor releases manganese to the air. ArcelorMittal releases manganese, chrome, iron and other heavy metals. Coal also contains trace metals, including mercury, which is highly toxic even at very low levels of exposure. Mercury is present in minute proportions but the massive scale of coal-burning by Eskom, Sasol and ArcelorMittal makes it significant.

Incidents – fires, explosions, leaks and flaring – occur with alarming regularity at many South African plants. As well as adding to the overall burden, incidents produce pollution spikes that result in intensive exposure. Even where the duration of such exposures is limited to a few minutes, the impacts on people's health are often severe and can be long lasting. Moreover, successive exposures have a cumulative effect that comes on top of the background exposure from normal operating emissions.

Incidents are not accidents. They are in principle avoidable and a sign of negligent environmental management. In 2005, the Sasolburg industries reported 86 significant incidents but there is no independent verification or guarantee that there were not in fact more. None of these incidents attracted sanction or any other visible enforcement from the DEAT. One example serves to illustrate the easy collusion between industry and government that has characterised the history of environmental regulation.

In July 2005, Sasol recorded 'a few exceedances of the proposed annual standard of 1.6 ppb [parts per billion]' for benzene.[8] This followed the release of 'cracker petrol' from the Sasol 1 plant. SAQMC reported petrol odours to Sasol on 26 July and took a bucket sample that showed benzene concentrations of 900 ppb. Exposure for one hour to this level of concentration results in serious symptoms. Sasol took its own sample 37 hours after SAQMC raised the alarm and after corrective actions had been initiated. It found 13 ppb and concluded that this did not warrant classification as a reportable incident. groundWork and SAQMC concluded that its sampling methodology was 'clearly flawed if not deliberate'. They sent all relevant information to the DEAT, which has regulatory authority for the large industries in the area, and called for Sasol to be prosecuted. The DEAT took no action to sanction Sasol.[9]

Poisoned waters

Intensive energy use is associated with intensive water use and pollution, which is ill-advised in a dry country. The water supply to Gauteng, the northern Free State and Mpumalanga is increasingly met by cross-watershed transfers from Lesotho and KwaZulu-Natal to the Vaal River and further transfer schemes are planned. At the upstream end, rural communities have been removed to make way for the large dams that supply the water and have lost their best land. At the downstream end of this massive 're-plumbing' of waterways, users across half the country – as far south as Port Elizabeth – will find themselves in competition with inland industry in years of widespread drought. Climate change greatly increases the likelihood of such droughts.

Eskom's national water use in 2005 amounted to 347 135 million litres, Sasol's global use was 163 203 million litres – most of it in South Africa – while Mittal South Africa used about 19 833 million litres in 2004. The Vaal River Eastern Sub-System Augmentation Project (VRESAP), opened in 2009, pipes additional water from the Vaal Dam to the eastern highveld and is particularly intended to secure the water supply to Eskom's Mpumalanga power stations and Sasol's plants in Secunda. In the Vaal Triangle, Lethabo draws 76 650 million litres annually, Sasol's Sasolburg plants draw 19 436 million litres[10] while ArcelorMittal's Vanderbijlpark and Vereeniging plants respectively draw 8 928 and 939 million litres.[11]

Much of the water used by industry is recycled or returned to groundwater or rivers but not necessarily cleanly. In 2005, Sasol produced 44 082 million litres of liquid effluent globally. Its Sasolburg operations produced 17 111 million litres that were treated and returned to the Vaal Barrage downstream. The returned water carries a heavy load of mineral salts into the river. Before the conversion from coal to gas feedstock in 2005, the salt loading was 30% higher but the water in the barrage must still be diluted with releases from the Vaal Dam to make it useable.

This logic extends upstream as the transfers from Lesotho are intended to compensate for the declining quality of the Vaal River water as much as to increase the supply. VRESAP adds a new twist as it delivers water back upstream. According to the minister of Water Affairs, it was built both to meet expanded demand and to compensate for 'an expected deterioration in water quality' on the highveld.[12] This expectation is well founded. The Olifants catchment, which drains northward across the Witbank coalfields, is degraded to the point that 'water in the Middelburg Dam is now no longer fit for human consumption for 40% of the time' (McCarthy and Pretorius undated: 15). The Vaal catchment drains south from the highveld watershed and includes Secunda, two power stations and new coalmines being opened up at the headwaters in the Mpumalanga lake district. Thus, clean water will be taken from the Vaal Dam but dirty water will be returned to it.

Nationally, according to the draft National Strategy for Sustainable Development, 82% of river systems are now threatened while half the country's wetlands are already destroyed. A number of small rivers and streams feed into the Vaal River through a complex of wetlands in the Triangle. The wetlands served both to regulate the flow of water and to filter it clean. These environmental services are now destroyed and the Rietspruit and Klip rivers carry a heavy load of pollution from the Reef mines and industries into the already polluted Vaal.

The industries of the Vaal Triangle itself have unerringly located their dumps and slimes on watercourses. Eskom's Highveld and Taaibos Power Stations are now demolished but five hills of ash still leach contaminants into the Taaibosspruit. Lethabo is on the bank of the Vaal while the New Vaal Colliery that supplies it is bounded by a bend of the river. A few kilometres downstream, Sasol's Wonderwater opencast coalmine, recently mined out and closed, occupies another bend of the river. Sasol's ash dump and effluent ponds, as well as Sasolburg's town dump, are all located above the Leeuspruit.

ArcelorMittal's massive site is located astride a ridge above Vanderbijlpark and contains the mountainous slag-heap and very large effluent dams. The dams have been in use since 1952 when Iscor started production but have never been lined to prevent effluent drainage into the groundwater. The ridge is a local watershed. In a detailed study, Cock and Munnik note that the site was purposely chosen 'to allow for waste water to drain away effortlessly' (2006: 12).

To the west, it drains through what was once the smallholder farming area of Steel Valley. More than five decades of unmitigated pollution has poisoned the groundwater with a toxic mix of heavy metals, dissolved salts and hydrocarbons derived from coal. It has also raised the water table. By 1996, the poison plume from the effluent dams was thought to cover up to seven square kilometres. It is supplemented by leachate from the slag-heap that rises darkly over Steel Valley and has not been capped. Farming is no longer possible: 'People and animals have been poisoned, crops have failed and lives have been devastated' (18). The area is now deserted. In 2006, two of 500 smallholders were still holding out. They lived behind high electric fences, recently erected by ArcelorMittal, which mark what is effectively an environmental sacrifice zone. One of them, Strike Matsepo, said it felt like his home had been turned into a prison. This western drainage flows on to the Rietspruit River and thence to the Vaal downstream at Lochvaal.

To the east of the ridge, water drains through the populous black townships of Boipatong and Sharpeville before emptying into the Vaal. This was previously a complex of wetlands and streams and local people say they once found freshwater crabs. Now it is stagnant and lifeless. An unlined canal drains water from the Arcelor-Mittal site and runs below the town dump but groundwater still rises to the surface in many places. Local people believe that a 'pollution plume is moving east, it is already in Boipatong and will soon be in Sharpeville' (quoted in Cock and Munnik 2006: 33). They have received no clarity on the status of the water and have seen no action from the national Department of Water Affairs and Forestry (DWAF).

Steel Valley residents challenged Mittal in a number of court cases and through direct action. The local and national media have also spotlighted their pollution. In ArcelorMittal's own words, 'because of legislation, legacy issues, legal action against the works and increased pressure from state departments during the late nineties, the need was identified to develop an environmental masterplan'.[13] Said to be 6 000 pages, the plan is secret so communities have to guess how much they have been polluted and trust ArcelorMittal to remedy it. ArcelorMittal says it has committed close to R1 billion for environmental mitigation. This includes what it claims is a 'zero effluent' water-treatment plant opened in 2005. There has been no indication that the plan includes cleaning up the Steel Valley aquifer. At the same time, ArcelorMittal plans to spend R8 billion on expansion. The end result may be more, not less, pollution.

As with air, business as usual pollution is supplemented by incidents. Arcelor-Mittal reported a 'serious incident' in July 2004 when it spilt a 'significant amount of "Spent Pickle Liquor"' – a hazardous waste composed of acid contaminated with heavy metals and sludge.[14] That this was reported represents some improvement on 1994 when Iscor failed to report a major spill of highly toxic chromium salts. ArcelorMittal said it had taken action to 'prevent future similar incidents' but made no mention of remedial action. Nor was there any reference to action or penalties by the regulator. Sasol reported three spills at the Sasol 1 site, including a serious spill of vanadium, in just the two months of February and March 2006.

South Africa is prone to floods as well as droughts, and the severity of floods will increase with climate change and so increase the likelihood of effluent overflows. Following rain, a white powdery substance – associated with sulphuric acid contamination – 'covered the veld' in Steel Valley and flooding in 1996 increased the general levels of contamination, according to erstwhile residents (Cock and Munnik 2006: 14). In January 2005, oil-contaminated water overflowed from Sasol's Secunda effluent dams into the Klipspruit River following heavy rain. Sasol reported the incident and took remedial action. What Sasol reports as 'very low volumes' overflowed from a Sasolburg ash dam into the Leeuspruit following rain in February 2006.

Floods also flush accumulated silts from river-beds and wetlands. In the Vaal area, these silts are heavily contaminated. During flooding in January 2006, sewage works overflowed and contaminated silt was flushed into the Vaal River, resulting in major fish kills and the virtual destruction of the river's ecosystem, according to *Beeld*.[15] Sewage was identified as the primary destroyer but it is most probable that there was also a heavy load of industrial pollution. This has not been investigated.

Poisonous work

Many of South Africa's industrial workplaces are highly polluted environments and workers are often not provided with proper protective clothing and masks. Workers who live near polluting industry thus get a double dose – at work and at home.

In 1999, medical tests were carried out on 509 workers at Samancor. The results showed that most workers, from all sections of the plant, suffered from manganese poisoning. This affects the mind – creating dizziness and confusion – as well as organs such as the kidneys. Ex-workers believe that a high proportion of their comrades have died as a result both before and after 1999. The medical report to Samancor recommended that workers should be informed of their individual results. The corporation did not do this. Instead, it proposed voluntary retrenchments and, when workers did not agree, implemented forced retrenchments. Samancor agreed to the redundancy deal with the National Union of Metalworkers of South Africa (NUMSA) but the union did not consult its members – it merely informed them and, indeed, put pressure on them to accept the deal. The retrenchments gave redundancy but not illness benefits.

In the meantime, the report was leaked to workers. Having lost their union membership along with their jobs and finding no support from NUMSA, they formed the Samancor Retrenched Workers Crisis Committee (SRWCC). This committee mounted a campaign demanding full reasons for their retrenchments and proper compensation for occupational illness from the corporation. In 2006, Samancor finally agreed to compensate workers found to be suffering from manganism by independent doctors. In 2008, ex-Samancor workers joined workers from Assmang's manganese smelter in Cato Ridge near Durban at hearings into the deaths and disabilities of Assmang workers from manganese poisoning. They marched into the hearings bearing a coffin to commemorate 'all our brothers killed by Assmang and Samancor'.[16] Just days before the hearing, five Assmang workers were killed when a furnace exploded. This followed an earlier blast that killed one worker in December 2007. NUMSA said that the furnace had been kept going against the advice of engineers.

Workers retrenched by Iscor before Mittal's takeover also observe that they received no compensation for occupational illness. The manganese from Samancor is just one of a number of toxic substances used or emitted at the plant at Vanderbijlpark and workers at the coke ovens, smelting furnaces and tapping floors are also subject to extreme heat. Coke-oven workers are typically exposed to a variety of VOCs, including benzene, and to hydrogen sulphide, carbon monoxide, ammonia and particulates. Furnace and tapping-floor workers are exposed to heavy metal fumes, carbon monoxide and particulates. Further down the line, they are exposed to vapours from solvents and acids (pickle liquor) used to clean metal surfaces and to various chemicals used to coat it. Workers say typical symptoms include 'high blood pressure, kidney problems, headaches, swelling feet, eye problems, ulcers [and] body swellings' (Cock and Munnik 2006: 41). Respiratory illnesses are also widely reported while cancers should be expected.

Several workers observe that they were retrenched when they showed signs of occupational disease. They are unable to corroborate this because the corporation says that their health records were lost in a fire. The story of the fire is both convenient and vague and workers suspect that it is more smokescreen than fire. More generally, they believe that Iscor used mass retrenchments to dispose of occupational health liabilities. The corporation agreed with NUMSA to retrench workers over 45, reversing the common practice of 'last in, first out' and enabling Iscor to rid itself of workers who were already sick or whose long-term exposure put them at risk. As at Samancor, workers say the union did not consult them on this policy change.

Throughout the Vaal Triangle, at Sasol as well as at ArcelorMittal and Samancor, workers and activists say that company doctors cover up occupational and environmental illness. Thus, it was said that company doctors always prescribe the same remedy irrespective of symptoms, that independent doctors give different diagnoses and frequently identify occupational and environmental causes, that most workers cannot afford such independent advice but those who do risk losing their jobs if they talk about it.

Back at Samancor, many of the jobs were outsourced to contractors. Ex-workers see this as the continuation of a strategy, already evident in the corporation's approach to retrenchments, designed to reduce the corporation's liability for worker health and safety and, conversely, limit workers' rights. They say that those still working at the plant know that the job may cost their lives but cannot afford to lose their livelihoods. At both Samancor and ArcelorMittal, it appeared that NUMSA had abandoned responsibility for health and safety, accepting it as a management prerogative.

Throughout the Vaal Triangle, people observe that outsourcing is now common practice. It takes two main forms with some variations in between. First, work is contracted out to small firms whose survival depends on the corporation and who must cut costs to meet the price terms dictated by the corporation. According to Mabuti Mlangeni, Sasolburg organiser for the Chemical, Energy, Paper, Pulp, Wood and Allied Workers Union, Sasol has now added to the price pressures by co-ordinating bidding for outsourced work and so intensifying competition between small firms. Cutting wages and health and safety standards are then made the basis for that competition. Alternatively, jobs are made temporary as individual workers are employed on fixed-term contracts and redefined as contractors. Such contractors are often supplied through labour brokers, who are supposedly responsible for benefits such as pensions and medical aid, and contracts are managed to prevent claims for permanent employee status.

The effects of these practices are devastating. Following a series of incidents in 2004 (see Chapter 6), Sasol undertook a major safety review and promised a makeover of its safety training programme and safety 'culture', including 'contractor management standards' (Sasol 2005: 45). But it ducked the central issue of using contracting to cut costs and limit liabilities. This results in the long-term erosion of institutional memory and intimate knowledge of complex plants. More immediately, it increases the likelihood of poor co-ordination and communication between different work teams. In June 2006, despite the safety makeover, nineteen people were injured in an explosion in Sasolburg. According to Sasol, an 'independent' contractor was clearing chemicals left by another contractor 'after vacating the premises'. Sasol's reporting is calculated to distance the corporation from responsibility although it clearly controls the site as well as the terms on which both contractors operated.[17] Globally, Sasol did reduce its reported number of 'fires, explosions and releases' from 32 in 2004 to 15 in 2006. But the number has risen every year since with 36 reported in 2009.

Enclosed economies
Jobs, income and poverty

Mining and heavy industry in the Vaal Triangle created a mass workforce largely made up of men. Workers faced the brutality of racist *baasskap* – which gave any white worker authority over all black workers irrespective of experience or position – and wages were below the costs of household maintenance.[18] This regime was increasingly challenged by union organisation but much of the cost of production was transferred on to increasingly stressed families both in the townships and in the rural homes of migrant workers. Faced with intense resistance, the apartheid government declared a State of Emergency in 1985. Brutal repression failed to subdue the anti-apartheid resistance but did enable corporate capital to crush a series of worker strikes that posed the most serious challenge to its power since the 1922 Rand Rebellion.

The workers' defeat enabled the corporations to impose a new round of industrial restructuring. Since the late 1980s, the masculine workforce has been torn apart and reconfigured. Across the Vaal area, people see the same pattern: massive redundancies have left a core of workers at the big plants while what the corporations redefine as 'non-core' business is contracted out. Between 1993 and 1998, 46 000 jobs, including 20 343 manufacturing jobs, were cut on the Gauteng side of the Vaal Triangle according to development researcher Wim Pelupessy (2000: 8). This followed the broader national trend. In the first decade of democracy, about two million full-time formal-sector jobs were cut in the name of competitiveness and productivity. Iscor alone cut 30 000 jobs nationally and Eskom cut over 10 000. Nor has the massacre of people's livelihoods been reserved for the urban areas. Many thousands of farmworkers in the Free State have lost both their jobs and their homes and moved into the towns of the Vaal Triangle to find somewhere to stay.

> ### Box 2.1   Retrenched and destitute
>
> Workers retrenched from the Vanderbijlpark plant believe they were cheated out of their full entitlement. Many have given up and returned to their rural homes but a group of men living at the KwaMasiza hostel have remained to fight for their rights and are engaged in a protracted court challenge to the legality of their retrenchments. It is a hard road. 'I am here in prison waiting for the money due to me,' said Ernest Sigaqana. He was offered a retrenchment package of R32 000 but refused to sign for it or to take it, believing that this was not the amount due after 22 years working at Iscor. The money nevertheless appeared in his bank account the next day.
>
> All the men were migrant workers and their families still live at their rural homes. Sigaqana came to work at Iscor from Qumbu in then Transkei in 1979. The men observe that the policy of retrenching older workers left them without hope of getting another job. Most of them are in their late fifties and early sixties. The retrenchment money has long since run out but they are not yet eligible for pensions. Their families no longer visit them and nor do they go home. 'There is no money here and no money there.'

What people see happening in the Vaal is what is evident at national and indeed global levels. Labour scholars Edward Webster and Karl von Holdt (2005) observe that the world of work is increasingly unequal and divided into three major 'zones': the core, non-core and peripheral zones.

At the centre is the core zone of permanent full-time workers, numbering 6.6 million nationally. Changes in the workplace regime have been highly uneven depending on the strategies and coherence of management and of unions at particular corporations and plants. Authoritarianism and racism remain entrenched in many plants and migrant workers are still employed, particularly in the mines including Sasol's Sigma Colliery.

In general, core workers' skills and wages have been upgraded and they have a degree of security both in their jobs and in benefits such as medical aid and pensions. They have access to legal rights under the post-apartheid labour laws and most are organised in trade unions. At the same time, they work under intense pressure to increase productivity and often in a dangerous environment. ArcelorMittal has two strategies for increasing productivity: multi-skilling to create a more flexible workforce, and retrenchments. In 2004, it agreed to a two-year moratorium on forced retrenchments with unions but reduced its permanent workforce by 9% and 8.5% in each of those years through voluntary retrenchments and by not filling posts.[19] Nevertheless, people in the Vaal Triangle observe that the big corporations find ways to get rid of workers who challenge them. Core workers are always at risk of being ejected from the inner zone.

Outside the inner core are the outsourced workers employed by contractors or employed as fixed-term contract labour, numbering about 3.1 million. They may be part-time or temporary workers, many are 'permanently temporary' and most are poorly paid. They, and the small contracting firms, are at the beck and call of the corporations – available when work picks up, dispensable when it falls off and vulnerable to arbitrary reductions in pay. Mostly, they are not organised, partly because unions have not come to terms with organising them and partly because they are threatened with losing their jobs, or their opportunities for work, if they join a union. Their insecurity is heightened by the knowledge there is a 'reserve army' of unemployed workers desperate to take their place. In Sasolburg, Sasol employed about 9 000 outsourced and contract workers as against 6 000 permanent workers in 2005.

The peripheral zone is made up of about 2.2 million informal workers and 8.4 million unemployed people. In Emfuleni Municipality north of the Vaal River,[20] Tielman Slabbert (2004) of the Vaal Research Group shows that about 9% of workers are active in the informal economy. They are regarded as employed however meagre or irregular their income. The garbage trucks arriving at Vanderbijlpark dump are met by more than 50 waste-pickers. They compete for recyclables such as plastic sheets and metal, packing them into large bags. The market in recyclables yields slim pickings but the pickers count as having jobs. Street traders set up their stalls mainly at taxi ranks but, because the people are poor, their income from trading is small. Unemployment in Emfuleni, using the broad definition,[21] was 51% in 2001 and rose to 54% in 2003 according to Slabbert. More than half the households in Emfuleni could not afford the costs of bare subsistence and 64% could not afford additional necessities such as school fees and medicines as well as subsistence.

South of the river in the Free State, Metsimaholo Municipality used the narrow definition of unemployment in its 2003 Integrated Development Plan (IDP). It put unemployment at just over 26% for the Sasolburg area. If the broad definition was used, the figure would be significantly higher. Further, this figure did not 'accommodate the outflow of farmworkers from farms to towns as well as growth in townships due to other reasons of urbanization' (Metsimaholo Municipality 2003: Section 5: 3). The way the figures were done, Metsimaholo's economically active population makes up only 42% of adults aged from 20 to 65. Local people believe something like 80% of all adults have no jobs and, according to the IDP, 63% 'of the potential labour force earn no income' (Section 5: 3). Many of these people were supported by others but more people depended on fewer incomes. Fully 9% of households had no income whatever.

Throughout the Vaal, more men than women have jobs and women are more likely to have low-paying work. Women are thus more likely to be poor. Poor households also tend to be larger than better-off households, so what money comes in has to support more people. Most of those supported are children. They are the majority of the very poor. Pensions and child support grants mitigate this poverty to some degree but do not reverse it.

In 2004, government claimed that over two million jobs were created in the first decade of democracy. Webster and Von Holdt note that most of these new jobs are either 'non-core' or informal. They conclude that 'the erosion of core jobs, the growth of insecure and low-wage non-core jobs, and the expansion of the peripheral zone have generated a widespread increase in poverty' (2005: 23). As with the apartheid work regime, the costs are transferred to households. Insecurity is taken home with alcohol abuse and conflicts over who spends what, and home becomes a fragile refuge, a place 'to hide one's poverty' (24).

By mid-2008, government was reporting substantially reduced unemployment – down to 23% on the narrow definition from 31% in 2003. Then commodities crashed and, by mid-2009, the economy had shed over a million jobs. Following a reprieve at the end of 2009, another 170 000 jobs were lost in early 2010. Statistics South Africa (StatsSA) put the official unemployment rate at just over 25% but the statistic was helped because hundreds of thousands gave up looking for work.

Enclave development

Clearly jobs are at a premium in the Vaal and local people say they want the big corporations to clean up, not to shut down. The corporations themselves see jobs as central to their 'social licence to operate'. Yet people are increasingly questioning the logic. They remark that those born in the Vaal Triangle are less likely to get jobs than newcomers to the region. One reason is that, having grown up in the bad air of the Vaal Triangle, locals tend to fail the pre-employment medical test.[22] The corporations, it seems, rely on the fresh blood of people they have not yet contaminated.

They also observe that these industries are capital-intensive and are making very substantial profits but few jobs. After so many rounds of retrenchments, ArcelorMittal is no longer seen as a source of work. The local economy may be heavily dependent on the corporations but more and more local people see less and less benefit. They are 'de-linked' from this formal economy and the money simply bypasses them. This is the logic of 'enclave' development.

The most extreme examples of enclave development are associated with extractive industries – mining and particularly oil – in poor countries with 'failed states'. ChevronTexaco's luxury Malongo compound in Cabinda, Angola, is separated from the surrounding poverty by several rings of security. Corporate personnel, all foreign, never step into the local community. The flow of oil money is equally divorced from the local economy but tightly integrated with the global money centres. It is repatriated to the US in corporate profits, it returns to global finance capital through Angola's endless repayments on national debt, and it lines the pockets – or international bank accounts – of the corrupt ruling elite. In 2006, Angola's GDP growth was turbocharged to 19% with windfall oil profits and massive foreign direct investment but no expansion of local jobs or income. As James Ferguson puts it: 'The movements of [global] capital cross national borders, but they jump point to point, and huge areas are simply by-passed' (2005: 379).

South Africa is not Angola and the Vaal Triangle is not Malongo. Yet the enclave logic is powerfully at work as the economy is increasingly integrated into the circuits of global production networks. If apartheid attempted to confine poverty to the homelands, townships and hostels, wealth is now securing itself within gated residential estates. The most extravagant is Sandhurst in Sandton where 640 houses, with an average value of R30 million each, are protected by kilometres of high-security fencing. The logic is actively extended through government economic and spatial planning and the ambition of the big metropolitan municipalities to create competitive 'world-class cities', on display to foreign investors in the glittering shop windows while the poor are packed off to the back of the shop.

Back-door delivery

At the edge of Gauteng, much of the Vaal looks more like the back of the shop than the shop window. As the jobs are swept out of the factories and neighbouring farms, 'delivery' – a catch-all term for the provision of jobs, housing, amenities and services – is the primary means for managing poverty. It is the supplement to enclave development, held out as the lifeline to many people and so also the measure of government legitimacy.

Since municipal restructuring in 2000, delivery has been increasingly devolved to local government. However, if South Africa is far from being a 'failed state' at the national level, there are many failed municipalities at local level. Emfuleni fits the bill. According to its 2005 IDP, its bureaucracy is over-staffed but under-skilled and consequently lacking accountability. Stories of corruption are common in every corner of the Vaal. Officials and politicians are seen to be making good with property development companies and shares in shopping centres and businesses. Local people see a repetition of the patterns of enrichment through office that defined the collaborationist black councils of the 1980s and provoked the Vaal Uprising.

The logic of enclave development supplemented by delivery is overlaid on earlier rounds of the Vaal's developmental history. What it gives rise to is not the single citadel of wealth surrounded by poverty but a variable pattern of inclusion and exclusion, of wealth and poverty, as the new logic is patched on to and into the old.

Housing is both the most visible evidence of this and at the core of delivery. For many people, a home is their first need and their last refuge. The history of apartheid removals has left many with a deep sense of insecurity and housing has been central to the conflicts that have escalated around the country in recent years. In the Vaal Triangle, the sense of vulnerability is palpable in some communities. And just as people are inhibited from criticising the corporations for fear that more jobs will be lost, so complaints about pollution have been muted in some areas for fear that they will result in removals rather than in the clean-up of pollution.

Housing delivery in the Vaal started with the conversion of hostels to family housing, and site and service schemes and the roll-out of Reconstruction and Development Programme (RDP) housing followed. In the first decade of democracy, government approved two million housing grants nationally but said housing demand was outstripping supply as people opted to live in smaller households. This resulted in 'an increase of two million additional households over and above that generated by population growth' and led to the proliferation of shack settlements (Presidency 2003: 26). However, this trend was already evident in 1994. In the old township of Bophelong, apartheid policies confined three or four generations in the same home. Despite the relative generosity of the original houses and the addition of backyard shacks, families had long since outgrown the space. In Evaton, many people rented backyard shacks densely packed into the same household. People moved out from these cramped conditions when they could and this movement itself was part of what cracked open the apartheid confinement. At the same time, just as farmworkers have lost their homes with their jobs, so too have many industrial workers who lived in 'tied housing' belonging to Sasol or Iscor. Many of these people have found no alternative but to house themselves in shack settlements.

The trend to smaller households has also been encouraged by government's own housing and service delivery programmes. Most RDP houses consist of two small rooms, including kitchen and washing areas, with limited space for extension while the free basic electricity and water supply penalises larger households because they get the same amount as small households. Houses are also poorly built and the basics of environmental design have been neglected. They are not energy-efficient and poor people must either pay for warming or cooling their homes or they must live with extreme cold or heat. People in Bophelong remark that the original houses built in the 1940s are more comfortable than new RDP houses simply because they have double-skinned outer walls. The saving on quality goes to government and contractors but is passed on as a cost to the 'beneficiaries'.

Harry Gwala is a new RDP housing estate built in Sasolburg beyond Zamdela. People forced to fit into the standard two-roomed houses have added shack extensions to accommodate their families. The local council is evidently not happy with this and wants the shacks taken down. Nearby is Iraq, set up by people who refused housing in Harry Gwala because their households did not fit into the RDP houses. It is termed a shack settlement because the houses are built of iron sheeting. Yet they are large by comparison with the RDP houses, well built and freshly painted with door and window frames picked out decoratively in the Sotho style. Plots are neatly fenced off and many have food and flower gardens.

In Sebokeng Zone 15, the five old Iscor single-storey hostels have been converted into 'family housing'. The residents pooled their housing grants to fund it but were told that the money had run out after just two of the hostels had been converted on the cheap. The result is that three or four beneficiaries have to share each of the apartments supposed to be family units. Most of them are unemployed following the Iscor retrenchments, their once-in-a-lifetime housing subsidies are used up and they are effectively trapped there. Local authority officials who administered the scheme could not account for spending. Residents believe that the job was not completed because officials pocketed the money. Whichever way – through negligently poor planning or corruption – they were dispossessed of housing rights at the very moment that these rights were created. Acting through the Vaal Working Class Crisis Committee, they have demanded that these rights should be restored.

For the most part, the enclave logic has drained money out of the Vaal townships along with the jobs. But those who have found a place within the 'core' labour zone are now highly visible. They are not part of the new elite settled on the river frontage, but full-time operatives – the new word for multi-skilled, flexibly-tasked, full-time workers – and professionals, mostly employed by the state, such as teachers. In Bophelong, isolated face-brick houses tower over their neighbours in the new RDP housing estates. Their owners claimed the housing grant to access land but immediately demolished the RDP house and rebuilt. In Zamdela, private developers built face-brick houses in the area now known as Success. They were bought by full-time Sasol operatives and middle-ranking professionals. Cars are parked or garaged in most yards behind the usual security gates but, after work, people are out strolling on the streets.

Corporate patronage is also visible. As company towns, Sasolburg and Vanderbijlpark received corporate largesse from the start. In the 1990s, the focus of patronage shifted from the white town centres to the townships and took the new name of Corporate Social Responsibility. In Zamdela, Sasol funded the well-appointed community centre that occupies a prominent and spacious block. It has also entered a public-private partnership with the Free State government to give the township's library a makeover. On the periphery of the Vaal Triangle, by contrast, Orange Farm remains remote from any benefits of the enclave.

Back-door services

Around the corner from Success and close to Zamdela's business centre in Joe Slovo, a shack settlement is squeezed into a lane between formal houses. On a winter's evening, men and women are in the street preparing for the night. The men are cutting up wood scrounged from old pallets while the women make balls of coal from a mixture of coal-dust, ash and water. The coal-dust is trucked in by entrepreneurs who get it free from Sasol. Like paraffin, it is sold in small quantities because people cannot afford to stock up – even if it costs more in the long term, the poor must budget daily. The braziers are lit in the street so that the worst of the smoke disperses into the outside air. They are then taken indoors to provide both for cooking and heating. The only ventilation is the gaps left by ill-fitting doors and the holes in the iron roofing and walls.

The variable pattern of enclave development supplemented by delivery applies to services as much as to housing. In old Bophelong, the houses were originally fitted with iron coal stoves ventilated through chimneys. Bophelong is upwind of white Vanderbijlpark and, in the 1970s, there was growing concern about pollution.

The old stoves were then replaced by newer models that could take smokeless coal. Both ordinary and smokeless coal is now available from Bophelong coal-merchants. These houses also had full services from the start in 1948 – electricity, water, sanitation and rubbish removal.

Total emissions from domestic coal-burning are comparatively minor. According to Scorgie, some 20% of homes in the Vaal Triangle rely on coal and they emit 2.8% of particulates compared to industry's emission of over 95%.[23] Just the low-level emissions from industrial combustion are 2.5 times greater than household emissions.[24] Nevertheless, household emissions have a significant impact on health because they are emitted where people live and are close to the ground. The effects of indoor emissions, where coal smoke or paraffin fumes are not ventilated, are many times greater for the household concerned. Indoor braziers and paraffin stoves also create fire hazards, particularly in crowded households where accidents are more likely, and in densely packed shack settlements where fires spread rapidly.

Noting that energy is necessary for life but that, for poor people, it has become hazardous to life, social movements have demanded delivery of clean, safe and affordable domestic energy. In practice, this has translated into a demand for electricity and it is twinned with the demand for clean water. In 2000, government responded with the 'free lifeline' supplies. The lifeline proved miserly: six thousand litres of water and 50 kilowatt-hours (kWh) electricity per month per household. Few can make it on this, least of all the poorest and largest households. In most distribution areas, the price rises sharply once the free allocation has been used and people end up paying as much as seven times more than industry per unit consumed.[25]

Prepaid meters and other technologies such as trickle-feeds, for both electricity and water, have come to symbolise government's insistence that delivery must be based on 'cost recovery' and are associated with privatised or commercialised service delivery.

Johannesburg contracted transnational corporation Suez to provide water management services in 2001. Anti-Privatisation Forum researchers remark that Suez was heavily indebted and saw Johannesburg as a good cash cow (Fiil-Flynn and Naidoo 2004).

Most households with electricity connections in the Vaal area are on prepaid meters and must cut themselves off before the end of the month. Indeed, many are cut off the moment their prepaid allocation is used up. In Emfuleni, the situation for water is anomalous because of the chaotic state of the council administration. In 2006, the history of water reticulation was evident in a patchwork of different systems. In the old Iscor hostels, for example, the council charged people collectively for bulk deliveries – including the leaks from poorly maintained water pipes. Since then, under the banner of the Campaign Against Water Privatisation, people have resisted plans to introduce prepaid meters.

Prepaid water meters were pioneered where people are most vulnerable – in Stretford, which is on the periphery of the Vaal Triangle in a part of Orange Farm administered by the Johannesburg Council. A high proportion of households are headed by women, all are poor and 30% have no income (Fiil-Flynn and Naidoo 2004: 16). The prepaid system replaced communal tap stands where water was free. It was justified as saving water and enabling the delivery of the free basic water supply. A year after it was installed, half the households ran out of water at some time because they had no money to pay for more units. Others ran dry because either the meters or the computers at the sales point broke down. Most had to restrict water use and local people remarked that children from Stretford were easily recognised because they 'go to school in dirty clothes' (20). Caring for sick people became more difficult and costly. For the most part, women carried the responsibility of managing water use, of begging from neighbours or of walking to find free sources. The system also provoked conflict in the neighbourhood over allegations of water theft and in households where people experienced increased domestic violence.

The system works well for the managers. The cost of providing the free basic supply is offset by the reduction in administrative costs, the need for billing systems and staff to read meters is eliminated as is the cost of cut-offs and the potential for conflict between people and utility staff – whether municipal or privatised. Beyond this, Greg Ruiters argues that the technologies of 'delivery' are also technologies of political and social control. They are part of a discourse that represents the 'empowered citizen as the customer . . . who pays for services, only uses as much as she can afford and makes wise, sovereign and informed choices with her limited means' (2005: 7). Discipline at the back of the shop is thus individualised as self-discipline divorced from social action: the poor must know their place in the order of the market; they must learn to be frugal consumers.

The system both reveals and hides poverty. Poor people must show their poverty to the authorities by registering to qualify for the free basic supply but prepaid meters erase politically embarrassing statistics on how many people are cut off because they cannot feed the meter. The effects of poverty are removed from the social realm and confined to the household. Yet the system computers still return a constant stream of information, creating banks of data that make 'customers' visible to managers while managers are largely invisible to the people.

Delivery of sanitation has lagged behind water but, in 2004, government promised to 'eradicate the backlog'. At Orange Farm, the promise of flush toilets was used to win acceptance of prepaid meters. People were given to understand that prepaid meters provided the means to obtaining flush toilets. They later discovered that the water to flush the toilets was more than they could afford. Water-borne sewage without water doesn't work so the pipes are constantly blocked. Johannesburg Water, however, has refused to accept that this is their problem. Instead it 'relies on the sheer necessity for hygiene and sanitation . . . to make residents provide this service for themselves' (Fiil-Flynn and Naidoo 2004: 17). Without proper tools, that means digging faeces out by hand.

Waste management has the lowest profile of domestic services. According to the Emfuleni IDP review 2008/9, 133 030 households have their waste collected once a week while 22 200 households in formal settlements and all households in informal settlements do not have their waste removed. Business waste is collected daily from 1 453 stands in Vanderbijlpark and Vereeniging. There are also street-sweeping services in business centres in Vereeniging, Vanderbijlpark, other outlying business centres, the Sebokeng taxi rank 'and all main roads'. But the service is visibly inadequate. Litter is commonly encountered, particularly around taxi ranks, and open ground is frequently used for informal dumps.

At the end of the waste pipe, sewage works and waste dumps are poorly managed throughout South Africa and have a major impact on rivers. The history of careless and profligate waste production combined with the gross neglect of waste management is clearly visible in the Vaal. None of the dumps has a permit. On paper they are illegal but in practice they are operated by the local authority and overseen by regulatory authorities. This semi-legal status is just one face of the greyness of waste and the shiftiness of definitions within the broader regime of negotiated non-compliance. The legacy costs incurred through decades of neglect are huge. The present waste-management system is not only unable to deal with them but is adding to them.

Emfuleni's planning ignores reduction at source and the waste managers must deal with what comes at them. As elsewhere, waste is a low priority and underfunded and managers are left to 'make a plan'. The Palm Springs dump across the Golden Highway from Evaton is a good example. Here the waste-pickers far outnumber the municipal team and Emfuleni's landfill manager responded by incorporating them into a benign management regime and supporting their capacity to earn a livelihood off recycling.

In Sasolburg, by contrast, the municipality more or less abandoned the dump and the waste reclaimers are practically in charge of it. They also established their own market, selling directly to buyers of recyclable materials. In 2006, with Sasol's support, the municipality awarded an exclusive right to recycle materials from the dump to Phutang, a local black-empowerment group. Having neither capital nor experience in waste recycling, Phutang soon merged with Remade, an established white-owned firm. Phutang-Remade simply inserted itself in the selling chain between the reclaimers and the original buyers. Its income depended solely on paying less to the reclaimers. The reclaimers resisted this but, having legally enclosed the reclaimers' livelihoods, the council literally enclosed the dump with a fence – 60% paid for by Sasol – to control access and prevent the reclaimers taking recyclables out to sell independently. Finally, in 2008, the council sent the police in to force the reclaimers to sign contracts with the company. In a report for groundWork, Melanie Samson observes:

> The combination of the police and the fence broke the reclaimers' ability to continue with their resistance. One reclaimer eloquently summarised the outcome of what she perceives as a hard-fought battle stating, '[w]e were chased away by the police on a Friday. We came back on Monday to surrender and sign the contract' (2008: 25).

ANOTHER VAAL

In 1996, democratic South Africa adopted a new Constitution. It is widely seen as a very progressive document but it provides two contradictory mandates for development. The Environment Right explicitly mandates sustainable development based, in the reading of *The groundWork Report 2004*, on environmental, social and economic justice. The Property Right implicitly mandates capitalist development. Since then, at least until the power went down (see Chapter 8), economic policy has scarcely deigned to mention the environment and development policy has been determinedly focused on securing the basis for the accumulation of capital as we shall see in the next chapter. The central idea of accumulation – that profits must be accumulated and reinvested to make more profits – is the basis of economic growth which is assumed to be a self-evident good.

This good is rather less self-evident to the people of the Vaal. Even as commodities boomed and growth topped 5%, most saw declining benefits and escalating costs. In contrast, the captains of industry did rather well. In 2005, Eskom's chief executive took home R10.4 million, Sasol's took R15.6 million including profits from share options, while Mittal South Africa's took R4.4 million. The bosses of the big mining houses with interests in the Vaal did even better. The CEOs of Anglo American and BHP Billiton respectively took R55 million and R30 million.[26] Starting in the 1980s, shareholder expectations for profit inflated alongside the expectations of executives and cost-cutting, starting with the wage bill, was made a core criterion for judging executive performance.

VEJA's first concerns are that corporations should clean up and compensate for the environmental harm done to people but, together with other fenceline organisations like the SDCEA, it is more broadly confronted with the nature of corporate production and the crisis in the lives of the poor. While the organisations that make up VEJA do not necessarily share a single analysis of their situation, a strong common language, critical of neo-liberal capitalism or of capitalism as such, is evident. At an air quality workshop in Sasolburg participants observed that the power of corporations is founded on their relationship with the state, both because the state creates the basis of corporate legitimacy and because corporations create the revenue basis for the state.[27]

Discussions of the purpose of struggle are located within this critical perspective. The immediate demands – clean up and compensate – are themselves ambitious in the current context. Beyond this, the desire for 'another Vaal' is very much part of the discussion with people raising the need to change the capitalist system for a new model of development, to socialise the economy and to create a new paradigm where communities and workers are central to the power system.

At a strategy meeting, people from the Vaal and other pollution hot spots observed that 'the corporations are like a brick wall while government is a wet blanket on people's action'. They see it as important to assert the freedoms and rights won through the anti-apartheid struggle and written into the Constitution. 'If we can't protect the achievements of the past, we cannot protect the future.' The tactics of struggle proposed by participants therefore include formal representations to parliament and other official bodies and engagement with formal processes of participation. And they are acutely aware of the potential for exclusion. 'Participation is not a favour from government but a right for us. When government did the EIA revisions, it consulted industry, not us.'[28]

Finally, there is a strong sense that popular mobilisation is the heart of struggle and that it must accompany whatever actions are taken through formal engagement. Action must always have its roots outside the bureaucratic wet blanket of state-sanctioned process. This mobilisation needs to link apparently disparate struggles that the organisations participating in VEJA are engaged in. This linking is at once an expression of solidarity and a recognition that these fronts of struggle are formed in resistance to the same processes of capitalist development.

There is a deep recognition in the Vaal that there is no blueprint for the future. What will be will emerge out of the process of engagement and struggle between people, corporate capital and government. This engagement takes place on many fronts and in many dimensions – on the streets, in the formal forums of participation, in battles for information, in the contestation of corporate control of knowledge and science, in the media and in struggles around the identity and purpose of the state.

These contending forces do not begin and end in the Vaal. The Vaal as an industrial and urban space has been produced within the old imperial economy of London and the new world economy of Washington as well as within the brutal politics and corporate rivalries specific to South Africa. It is the product of the racist divisions of labour that took particular forms in South Africa but were also part of the structuring of a global division of labour. The story opened with the dispossession of people to force them into mining and industrial work. It has arrived now at the dispossession even of that labour. It opened with an industrial regime indifferent to its environmental destruction. It has arrived at the prospect of the wholesale junking of the basis of life. The story does not close here.

The crises into which the world is now led will send successive shocks through South Africa and the Vaal. Those who direct the processes of accumulation that produce the crisis will seek to save themselves by managing the displacement of its violent effects on to those made poor. Yet crisis is also opportunity. Resistance to the orders of accumulation is gathering in all corners of the world and resistance itself becomes part of the crisis of the ruling powers. And just as these struggles shake the Vaal, so too will the struggles of the Vaal people reverberate around the world – as they have done in the past.

3

New South Africa

INDUSTRIAL DEVELOPMENT CASTS a deep shadow of environmental injustice but is still widely assumed to be at the heart of economic development and to hold the prospect of adding to the sum of social wealth. This reflects a more global and deeply-rooted assumption of modernity: in official and academic language, development and industrialisation are virtually synonymous – 'developed countries' and 'industrialised countries' have become more or less interchangeable terms. Development is about how societies are ordered. It is deeply political. That developing countries should aspire to become developed, that they should see the image of their future in the rich world and the means of realising it in the process of industrialisation is thus naturalised in the language of international and national institutions. It is taken not so much as the matter of politics but as its foundation, as the unexamined assumption on which politics rests. What development is about is fiercely contested but development as such works as an imperative.

Chapter 2 shows how the developmental process has shaped people's lives and reproduced environmental injustice on the ground in the Vaal Triangle. The enclave development described there has been actively produced through the so-called 'post-Fordist' strategies of corporations enabled by the neo-liberal policies initiated by the apartheid government and entrenched by the government elected in the first democratic elections of 1994. The logic of the enclave has been patched into the spatial orders produced in previous rounds of development – the colonial free market ruled over by the British Empire and the racial Fordism of the apartheid era tied into the Cold War US hegemony. This chapter looks first at the policies that have broadly shaped development since 1994. This sets the context in which environmental policy developed. The second section shows how it has been driven by fierce contestation on industry's polluted fencelines. In the last section, we open the lens on the wider world to interrogate the promise of green capitalism.

NEGOTIATING THE FUTURE

South Africa's democratic constitutional order was won at considerable cost through a struggle that mobilised people across the full spectrum of social relations: through labour and civic organisations, religious bodies, women's organisations, human rights groups and in struggles for land. It terminated the brutal repression and authoritarian racism of apartheid and so represents a victory for the majority of people. As is so often the case in struggles against oppression, what has been won is very far from what many South Africans imagined they were fighting for. For the majority, apartheid was not just about political exclusion from decision making: exclusion was the means by which they were dispossessed and impoverished to secure cheap labour for capital. Similarly, democracy was not an end in itself but the means by which the majority would gain freedom from want through their control of production. Thus, the Freedom Charter called for the return of the land to those who worked it and the people's control of the 'commanding heights' of the economy. These demands picked up strands of egalitarian and socialist thinking within the liberation movement but were increasingly marginalised in the course of the transitional negotiations.

The crisis of apartheid created the conditions for the transition to democracy but bequeathed a tattered economy. Commonly identified problems included that the local market was too small for large-scale production that could compete with imports; investors were therefore not interested; capital goods (manufacturing plant) had to be imported at huge expense; and the local skills base was constricted by the perversity of apartheid education. Many analysts saw these problems as symptoms of the exhaustion of industrial strategies, originating in the 1920s, and founded on import substitution.[1] The Macro-Economic Research Group (MERG), by contrast, found little in the way of coherent industrial strategy in South Africa's history. It identified the centrality of the minerals-energy complex in defining South Africa's development or, in Ben Fine's words, 'the concrete form of accumulation of capital taken in South Africa' (2008a). This implied that industrialisation in South Africa was founded on the power of the minerals-energy complex which was, from the beginning, geared for export of resource commodities demanded by imperial capital and indifferent to local needs. The constriction of the local market and the failure of investment flowed from this. Ad hoc responses to demands for protection meanwhile allowed the development of consumer goods industries to supply the top end of a stunted (white) market. The overall result was an industrial structure developed at two extremes with nothing in between to link the dominant core created by the minerals-energy complex and the rest of the economy.

South Africa's negotiated transition was not just a compromise with apartheid's ruling elite or even with the representatives of racial capitalism. Negotiations took place in the context of change in the global order of power. The collapse of the Soviet Union marked the end of the Cold War and left the US as the only superpower. The victors proclaimed the triumph of capitalism and aggressively redefined 'development' in line with neo-liberal ideology. Behind the scenes at the South African negotiations, the institutions of global capital profoundly influenced the terms of the economic debate.

During the early 1990s a range of experts, including the World Bank, business leaders and economists of various political leanings, said that South Africa's economy would need to grow by at least 6% a year to create enough jobs to reverse poverty. There were intense arguments about how this was to be achieved and what the role of the state should be. The trade unions called for 'growth with redistribution' and an interventionist state. Their thinking was strongly represented in the Reconstruction and Development Programme (RDP) published as the ANC's manifesto for the first democratic elections in 1994.

Capital, on the other hand, argued that the economic failure of apartheid was precisely the consequence of state intervention and that the 'market' should regulate itself. This view was endorsed by the international players. They emphasised that 'there is no alternative' to an export-oriented economy integrated into global relations of production and marketing, and this message was strongly supported by those sectors of South African industry that saw profits in the world market.

In 1993, MERG was specifically tasked by the ANC to prepare an alternative to these prescriptions. On Fine's account, however, 'both the substance of economic policy and the way it was produced (from on high . . . as opposed to organised root and branch discussion in the earlier period leading to the RDP) had changed dramatically' and the ANC leadership disowned MERG's report even before it was published (2008a). It thus accepted South Africa's place in the global orders of production as prescribed by the Washington Consensus.

Consuming the liberation

Shortly thereafter, the Melamet Commission recommended that those holding public office should be paid 'market rates', effectively taking the acquisitive corporate executive class as the market benchmark. As veteran anti-apartheid activist Neville Alexander remarks, this was accepted without question. During the struggle years 'everyone, including your "Comtsotsi", was seen to be and treated as an equal, whereas after 1994, there was this sudden and very visible divide between those

81

who were deemed to have been "successful" . . . and the veritable underclass, victims of apartheid before 1996 and of neo-liberalism thereafter' (2009). Conspicuous consumption as the reward of office has since become the norm as the parade of million-rand ministerial cars testifies.

South Africa thus joined the world market defined by Ponzi capital's globalisation strategy. This strategy produced growing inequality both globally and within individual countries while trumpeting the promise of individual fulfilment through consumerism. For those on the wrong side of the wealth gap, this is a promise broken as it is made. Anticipating Alexander's critique, media analyst Eve Bertelsen observed: 'The heroic figures (role models) of the anti-apartheid narrative were . . . principled political and union leaders who lived frugally and stoically battled for the cause. In the organising narrative here collaborative endeavour ensures the betterment of the whole community' (1998: 232). Since the 1994 elections, however, the political leadership 'enthusiastically embraced the philosophy of the late capitalist "free market"' (221). Commercial advertisements, meanwhile, played on the symbolism of struggle by emptying its political meaning and 'recoupling its potent signifiers (of black pride and accomplishment, democratic choice, civil rights, the right to a better life) with the discourse of consumerism . . . In the process, the desire for a shared social good is replaced by the desire for consumer goods or commodities' (231).

The values that stitch together status, power and consumption have trickled down through the layers of government and bureaucracy. It may not be that the values of liberation have been entirely consumed on the gravy train. Yet the many people who continue to serve their communities with dedication – whether as professionals or volunteers working in such fields as health, education and development – are confined to the back of the shop where the narrative of poverty alleviation offers small rewards when compared with the promise of liberation.

Consumer gratification no doubt compensates for the disappointment of that promise. Thus, Thokozani Xaba recounts the tragic transition of struggle hero to community pariah in the Durban townships and notes that 'comrades' and 'exiles' without employment 'could not legitimately get money to buy what they considered to be "the best things in life" – expensive cars, clothes and jewellery (for their girlfriends)' (2001: 112). He concludes that such luxuries are won by those who are 'more daring and brutal' and that the risk of being killed is preferable to the benefits promised by government's development programmes (119).

Jacklyn Cock observes that a certain masculinity is linked with particular icons of consumerism. She quotes a research respondent from Lenasia: 'If you have a

BMW, a cell phone and a glamorous woman you've got a lot; if you've got a gun you've got everything' (2001: 14). This masculine consumerism is not restricted to the criminal fringes. Indeed, criminal activity seems very much in tune with 'core' development. Hein Marais argues that syndicates not only demonstrate the entrepreneurial virtues promoted by government but also provide start-up capital for many legitimate businesses. Property crime reinforces 'the values of consumerism and . . . helps sustain consumer capitalism in a society that deprives the bulk of society from sharing in its ostentatiously advertised fruits. In many depressed communities, circuits of illegal accumulation have become integral to social and economic reproduction' (2001: 196). Corruption, whether of politicians or officials, similarly reflects the values of consumerism and provides a link between criminal activities and state institutions.

The gravy train did not start out in 1994. These values were already anticipated in the colonial and apartheid values of racial capitalism and they are now refigured in the conspicuous displays of global consumer capitalism. The corporate leaders who are, to paraphrase Bertelsen, 'the heroic figures (role models) of the narrative' of capitalism-as-development are themselves none too precious about the getting of wealth as the fraudulent dealing at the centre of global capitalism demonstrates.

Geared

Two years into the life of the ANC-led Government of National Unity,[2] international finance capital demonstrated the material force behind the World Bank's arguments. Trevor Manuel, just appointed as the first ANC (and black) minister of Finance, made a mildly sardonic comment about 'the market' and foreign capital was instantly withdrawn from the economy, putting the value of the rand on the skids. According to economic policy insider Alan Hirsch (2005), 'restoring credibility' with international capital was then made government's top priority. Macro-economic policy, given the title Growth, Employment and Redistribution (GEAR), was put together in a matter of months and adopted in 1996. It was the culmination of the process that began with the business delegation visits to the ANC in exile. In that process, 'the working class [was] forsaken as the agent of the National Democratic Revolution in favour of the state and a black bourgeoisie' (Chipkin 2003: 35). Government's agenda for transformation thus came to centre on Black Economic Empowerment (BEE) aimed at creating a black capitalist class and, in Moeletsi Mbeki's view, represented the deal done between the new black political elite and the old white business elite.[3]

With some variations on the theme, GEAR replicated the development model prescribed by the neo-liberal Washington Consensus. Indeed, several World Bank staff members were on the team that developed the policy. While the bank, effectively representing the 'virtual senate' of global investors, was given an inside track, the policy was developed in secret and presented as a fait accompli to the South African people and even to the unions who were formally in alliance with the ANC.

What remained consistent through the policy shifts that led to GEAR was the vision of a modern industrial economy aiming for the magic number of 6% growth. The neo-liberal strategy accepted global, rather than national, capital as the agent of modernising development. The state's own strategies were then geared to attracting private capital, and particularly transnational corporations, through privatisation or through state infrastructure investments in Spatial Development Initiatives (SDIs) and Industrial Development Zones. It was claimed that benefits to workers, to women and to the poor would then flow from economic growth. The effects, however, were to expose these supposed beneficiaries to the full blast of globalisation and GEAR met with resistance from the start.

Government defended GEAR on two grounds. First, South Africa had to deal in the real world. The political transition had coincided with globalisation, the world economy was expanding rapidly and, in contrast to the economically isolated apartheid regime, democratic South Africa would ride it. Higher rates of investment into South Africa and access to global markets would provide for growth and growth would provide jobs. The domestic market would then grow to create a virtuous cycle. Foreign investments would usher in new skills and clean and green technologies to replace South Africa's dirty old industrial plant. Second, government argued that the RDP remains the development programme but GEAR provides the economic means through which the necessary resources can be mobilised. Fiscal conservatism and the proceeds from privatisation would reduce South Africa's debt, the country would avoid the debt trap and the consequent dictation of policy by the International Monetary Fund (IMF) and World Bank, and the money saved on interest repayment would be freed up for redistributive development.

Contrary to the words of its title, and as its critics predicted, GEAR produced little growth; formal sector employment shrank with the loss of two million full-time jobs in the first decade of democracy, and inequality grew as wealth was indeed redistributed but from the poor to the rich. Resistance also grew as the effects of GEAR were trickled down through cost recovery on services, inadequate housing provision, the collapse of already pitiful environmental regulation and the lacklustre performance of land reform. The value of pensions, on which many poor households depended for survival, was eroded and the very principle of welfare grants came under sustained assault.

At the same time, the private corporations at the heart of the minerals-energy complex were allowed to list offshore in the global financial capitals. This came on top of large and illegal capital outflows and represented a huge disinvestment – appropriating resources produced in South Africa to finance the corporations' global ambitions. Fine (2008b) observes that the process was facilitated by government. Treasury effectively lined the corporations into a queue for permission to list in London and New York. The intention was to prevent too much money leaving the country at once as this would have devalued the rand and hence also the value given to the corporations' global listings by the market. State-owned corporations were meanwhile instructed to restructure in preparation for privatisation.

Resistance was deeply resented, particularly when it threatened the ANC government's claim to the mantle of emancipation. Critics, including the ANC's alliance union partner, were castigated as 'ultra-left' or 'infantile'. More insidiously, resistance was associated with a supposed right-wing reaction against black rule or said to be manipulated by a 'third force'. As it cracked down on dissent, government upped the rhetorical emphasis on 'poverty alleviation' and turned up the volume to legitimate a priority for 'development' over 'environment' at the 2002 World Summit on Sustainable Development (WSSD) in Johannesburg. Instead of submitting to the ANC's leadership, however, people took to the streets outside the approved orders of participation. There they were confronted by armed force. At the WSSD, police violently disrupted disapproved protests. Activists from the Landless People's Movement and the Anti-Privatisation Forum were arrested following one demonstration. The following week, police used percussion grenades to break up a march of activists demanding their release. Several people were injured including international visitors. At the same time, government moved to ban a mass march under the banner of 'Social Movements United' but simultaneously promoted a march organised by the ANC. In the event, it backed down on the ban as the political costs became clear. The social movements' march, involving thousands of international activists along with the local movements, substantially outnumbered the ANC's march.

While losing the battle on the streets, government was winning in the official negotiating halls. The top environmental official told parliament that the United Nations Conference on Environment and Development had sub-ordinated development to environment and the WSSD would put this right. In the name of 'poverty alleviation', the Johannesburg summit firmly prioritised development and held up government 'partnerships' with corporations as central to the means of delivery. Far from addressing poverty, however, development was producing it along with environmental degradation. Government's policy was really about creating the conditions for capital accumulation. And even that was not working. Rather than make direct investments in local enterprises, the global managers of capital preferred to speculate on the South African economy and, in 2002, they were taking their 'hot' money out. The value of the rand sank dramatically, exposing the country to the vulnerability caused by the GEAR policy of opening capital markets. Food prices ballooned, most spectacularly for maize meal and bread, the staples of the poor.

The developmental state

If government resented being challenged, GEAR was also failing in areas that were of concern to it as the manager of a capitalist economy. On its own terms, GEAR was successful in imposing spending constraints and reducing inflation and the national debt, but it completely missed its stated goals on economic growth and job creation. In government's view, the key problem was that GEAR failed to attract private sector investment – and particularly foreign direct investment. Finance minister Trevor Manuel expressed his frustration at an international Financing for Development Summit in March 2002: 'You can subject South Africa's policies to the tests of salt water and fresh water economists, and we will pass those tests. But that has not translated into a great flow of investment'. In other words, the economic fundamentals required by the neo-liberal Washington Consensus were in place but the development story did not go according to the script.

The script itself was beginning to change, however. The World Bank backed off some of its more extreme free-market positions and allowed an economic role for the state beyond macro-economic rectitude. In 2000, World Bank president James Wolfensohn made 'a strong plea' to President Thabo Mbeki for South Africa to get its 'micro-fundamentals' right (SALB 2002: 8). This was a green light for more active state intervention in the economy but for the purpose of extending the logic of the market into the day-to-day working of the state and into the fabric of social life.

The Micro-Economic Reform Strategy and the Integrated Manufacturing Strategy followed in 2002. They were welcomed by labour as signals of a more interventionist approach to economic growth but were cast in the mould of GEAR: they were premised on an open, export-oriented economy tied into the world economy through global production chains; and they were formed from an imagination of development as produced through market competition based in high-tech, high-capital and high-energy enterprises. This excluded the majority of

South Africans from the core of the economy while subordinating this economic core itself to the needs and profits of global capital.

Government subsequently admitted that its development policies excluded the poor. In 2003, Thabo Mbeki introduced the metaphor of the two economies: the First World market economy and the second economy, a Third World survivalist economy. Wealth, he said, would not 'trickle down' from the first to the second economy so we should 'not assume that the interventions we make with regard to the "first world economy" are necessarily relevant to the ["third world economy"]'. A distinct development strategy was needed 'to transform this economy... so that the "third world economy" becomes part of the "first world economy"' (quoted in Hirsch 2005: 233).

The metaphor produced a dissonant echo of the apartheid description of South Africa as two worlds – First and Third – in one country. The problem of the second economy was held to be that it lacks development rather than that it is the product of development. The process that produces wealth was thus made to appear separate from the process that reproduces poverty. This justified a dual development strategy: core development promoted capital accumulation while development-as-delivery contained the fallout from the enclosure of common and public resources and the costs of externalisation at the 'back of the shop'.

This dual strategy came to be named the Accelerated and Shared Growth Initiative for South Africa (ASGISA). The 'first economy' strategy focused on the magical 6% growth target. Government put the privatisation of key state industries, notably Eskom, Transnet and the arms group Denel, 'on hold' and channelled very substantial investments into infrastructure, primarily designed to service the needs of capital, through these parastatals. Asserting that the poor lacked development, government focused the 'second economy' strategy on turning them into budding entrepreneurs who could be linked into the 'first economy' while also providing 'work opportunities' – that is, badly paid temporary jobs – through the Expanded Public Works Programme (EPWP). Outside of the ASGISA framework, government expanded social spending – on grants and housing, etcetera – and belatedly acknowledged that this was the only policy that had any effect in alleviating poverty. Calls for a Basic Income Grant payable to all adults were nevertheless brushed aside on the rationale that it would create a 'culture of dependency'.

These second economy initiatives followed the World Bank prognosis that, throughout the developing world, the informal sector will now provide the jobs that the formal sector no longer offers. It also provides the most flexible of labour markets. Nevertheless, it is commonly recognised that the informal sector relies on formal-sector wages for its market. As the number of jobs shrinks, so too will the opportunities for making a living. Further, informal trading is increasingly saturated. There are only so many shebeens that can survive competing for the same shrinking market. In 2003, UN Habitat concluded that the rise of the informal sector, together with the slums that now house close to one-sixth of the world's population,[4] was itself 'a direct result of liberalisation' dating from the 1980s and enforced by the IMF and World Bank (quoted in Davis 2004: 23).

Government now argued that GEAR was always intended to create the basis for a more expansionary programme of intervention.[5] Yet the programme outlined in ASGISA clearly marked a radical shift from GEAR in respect of the role of the state in development. GEAR explicitly relied on the market to drive growth and create jobs once the economic fundamentals were put in place and it assumed that state investments would crowd out private-sector investments. If the story had gone according to script, ASGISA would be superfluous. Instead, ASGISA made the opposite assumption, that public investment would crowd in private investment and the state would lead the drive for growth. In short, it is an expansionary programme where GEAR was not. The shift is most obvious in energy policy described in Chapter 8.

Nevertheless, government had considerable justification in claiming that GEAR still lived. If the role of the state changed, government's imagination of development did not: modernising development – catching up with the industrialised nations – would be achieved through an export-oriented economy driven by international competitiveness and by infrastructure investments intended to create a competitive advantage for South African industry. These investments, now in progress, are capital-intensive and create significant numbers of jobs only during construction. The creation of a black capitalist class through BEE remained at the centre of government's project for transformation and was promoted with increasing vigour through BEE charters for most economic and industrial sectors. The charters are intended to ensure that business does not renege on the transitional deal, but also have the effect of intensifying the networking between the new political elite and the old economic elite.

Indeed, these interventions were to give greater depth to that imagination, taking it into the details of economic life and attempting to recreate people's subjectivities – their sense of themselves in the world – in its own image. The first-economy interventions were to address the constraints to international competitiveness and accepted that it is defined by global capital and the major powers. Box 3.1 looks at the implications. Industrial policy remained fixated on 'knowledge intensity' and finding a place in the global production networks presided over by the leading transnational corporations who determine the terms of production and access to markets. By this means, government aimed to move South Africa up the value chain, to higher, value-added production and therefore higher GDP growth, but it also signalled that South Africa 'knows its place' in the global order, just as it expects the denizens of the second economy to know their place.

Box 3.1   Competitive regulation

The constraints on international competitiveness are not just about technologies of production. ASGISA mandated 'a system of regulatory impact analysis [which] will add well-designed procedures (first developed in the United Kingdom) to reduce or eliminate the negative unintended consequences of laws and regulations, especially on job creation' (SAG 2006: 12).

It expresses particular concern about local and provincial planning and zoning and the Environmental Impact Assessment (EIA) system. Revised regulations 'streamlining' EIAs were introduced in 2005 but government ministers kept up a verbal barrage to the effect that they were holding up development. Delays were attributed to the constraints of bureaucratic capacity and to 'special interest' groups – read local activist groups concerned about the impact of proposed development on their communities. What was not, and is not, acknowledged is that business itself contributes significantly to delays. Brent Johnson observed that environmental and social impacts are peripheral to the concerns of most project developers who 'often fail to plan or provide adequate resources' to address them.[6] Even as ministers reiterated their commitment to environmental integrity, the discourse of development – the framing that defines the issues – made the environment, not to mention environmental justice, a point of resistance in the heroic narrative of accelerated growth.

ASGISA also called for 'a review of labour laws, including their impact on small businesses' (10). The phrasing suggested that the legal endorsement of a dual labour market, consistent with a dual development strategy, was under consideration. It also suggested that 'labour market rigidities' in the first economy would be subject to hostile scrutiny. These were not novel proposals but confirmed that the 'developmental state' was indeed the child of GEAR. The unions once more rallied in opposition.

> This regulatory agenda confirmed the subordination of the 'developmental state' to global capital. French legal scholar Alain Supiot remarks on the influence of the World Bank's annual *Doing Business* reports which 'provide a systematic evaluation of every feature of national legal systems that have a bearing on economic efficiency' (2006: 115). They provide a supposedly objective benchmark against which international investors and governments can measure competitiveness – or profitability. In respect of labour regulations, 'a "rigidity of employment" index penalizes countries that recognise too many workers' rights: social insurance for part-time employees; excessive minimum wages ($20 a month is deemed too high for an African worker); a working week limited to under 66 hours; the requirement to give third parties (for example, a union) notice of dismissal; programmes to fight racial or sexual discrimination' (116). *Doing Business* is both a symptom and an instrument of a global economic system in which 'it is no longer products that are in competition but the normative [regulatory] systems'. The obvious 'consequence is a race to the bottom in fiscal, social and environmental deregulation' (119).

President Mbeki was ousted by his own party just as the effects of the global financial meltdown hit South Africa. While government denied that the economy was in recession,[7] corporate capital cut 20 000 formal jobs in mining and manufacturing in the fourth quarter of 2008. Over a million more jobs followed in 2009. In February, Manuel finally admitted that the South African economy would not escape the storm but claimed that the wisdom of past policy had created the resilience to weather it. Government's infrastructure programme, along with the 2010 football World Cup spend, was then repackaged as a 'countercyclical' stimulus to the economy. Hot money meanwhile flooded out and then back into the economy – more or less tracking the volatile fortunes of commodities – as global capital looked either for safe havens or piratical profits.

President Jacob Zuma's administration took office in May 2009. The wholesale restructuring and enlargement of cabinet portfolios appeared as much about rewarding those whose loyalty sustained Zuma through numerous court appearances on corruption charges as about executive effectiveness. Policy hung in suspension between the conflicting interests of his supporters: the 'left' alliance partners; a raft of business opportunists; and the moderates representing the interests of global capital.[8] ASGISA is rarely invoked but policy is more or less defined through the revision of its elements. Putting snigger marks on 'the second economy', the ASGISA Annual Report 2008[9] rejects the dual economy paradigm, ditches the fantasy of poor entrepreneurs innovating their way out of poverty, recognises social spending as an instrument of economic policy in response to inequality, and calls for an expansion of public employment. In practical terms, following Zuma's promise of 500 000 'work opportunities', the EPWP is once more expanded but is still a stopgap that does not address long-term structural unemployment. The countercyclical initiative as a whole remains framed by the market and 'decent work' – adopted as a central plank of the alliance platform but dependent on an unlikely expansion of a formal labour market with little history of decency. The infrastructure spend meanwhile ran into funding limits while the big 2010 construction jobs were being done. The future of the new stadiums is now in question as the municipalities are confronted with unpayable maintenance bills. It may be that the demolition jobs will provide some relief to construction workers.

ENVIRONMENTAL POLICY

Whether left or right, the participants in the transitional debates on economic development ignored the environment. GEAR did not mention it and the 2002 Integrated Manufacturing Strategy mentions it once and then only in reference to the King Commission[10] on corporate governance and its recommendation for 'triple bottom line' reporting (DTI 2002: 19). This purportedly gives the social and environmental practices equal weight with profits in corporate accounting. This implied that environmental management be left to the self-regulation of the market and/or that the Department of Trade and Industry (DTI) would leave the environmental consequences of industrial development to the poorly resourced Department of Environmental Affairs and Tourism (DEAT).

South Africa's Constitution, as noted in Chapter 2, provides contradictory mandates for development in the Environment Right and the Property Right. While economic and industrial policy took its cue from the latter and barely mentioned the environment, environmental policy initially took its cue from the former. At the same time that GEAR was peremptorily announced, government initiated the Consultative National Environmental Policy Process (CONNEPP) widely seen as a highpoint of open and participatory governance.

CONNEPP was fiercely contested. Under apartheid industry dominated the DEAT and enjoyed a virtual monopoly on policy inputs. As business repositioned itself in the early 1990s, two powerful lobbies were formed to represent industry's environmental interests: the Industrial Environment Forum and the Chemical and Allied Industries Association. They called for self-regulation in the place of apartheid's 'command and control' approach. Neither command nor control were much in evidence but the phrase was intended to associate apartheid with authoritarian socialism and contrast both with free-market capitalism. They thus echoed capitalism's triumphant bray following the collapse of the Soviet Union.

This agenda reflected that of international business which was repositioning itself as 'part of the solution' to the growing environmental crisis. The self-regulatory manifesto was taken into the 1992 United Nations Conference on Environment and Development at Rio de Janeiro by the World Business Council for Sustainable Development (WBCSD) that was formed for the purpose. It subsequently spawned offshoots such as Business Action for Sustainable Development created largely to co-ordinate business interests at the WSSD in Johannesburg ten years later.

Industry's environmental agenda was confronted by an alliance of trade unions and local community and environmental groups who found common cause in the idea of environmental justice. Initially under the banner of 'green politics', the environmental justice movement emerged in the early 1990s. It saw the environmental destruction of apartheid in explicitly political terms and challenged the dominant view that reduced the environment to wildlife conservation.

It also responded to the peripheral place of the environment within the imagination of liberation. For many black people, the environment was associated with conservation and conservation with forced removals. It was a middle-class white concern that put animals before (black) people and 'not relevant to the urgent needs of the country for development and social justice' (Whyte 1995: xviii). Nevertheless, many of the demands articulated during the 1980s responded to environmental injustice: unions demanded health and safety at work; civics demanded water, energy and waste services; and everyone demanded the total transformation of South Africa's spatial regime – an end to pass laws and urban influx controls and comprehensive redistribution of land. So in many ways, the struggle against apartheid was implicitly also an environmental struggle as was first recognised by the National Environmental Awareness Campaign, founded in Soweto in the aftermath of the June 1976 uprising.

The environmental justice agenda was given visceral meaning through a series of struggles around waste and air pollution.[11] Key campaigns related to the illegal dumping of toxic waste on open ground, invariably in poor black neighbourhoods, the appalling state of formal dumps, waste incineration and trading in toxic waste.

The Thor campaign, co-ordinated by environmental group Earthlife Africa, was emblematic. Thor Chemicals, a British transnational corporation, traded in toxic mercury waste from the US and used an incinerator to separate out the mercury for 'recycling' at its plant at Cato Ridge outside Durban. It had a similar plant in Britain that was under investigation by British regulators. They found mercury in the air up to twenty times the legal limit and, in 1987, told Thor to clean up or face court action. Rather than do either, it closed the British plant but expanded at Cato Ridge where it could draw on a pool of cheap labour from the impoverished Inchanga area. The consequences were stark: four workers are known to have died from mercury poisoning; many more suffered chronic poisoning; the site itself was saturated with mercury; a stream used by local people was heavily contaminated; and mercury emissions into the air were unknown because they were not measured.

While apartheid South Africa liked to represent itself as a First World state, the case highlighted its habitual collusion with industry and the Third World state of environmental regulation: Thor ignored air-pollution regulations and was not penalised by the DEAT; the Department of Water Affairs and Forestry (DWAF) did not act on evidence of water pollution until it became a public scandal; the Department of Manpower inspectorate failed to identify health and safety issues until forced to do so. Next, prosecutors were reluctantly forced to bring Thor to court but botched the case and created the impression that legal action against corporate interests would fail in South African courts. Under the new government, the Davis Commission was set up to enquire into the case and found that government shared responsibility with Thor for the disaster because of the failure of regulation.[12]

The Thor campaign also set the pattern for coalition campaigning. Earthlife made a point of engaging with organisations representing people directly affected by pollution as well as networking internationally for information and support. This principle was given organisational form at a conference organised by Earthlife in 1992. The conference emphasised the connection between relations of power and environmental degradation and aimed to connect South African civil society with international debates (Hallowes 1993). The concept of environmental justice, introduced by US activist Dana Alston, resonated with the experience of South African delegates. News received during the conference that an Italian corporation was dumping toxic waste in war-torn Somalia demonstrated its pertinence and urgency. Delegates adopted environmental justice as the core idea capable of linking disparate struggles – struggles for land, housing and services, and struggles against pollution, dispossession and exclusion – to a common movement. They also mandated the formation of the Environmental Justice Networking Forum (EJNF) to give shape to the movement. EJNF grew rapidly to provide a shared forum for over 600 organisations representing workers, local community groups, religious bodies and women's groups as well as non-governmental organisations (NGOs) active in a range of sectors. It took up the Thor campaign amongst others and led civil society participation in the debates that then seemed to promise a wholesale transformation of policy.

Spurred by controversy and bad publicity, the state began to fashion a response from the early 1990s. The Integrated Pollution Control (IPC) process, initiated in 1993 and jointly managed by DWAF and DEAT, reflected the interests of the state and corporate capital in minimising the costs of environmental management. It was criticised for its exclusion of labour and civil society, and even industry agreed that it was ill-conceived and ineffective. Rising environmental activism, linked to the prospects of a democratic government, appeared to be driving change. Earthlife adopted the waste hierarchy to reduce, re-use and recycle waste but pointedly omitted disposal. The ANC and its alliance partners meanwhile commissioned the International Mission on Environmental Policy, which adopted the framing of environmental justice (Whyte 1995).

In the context created by local struggles to close existing dumps and oppose new ones, the IPC process lost all credibility and was terminated. Government launched CONNEPP, which was to produce an environmental policy framework within which a coherent policy on pollution and waste could be developed from scratch and with the involvement of all sectors. Most of the principles that civil society had been calling for – including sustainable development, environmental justice, the waste hierarchy, the polluter pays principle and the precautionary principle – were adopted as policy and subsequently incorporated into the National Environmental Management Act (NEMA) of 1998.

These fine principles notwithstanding, NEMA was compromised. During CONNEPP, industry argued for 'environmental management co-operation agreements' (EMCAs), that is, for self-regulation dressed up as 'partnerships' with the state and made into law. EJNF and the unions countered that this was intended to avoid effective regulation and called for legally binding pollution standards with stiff penalties for non-compliance. They eventually agreed to EMCAs only on the condition that they would be additional to rigorous regulation based on enforceable environmental standards and would be designed to take industries beyond compliance. In return for this concession, NEMA would enable communities to take industries to court if they broke the law.

At the national conference closing CONNEPP in January 1997, the minister agreed that state regulation should precede EMCAs because industry could not be trusted to regulate itself. With the formal consultation process closed, however, industry went to work in the back-door lobbies. The first draft of the NEMA placed clear conditions on the development of EMCAs, including that all stakeholders should participate in their design, that they should include quantifiable targets with independent monitoring and that there should be penalties for non-compliance. These conditions were diluted or deleted in the final version. The agreed purpose of improving on standards was made optional. EMCAs would merely 'promote' compliance with the principles of the Act.

Moreover, NEMA had no teeth. Environmental organisations found that the provision that enabled them to take corporations to court had little meaning both because environmental cases are generally difficult to prove, and particularly so where there is no credible information, and because specific laws and legally binding standards for air emissions and other wastes were not developed.

Follow-up processes meant to establish the basis for implementing NEMA were first rushed and then stalled. Thus, policy on Integrated Pollution and Waste Management was to be implemented through a National Waste Management Strategy (NWMS) within the legal framework provided by NEMA. The NWMS was finally published two years late but without consultation. In theory, it marked a paradigm shift in the approach to waste management. Founded on the waste hierarchy, reducing waste generation is central to the stated objective. The objective, however, faded from the practical strategies: Waste prevention was altogether lost; minimisation through 'cleaner production' was flagged but with no real means of implementation; incineration, particularly for hazardous waste, was retained as a disposal option despite the fact that it contradicts reduction; recycling was reduced to a symbolic cipher without adequate funding or any requirement for producers to use recycled materials and so create a viable market. Further, while NWMS prioritised development of a waste information system, it did not require industries to report on waste. It thus created no credible basis for waste information but rather sustained the wilful ignorance that allows producers to disregard their own waste and the externalised costs.

In short, the NWMS had no purchase on the production system presided over by the 'senior' departments. For government as a whole, the environment was scarcely a priority. The intergovernmental Committee for Environmental Co-ordination, established in law by NEMA and chaired by the very junior DEAT, was all but ignored by senior departments. Meanwhile, environmental budgets were squeezed, waste management was abandoned without adequate resources in the flood of waste, and environmental regulation for air and water, such as it was, collapsed. There were just four air pollution control officers left by 2002, down from five in 1994 and seven in the 1980s. In KwaZulu-Natal, the regional DEAT office was abandoned without so much as a forwarding address. Industries required to submit reports in terms of their scheduled process permits had them returned unopened. The environment was left to the mercy of the market constrained only by the resistance of local environmental groups.

Bad air

Local-level environmental struggles were most intense in south Durban. Under the 1965 Air Pollution Prevention Act, and protected from public scrutiny by the Key Points Act, apartheid industry was effectively self-regulating. All industrial centres were, and still are, pollution hot spots. Situated next to Africa's busiest port, south Durban houses some 600 industries including the two largest oil refineries and Mondi's large paper mill. Until 1998, the Engen refinery's permit allowed it to emit 72 tonnes of sulphur dioxide a day while Sapref, jointly owned by Shell and BP, could emit 50 tonnes a day. Both said that they operated well below these limits and the combined emissions from south Durban's top eight polluters was put at over 95 tonnes a day.[13] However, Sapref had under-reported emissions for five years and was well over its limit so the combined emissions were actually around 107 tonnes a day – assuming no one else was under-reporting.

It was the bad air that brought the communities of south Durban together. They have a long history of environmental concern and the transition to democracy enabled its more robust articulation. In 1993, a group of community organisations and NGOs formed the South Durban Environmental Forum to co-ordinate civil society action on air pollution. This forum was the forerunner to the South Durban Community Environmental Alliance (SDCEA) that was constituted in 1997.

Engen, meanwhile, anticipated that the democratic transition would create the need for a 'policy of transparency'. With President Nelson Mandela due to visit the plant in 1995, management told community representatives that they wished to announce the formation of a Community Awareness and Emergency Response (CAER) committee ahead of his visit. CAER is central to the 'Responsible Care' initiative administered by the International Council of Chemical Associations. It was developed by the industry following the Bhopal catastrophe to 'build trust' with people neighbouring chemical plants. CAER gives industry a tool for managing local participation and shifting responsibility to communities while not taking on any obligations.

The local organisations rejected this approach and proposed instead a detailed 'Good Neighbour Agreement' modelled on community-initiated agreements with companies in the USA. This would commit Engen to reporting impacts, and to workplace and environmental planning including pollution reduction, improved emergency planning and active affirmative action. The company declined the agreement and members of the Wentworth community demonstrated at the plant during Mandela's visit. Mandela stopped to talk to them and subsequently called community leaders to a meeting with cabinet ministers and Engen executives where the latter pledged that they would address the problems of pollution.

Talks on a Good Neighbour Agreement resumed but deadlocked for two reasons. First, Engen attempted to intervene in the selection of community representatives and, second, it argued that its emissions were within the legal limits set by its DEAT permit, and that health impacts were unproven. Wiley, Root and Peek (2002) argue that the company would have found reason for holding out on pollution reduction both in government's patent lack of interest in pollution control and in the pro-corporate context created by GEAR. But community activism, now co-ordinated by SDCEA, intensified and resulted in bad publicity while civil society demands for national air quality standards were gathering momentum at CONNEPP. In 1997, Engen re-opened negotiations on an agreement.

The Environmental Improvement Plan agreed by Engen and SDCEA depended heavily on switching from heavy furnace oil – with a high sulphur content – to gas to fuel the refinery. As it happened, this fitted well with government's agenda. Sasol was planning to pipe gas from its Secunda plant to Durban and this linked with the development of Mozambican natural gas, a project to which government was committed in the name of regional development. Further, the DTI was shortly to promote the development of a 'world-class chemicals cluster' in south Durban. Gas then appeared as the means of legitimizing a new round of industrial expansion within the discourse of ecological modernisation. Whatever undeclared agenda may have lurked behind Engen's decision, SDCEA was effectively put in the role of surrogate regulator as the DEAT abandoned its responsibility.

Contested knowledge

As the energy around CONNEPP dissipated, EJNF lost focus. For six years, the forum had co-ordinated the emergence of a remarkably vibrant environmental justice movement linking disparate struggles over a very wide range of issues including land and labour, municipal services and waste, and air pollution and climate change. In 1998, however, internal tensions opened into painful divisions and the organisation suffered a collapse of capacity.

groundWork emerged from the fallout to focus on working with local activists mobilising against industrial pollution and challenging the power of large corporations. It introduced the 'bucket brigade' to the refinery fenceline organisations with electrifying effect. Using a low-tech air sampling method developed by Communities for a Better Environment on the fencelines of US refineries, the bucket brigade showed a cocktail of chemicals in the air at all sites: sixteen compounds on the official US list of hazardous air pollutants in Sasolburg; fourteen on the fenceline of the Chevron refinery in Cape Town; and nine at Engen in south Durban. Readings for benzene were extraordinarily high at all sites. The documented health effects of these chemicals show harm to all the bio-physical systems of the body as well as to individual organs. Many are carcinogenic. Many cause death following high-level exposure and many eventuate in death following prolonged exposure at lower levels.

Under the Air Pollution Prevention Act, air pollution control officers negotiated permits in secret with industry and relied on what industry told them. They were also concerned exclusively with sulphur dioxide emissions. They produced no credible information on pollution but both industry and the regulators used this lack of information to dismiss the concerns of neighbour communities as uninformed. Industry, left to monitor its own emissions, represented itself as the only reliable source of knowledge but ensured that its knowledge was untroubled by some very basic questions. The bucket brigade upset this purposeful ignorance.

Sasol immediately contested both the findings and the bucket method. It commissioned its own sampling programme, undertaken by the South African Regional Science Initiative (SAFARI) in 2000 and Leeds University, but the results confirmed the bucket findings and hence also the credibility of the method. The campaign thus discredited industry claims to superior scientific information along with its assurances that it could be trusted to monitor its own emissions. It also discredited government's reliance on industry figures and exposed the paucity of official information.

Meanwhile, a string of incidents across the country and a series of articles in a local newspaper on health impacts in south Durban provoked an intense public reaction.[14] In April 2000, groundWork and SDCEA organised a mass protest in the city centre on Earthday. Calling for a clean energy future, the protestors asserted that 'clean energy is our constitutional right' and called on the city authorities to act on pollution in south Durban. The march was notable for the number of children participating. Many had been gassed out by toxic blowouts from local plants. An 'irregular discharge' of sulphur dioxide from Sappi-Saiccor's Umkomaas plant engulfed the Naidoo Memorial School. On three successive occasions, chlorine clouds blew across the Strelitzia School from Sasol's polymer plant at AECI's Umbogintwini complex. At Settlers School, situated between the two refineries, children had difficulty breathing even when the plants were operating 'normally'.

Civil society's perception that government had lost – or abandoned – regulatory control of polluting industries was fast becoming the public perception. In December 2000, the national and provincial ministers responsible for the environment, for health and for industrial development, together with the Durban city mayor, finally responded. At a stakeholder meeting they jointly announced a 'multi-point plan' for environmental management to be piloted in Durban. The plan included: setting national sulphur dioxide standards; introducing new legislation to replace the moribund Air Pollution Prevention Act; banning the use of dirty fuel by industry in south Durban; improving air pollution monitoring; identifying and minimising fugitive emissions; and assessing community health impacts. This reflected, point by point, the demands of civil society. At the same time, regulatory authority was to be devolved to the municipal level.

Local industry mobilised against the ban on dirty fuels, saying it would put them at a competitive disadvantage and deter local investment, and government ditched this promise almost immediately. Much of the rest of the multi-point plan has since been implemented albeit unevenly, slowly and often grudgingly. Under sustained pressure from SDCEA and in the face of considerable resistance from the refineries, the Durban regulator opened up the once secret permit process to public scrutiny and implemented a more credible monitoring regime for ambient air quality. At the same time, the implications of devolution were stark. Outside the metropolitan areas, local government capacity varies from minimal to zero. Even in metropolitan areas, effective regulation depends on the relation of power between local organisations, industry and local government.[15] This lent urgency to the demand for effective national legislation backed by stringent standards.

While government outsourced much of its environmental responsibility to civil society, it simultaneously disarmed them. Participation was increasingly confined to forums where the outcome was predetermined or could be ignored while the 'partnership' of government and corporations appeared focused on outlawing dissent. In 2003, clauses in NEMA guaranteeing rights of access to environmental information were removed and access made subject to the more restrictive Promotion of Access to Information Act. Noting that information was critical to environmental struggles against polluting industry, groundWork observed that 'industry and government [are] working hand in hand to ensure that environmental information is kept away from the very people that are living on the fenceline of polluting industrial development'.[16] The Key Points Act was revived by the Ministry of Defence, which told south Durban industries that environmental information should be treated as 'extremely sensitive'. The Mondi paper mill managers saw the point. They sought to restrain SDCEA from publicising information on worker injury and death at the plant and threatened to use Mondi's influence with the media to block stories highlighting its pollution. Steel giant Iscor similarly sought a gagging order against community members who had taken legal action in an attempt to hold it to account for the pollution that destroyed smallholder farming in Steel Valley. And, while the DEAT was simply not producing credible information, both government and industry worked to discredit those who claimed that their health was affected by pollution.

Health impacts

The Air Quality Bill presented to parliament in February 2004 did not recognise the protection of people's health as an objective. For communities on the fenceline, the relationship between health and pollution is central and civil society representatives vigorously objected to the omission, pointing out that the Environment Right in the Constitution emphasises health. The final version of the Bill draws its objectives directly from the Environment Right.

The original omission, whether intentionally or not, appeared to play to a corporate agenda that works to dissociate health and industrial pollution on the grounds of 'scientific uncertainty'. Scientific certainty is in fact the twin of wilful ignorance. As industry uses it, certainty must be absolute: the link between pollution and ill-health must be demonstrated in each case. Medical studies on the causes of ill-health, however, work on the basis of statistical probabilities and are not compatible with absolute certainty. Industry thus demands a standard of proof that it knows is impossible. The strategy is to invalidate statements linking pollution and ill-health and so exclude them from public debate and make the relationship invisible. It puts the onus of proving harm on to those who suffer it and simultaneously raises the costs of doing so.

The relationship is also made invisible because the Department of Health does not collect relevant health statistics. Nevertheless, groundWork was able to access clinic records for its 2003 report on air pollution. In Sasolburg, it found that 'respiratory illnesses can account for up to 40% of all illnesses treated at the clinics' (2003: 27). Subsequent to this, clinic records were withdrawn from local scrutiny as the Ministry of Health decided 'that one cannot access information on health without pre-approval from the provincial offices' (Fiil-Flynn and Naidoo 2004: 20).

Fenceline activists see the impact on people's health every day and the Vaal Environmental Justice Alliance (VEJA) participants note that independent doctors regularly tell patients that they will not get better unless they leave the area. Their struggle is against official silence and the wilful ignorance that serves to frustrate their core demands that industry must clean up and compensate those it has harmed. It is a struggle to have what they know substantiated by medical science so that it can no longer be excluded from public debate and ignored. Their belief that the health impacts are pervasive is in fact supported by a massive international literature, which shows that exposure to specific pollutants results in specific health effects. It is also widely acknowledged that the cumulative effect of exposure to many pollutants is probably greater than the sum of impacts from individual pollutants but this is not well studied due to the limitations of scientific procedure.

Faced with denial of health effects from industry and government, SDCEA campaigned over several years for a health study to corroborate what they knew to be a heavy toll of death and disease from pollution. In 2000, government finally agreed. An initial study focused on respiratory effects at a local school situated between the two Durban refineries. After numerous delays in getting started, a larger study compared south Durban with other sites in eThekwini, including Warwick next to the major city traffic intersection, and sites that are remote from industry and traffic. The *South Durban Health Study* (Naidoo et al. 2006) corroborated local people's perceptions and its recommendations echoed their demands. It comes in two sections: a health risk assessment based on monitoring people's exposure to air pollutants; and an epidemiological study that looked at the actual status of people's respiratory health by examining children at selected schools and their parents.

The risk assessment concludes that the number of cancers caused by pollution will be very high in south Durban compared both with the other sites and with figures from studies in other places in the world. It estimates 25 cancer cases for every 100 000 people, which is 250 times the accepted norm, but notes that this figure is conservative for three reasons: air monitors were not necessarily located at hot spots; estimates of people's exposure were based on averages for pollutant concentrations and actual exposure may have been much higher; the amount that people actually breathe in was conservatively estimated. Real cancer rates in south Durban are thus likely to be higher than the study estimates and may be much higher. The study identifies benzene, naphthalene, and dioxins and furans as the pollutants responsible for most of the cancers, but ethyl benzene, polychlorinated biphenyls (PCBs) and styrene and the metal particulates of nickel and chromium also contributed to cancers.

The epidemiological study finds that respiratory ailments in south Durban are high by comparison with other sites. In particular, it notes that previous exposure increased people's vulnerability. Further, relatively modest increases in pollution affected vulnerable children. The key pollutants linked to respiratory ailments were sulphur dioxide, nitrogen dioxide, nitrogen oxide and particulates.

The study recommends tighter regulation with further improvements to the air quality monitoring system and of emission controls. For the 'conventional pollutants', it notes that sulphur dioxide emissions have been reduced from historically very high levels but further reductions are needed, particularly if there is going to be more development. As it is, daily and ten-minute limits are frequently exceeded by pollution peaks although the annual limit is met.[17] Particulates (PM_{10}) and nitrogen dioxide exceed the annual limit and very frequently exceed short-term limits. Limits for fine particulates ($PM_{2.5}$) – which penetrate deeper into the lungs – have not been set but limited monitoring specifically for the study showed that concentrations exceeded the World Health Organisation annual guideline and regularly exceeded its daily guideline.[18] It recommends that a 'strategy and timeframe for attaining compliance with standards, guidelines and targets should be developed for each pollutant' (Naidoo et al. 2006: 202). For other pollutants, it recommends expanded monitoring for a wider range of volatile organic compounds (VOCs)[19] and for metals, dioxins and furans, and total reduced sulphur compounds. The sources of all these pollutants should be tracked down and rigorous emission controls introduced. Finally, it calls for a programme of asthma education in Durban communities and for better health information, including long-term monitoring of respiratory diseases and setting up a cancer registry.

Framing the law

Although the environment still ranked at the bottom of government's priorities, the DEAT budget for environmental management, supplemented by donor funding, expanded. In 2003, the DEAT started setting up the Environmental Management Inspectorate – or Green Scorpions – and announced its presence by stinging the operators of an illegal toxic dump. The inspectorate was formally established in 2005 and is working, according to the minister, 'to change the common perception in South Africa that government lacks the will to enforce our environmental legislation'.[20]

The DEAT also started actively preparing long-promised laws. It finally brought the Air Quality Bill to parliament in 2004 while the Waste Bill started going through the parliamentary process in 2007 – seven years later than originally promised. Civil society participation in these legislative processes was generally restricted to formal parliamentary hearings required under the Constitution. Business, on the other hand, appeared to have prior access during the drafting stage.[21] Fenceline groups put together joint submissions arguing that people's right to an environment not harmful to their health should be written in as the primary purpose of the Bill; it should focus on reducing pollution at source and should therefore mandate emissions standards rather than rely exclusively on ambient air regulation; and that various measures should be made mandatory rather than left to the discretion of the minister. The department accused them of delaying the urgently needed Bill but they argued that, having waited ten years for it, they would wait a few months more rather than accept a flawed Bill. The amended Bill included several significant changes demanded by communities and was passed in August 2004.

National environmental laws take the form of framework legislation, outlining broad objectives and either requiring or allowing the minister to promulgate more specific regulations required for implementation. As with the laws themselves, these processes are contested and vulnerable to delay, intentional or not. Thus, ambient air quality standards required by the Air Quality Act were developed four years after the Act was signed into law and nine years after they were promised. Industry was particularly resistant to meaningful emission standards. Through the long process from the 2000 multi-point plan announcement, fenceline groups kept them on the table, insisting on their inclusion in the Air Quality Act and refusing to allow their neglect in the standards-setting process. They were finally promulgated in 2010 but, for most industrial processes, regulate only for the conventional pollutants and not the more exotic substances noted by the Durban health study.

The Waste Bill followed in 2008. Civil society contestation focused particularly on incineration and the DEAT's refusal to recognise waste-pickers. The DEAT then rallied industry in defence of incineration. Ahead of the parliamentary hearings, it flew officials to all the cement kilns lining up to burn toxic waste to consult with management but not with neighbouring communities. This disregard was highlighted during the hearings and it belatedly visited three communities. The Waste Act codifies the current state of play in struggles around waste initiated in the late 1980s. Understood this way, it makes sobering reading: waste is generally to be defined by the market; the waste hierarchy is invoked but effectively inverted to prioritise disposal over avoidance and minimisation; incineration is permitted although, in a nod to civil society objections, subject to parliamentary approval rather than ministerial fiat; and toxic waste trading, so long resisted through successive campaigns, is regulated in order to allow it. Over the objections of the department, the Act does recognise waste-pickers and so creates a political opening for them to engage with the official processes that define their work.

More broadly, the DTI's view of the place of environmental management in the developmental state was confirmed by DEAT officials at the parliamentary hearings: 'The bottom line [is] that South Africa [is] a country in need of economic growth and development. DEAT [is] thus trying to manage the negative effects of dealing with waste'.[22]

PAINT IT GREEN

Despite having a record of supporting dirty projects, the DTI's endorsement of 'triple bottom-line' reporting, together with its promotion of knowledge intensity, gestures towards an assumption that a new round of globally integrated development is accompanied by cleaner production driven by the market. In the last decade South Africa's major corporations have shown increasing concern over the representation of their environmental and social records. They have introduced sustainability reports and signed up with international organisations such as the Global Reporting Initiative and the WBCSD, which promote triple bottom-line reporting and proclaim a new age of clean and socially responsive development founded on networked production (WBCSD 2010).

Networked production has its origin in the East Asian economies and Japan in particular. It introduced a range of innovations to cut production costs such as 'just in time' delivery of inputs and 'total quality management' aimed at 'zero defect' in goods produced. The concept of zero waste, according to industrial economist Robin Murray, is an extension of zero defect and derives, on the one hand, from the pressures exerted by the environmental movement and, on the other, from 'the world of industry and its rethinking of production' (2002: 19). Zero waste, he argues, is central to a new 'wave of industrial development . . . centred on electronics' and 'marked not so much by a new material . . . as by the pressure to reduce materials and their toxicity. . . . We live in an age [that] speaks of "dematerialisation", of finding ways of avoiding production, of making more with less' (69). And he goes on to applaud the WBCSD's leadership in promoting 'eco-efficiency'.

Murray emphasises the role of social movements and government regulation in ushering in a new paradigm of 'post-industrial' production with design for inbuilt re-use, upgrading and recycling, etcetera. But finally it is corporate capital that leads this wave of development and shapes the new world of clean production. 'In the words of Edgar Woolard Jr, former chairman of DuPont, "The goal is zero: zero accidents, zero waste, zero emissions"' (71).

This representation of green capitalism, in a book written for Greenpeace, could not be further from the experience of actual networked production and it fundamentally mistakes the nature of capital. First, the new wave of development has been accompanied by a new wave of waste precisely from the cutting-edge sector of electronics, as shown in Box 3.2.

Box 3.2   The most (post)modern waste

A growing and toxic electronic waste stream flows from the so called 'post-industrial' and 'resource-light' economies of the North. The major components of e-waste are discarded personal and mainframe computers, printers, copiers, faxes, cell phones, telephones, televisions and high-end telephonic equipment. In Europe it is growing three to five times faster than municipal waste as a whole. In the US, where around half of all households own a personal computer, the Environmental Protection Agency estimated in 2001 that e-waste in US landfills would grow four-fold.

This rapid growth results from the purposeful design of inbuilt obsolescence. From the 1950s, as Annie Leonard observes, industrial design journals 'actually discuss how fast [designers] can make stuff break and still leave the consumer with enough faith in the product to go buy another one' (2008). Electronics take obsolescence to new heights. Rapid technology change is part of the arsenal. Computers are made to become incompatible with evolving information and communication technology systems. They could be designed for upgrading but, says Leonard, the 'piece that changes' is given a different shape so it won't fit and 'you gotta chuck the whole thing and buy a new one' (2008).

E-waste is toxic, yet most of it enters the municipal waste stream. In 2001, e-waste was reported to be the source of 70% of the heavy metals in US landfills, including mercury, cadmium and hexavalent chromium (the biologically absorbable form of chromium). Computer monitor screen glass contains lead to stop radioactive gamma rays from the display cathode from reaching user's eyes. This contributes 40% of the lead now in US landfills. Computers also contain polyvinyl chloride (PVC), which generates dioxins and furans during production and disposal by incineration, and other toxic compounds.

> By 1999, only 11% of discarded computers in the US were recycled. The task of recycling is dangerous to workers' health, especially in informal or semi-formal conditions. E-waste is moved to the South as 'donations', much as expired pesticides are 'donated', where they become toxic pollution sources. In May 2009, the Basel Action Network uncovered just such a scam. US company EarthECycle staged charity events to collect e-waste supposedly for recycling in the US but then exported it to Southern countries. One load was destined for Durban. groundWork alerted DEAT officials who took no action. The Green Scorpions proved more responsive. They eventually caught up with the container in Johannesburg and, two years later, were still trying to negotiate its return to the US.
> Sources: Pichtel 2005; Leonard 2008.

Second, global production networks have located the dirty end of the production chain in the global South, giving the North the appearance of clean production. This is an uneven process but, schematically, what has emerged is a triangular ordering of the global economy. Raw materials from Africa and Latin America are taken to the Asian factory to produce goods consumed in the North. This flow of resources is largely managed by Northern transnational corporations who also determine the technologies of production, control product development and allocate 'value' – or profits – through the network. The global concentration of control in the hands of transnationals is a striking feature of the global restructuring of production and this intensified following the financial meltdown (Nolan and Zhang 2010). Heavy pollution in China, and recent scandals involving the contamination of goods produced there, has as much to do with cost cutting imposed by Northern transnationals as with cowboy development in the wild East. Wolfgang Sachs observes that 'self-poisoning is the price [newly industrialising nations] have to pay for a greater share of value creation' while producers of raw materials, at the bottom of the industrial supply chain, face the wholesale destruction of their environments (Sachs and Santorius 2007: 66).

Third, the management of production networks is counted as 'services' rather than 'industry'. The transition from high-energy industrial to low-energy service economies is generally represented as inherent to the trajectory of development: where the (post)industrial developed world leads, the developing world will follow as they 'catch up'. But first they must pass through the stage of industrialisation. To the contrary, however, the service economies are possible only on the basis of the global structuring of production described above and they rely on the unequal global division of labour. This brings us to the fourth problem – the wasting of people. As described in the chapter on the Vaal, the world of work is increasingly unequal and divided into three major 'zones': the shrinking core of permanent workers; the growing 'non-core' of insecure casualised workers; and, in the 'peripheral zone', the vast pool of informal workers and unemployed people made surplus to the requirements of networked capital.

Fifth, the age of globally networked capital is integrally bound up with the neo-liberal policies given global force through the Washington Consensus. A critical aspect of this revolution from above was the financialisation of capital resulting from the crisis of over-accumulation. The series of financial crises devolved to Southern countries since the 1970s have yielded high returns to global capital that could appropriate assets at fire-sale prices. This was just one aspect of 'accumulation by dispossession' through which capital managed the spectacular transfer of wealth from poor to rich globally and within most countries, South and North. Finance is, of course, also a service sector and it is financialisation, rather than reduced materials intensity, that has 'dematerialised' economies. In South Africa, the finance sector now accounts for 20% of GDP but, as Fine (2008a) argues, this was not a contribution of 'value added' to the economy but rather the finance sector's appropriation of value from the economy.

The crisis was also passed on to workers and the environment as indicated in the first and third points above. Beyond e-waste, however, the WBCSD is very much part of the neo-liberal moment, promoting 'flexible business solutions' in opposition to mandatory regulation and precisely to deflate pressure for such regulation. Murray (2002) provides a seductive account of initiatives by this or that corporate. Many of the same corporations, however, operate by other standards in other parts of the world. And, as lead corporations in global production networks, their demands on subordinate firms ensure practices that are directly contrary to those advertised to consumers. This is the flexibility that corporations seek to protect and it is enhanced by corporate advocacy in other forums such as the World Trade Organisation. It is this advocacy that has subjected national regulatory systems to international competition. And it is this advocacy that creates Byzantine market responses such as carbon trading: after a great deal of mathematics, and profit, the carbon credits traded still have no relationship to actual carbon emissions.

4

The toxic cradle of production

CAPITALISM IS NOT ONLY a 'gigantic accumulation machine' (Kovel 2002: 59); it is also a gigantic waste creation machine. Its logic is to turn more and more raw materials and energy into sellable commodities, commodities into accumulated profit and profit into investments which then expand the system as a whole. Its restless need for never-ending accumulation and expansion means that it must keep on consuming resources and creating ever-growing wastes. Behind the product on the shop shelf lies the 'value chain' of production that is shadowed by a vast chain of waste and destruction. This shadow leaves a deep toxic stain that spreads through air, water and land across the face of the earth and across time into a poisoned future. This chapter focuses on minerals, looking at mining, the first link in the waste chain, and then at selected industries further along the chain: iron and steel, aluminium and cement.

THE SACRIFICE TO MINING

The post-war period in South Africa saw a rapid expansion of industry centred on the minerals-energy complex. The apartheid state's massive investments in Iscor, Sasol and Eskom enabled the vertical integration of production under the control of the corporations. This meant controlling the entire chain of production from raw material inputs to the marketing of products. More broadly, this process of industrialisation created the giant corporations, private and state-owned, necessary to manage vertical integration and concentrated economic power in their hands. By the 1980s, Anglo American and the state each controlled 25% by value of South Africa's top 50 corporations. 'The picture is essentially one of a relatively small economy with three main pillars: the state, the three insurance-based groups,[1] and Anglo' (Pallister, Stewart and Lepper 1987: 38). If anything, this understates the degree of concentration since the finance houses were themselves tied in with the big mining houses (Fine and Rustomjee 1996) and Anglo's influence extended well beyond its control of over half the capital value of listed companies.

Anglo had gained control of the De Beers diamond monopoly and emerged as the top gold-mining house during the 1930s. It also started diversifying into coal, base metals and industry. Following the war, it too established vertical control of production relating to mining – mining inputs and minerals and metal industries – as well as in other industrial sectors such as the auto industry, forestry, timber and paper, and agriculture, food and beverages. Anglo became Iscor's main partner in upstream and downstream businesses and also both the main supplier of coal to Eskom's power stations and the power utility's largest customer. The interests of the minerals-energy complex were thus consolidated through tight institutional relationships across the state and private sectors. They were also given representation within the state through the Department of Minerals and Energy (DME), which was dominated by the Chamber of Mines on the one side and Eskom on the other.

With the political transition of the 1990s, Anglo moved considerable assets out of the country so as to pre-empt the possibility of nationalisation. Thus, it transferred substantial shareholdings to Minorco, an established subsidiary registered in Luxembourg, which in turn invested heavily in international acquisitions, particularly in North America. At the same time, it began a process of unbundling, selling off non-strategic assets and using the process to initiate the first black-empowerment deals and so consolidate relationships with the new political elite.

The offshore listing in London followed in 1999. Anglo was then level pegging with Rio Tinto for the top spot as the world's largest mining corporation. Starting with the merger of BHP of Australia and Billiton – to which we return below – the 2000s have seen a major concentration of ownership at the global level. By 2005, Anglo had been knocked down to fourth place behind BHP Billiton, Rio Tinto and Brazilian corporation Vale.[2] It remains South Africa's largest mining company and massively influential in the broader economy.

Spoils

Mining is literally an extractive industry, clawing materials from the ground and generally impervious to the environment and people around the mines. Solid mining waste is rarely managed beyond being piled into heaps or dams next to where it has been excavated. For the most part, mining slurry has simply been dumped into rivers, lakes or the sea although mining engineers say this is changing:

> Historically the easiest and most economical solution was to discharge tailings slurry by gravity to the nearest body of water and let nature take

care of the problem. However, as communities and farming activities have encroached on mining areas, and fishing industries and interested individuals have applied pressure to government regulatory bodies, the need for properly engineered tailings disposal areas has become apparent (Robinson and Toland 1979: 782).

In fact, it was mostly the mines that 'encroached' on farming, fishing and communities, but the idea that the land was empty made its enclosure easy, especially if it belonged to indigenous people, North or South, without capitalist property rights. The term 'sacrifice area', reports mining activist and researcher Roger Moody,[3] was first officially attached to the Four Corners region of the US Midwest by the US Academy of Science in 1973, after it had been trashed by uranium, coal, oil and gas mining. In July 1979, a tailings dam in the area burst to release 1 100 tons[4] of milling waste and nearly 100 million gallons[5] of radioactive liquids into streams on Native American (Navajo) territory. According to Native American activist Winona La Duke 'at least one member of every Navajo family has likely died from lung cancer and other diseases resulting from uranium mining' (quoted by Moody 2007: 127).

In Papua New Guinea, Rio Tinto insisted on the right to dump wastes from its very lucrative Panguna mine in Bougainville into a nearby river and so provoked a civil war:

> By 1988 a few of the Panguna indigenous landowners, led by a former Rio Tinto mineworker, Francis Ona, demanded US$10 billion compensation for the ruination of their gardens, forest and waterways. The company jeered at the claim and refused to negotiate. Ona set up a nucleonic 'Bougainville Revolutionary Army', declaring independence from Papua New Guinea. Backed by Australian helicopter gunships, troops from the mainland invaded the island. In the bloody civil war that ensued up to a fifth of the island's population (between 15 000 and 20 000 villagers, many of them women and children) were to die before peace was reached in early 1998 (Moody 2007: 2).

Active mines pollute water in two ways according to a textbook on coalmining. First, water used for mining processes 'is often seriously polluted and cannot be returned directly to the hydrological cycle without prior treatment'. Second, 'a large volume of water . . . is casually affected' by surface run-off, acid mine drainage, pumped mine water and groundwater flows. 'It is not possible to apportion the damage among the "process" and "casual" categories, but the latter is probably the more important' (Down and Stocks 1977: 91).

Acid mine drainage results when sulphates in rock are exposed to oxygen, on mine dumps or underground, to produce sulphuric acid. The acid then dissolves and mobilises heavy metal toxins. Millions of litres are pumped from South Africa's mines daily and 'partially treated' with chemicals to neutralise the acid, but not the metal toxins, before being released into the surface water. If it is not pumped, it fills to the surface and decants untreated. The Johannesburg conurbation is now sitting on a rising tide of toxic water as the old mines fill. The first large-scale decant occurred on the West Rand in 2002. The water flowed out through mine shafts and boreholes and through springs that had dried up when the mines originally ruined the aquifers that fed them. That decant has been 'managed' but scarcely contained. The year 2010 opened with a new round of unmanageable decanting on the West Rand as several mining corporations stopped pumping. The Central and East Rand basins are also filling fast and will reach critical levels in the next year or so. The Witwatersrand is a major watershed, draining south and west to the Vaal and north and east to the Limpopo. Large-scale decanting will flow both ways and threatens the ruin of life over an immense stretch of the country. While the mining houses ducked for cover from liability, government ignored the problem for over a decade. In August 2010, it put together an inter-ministerial team chaired by Water and Environmental Affairs minister Buyelwa Sonjica who said they were looking for 'a cheap, effective and sustainable' solution.[6] Whether these aims can be reconciled remains to be seen.

Abandoned mines

Mining corporations arrive brazenly, but leave furtively when the profits dry up. In North and South alike, 'abandoned and ownerless mines' litter the landscape. In the US, there are half a million such mines (Moody 2007: 129ff.). In South Africa, the list is not complete but is estimated at 6 000. The mine owners simply abscond, or slip out of one corporate skin into another, taking their wealth with them but leaving toxic liabilities for others to clean up.

The Transvaal and Delagoa Bay coalmine near Emalahleni (formerly Witbank) tops the list of abandoned mines in South Africa. It operated from 1896 to 1953 but, more than half a century later, its waste is still producing an ongoing ecological catastrophe. Underground fires still smoulder, releasing sulphur dioxide, methane and carbon dioxide. Acid mine drainage seeps from various cracks and covers the area with sulphate salts that kill all vegetation they touch.

Near Maguqa, one of Emalahleni's townships, local children use a warm pool to swim in. It is filled with acid mine drainage water heated by the underground fires and likely to contain carcinogens, including benzene and toluene, which have been detected in the gases from the fires by Pone et al. (2007). The pond is one of a series constructed by the Department of Water Affairs and Forestry (DWAF) when it took responsibility for the abandoned mine. The ponds collect acid mine drainage water that is then supposed to be pumped to a DWAF treatment plant built in 1997. Although only ten years old, the plant was taken out of commission in late 2006 for want of staff and fairly minor repairs (Hobbs, Oelofse and Rascher 2008) and the acid mine drainage just ran into the Brugspruit which flows past Maguqa. In 2008, the sulphate salts were so thick on the water that the stream looked like it was snowed over. The toxic water then ran into the Olifants River, past fruit farmers and into the Loskop Dam. Over the past few years, officials at the Loskop Dam nature reserve have reported thousands of fish deaths as well as the deaths of crocodiles and water turtles.

The acid mine drainage degradation seems to have encouraged other factories – Highveld Steel, Vanchem, Samancor – to release their untreated waste water into these streams. The Emalahleni municipality similarly releases raw sewage into local streams and this too arrives in the Brugspruit. The stream is surrounded by townships in a busy valley. Children play in it, people cross it on their way to work, herders graze cattle and coal-pickers work over heaps of discarded coal.

Four mines in the Witbank area, belonging to AngloCoal and BHP Billiton, have constructed an acid mine water treatment plant. They show that acid mine drainage water can be treated but the price tag of R300 million deters hundreds of other coalmines. The externalised cost from untreated acid mine drainage is far greater. It is imposed on the environment and the people living there. Finally, the costs from working and abandoned mines are imposed on the public purse – except, of course, that DWAF also abandoned the responsibility. By the time the coal is mined out, both ground and surface water will be severely contaminated and 'the region could become a total wasteland', according to McCarthy and Pretorius (undated: 16). There is no plan to prevent this.

Gold's wasteland

In more than a century of mining, South African gold mines have covered an estimated 180 square kilometres under more than 200 tailings dams. These areas are now permanently contaminated. In 2001, Roesner et al. estimated that treating just the polluted topsoil (top 30 cm) would cost \$550 million. Mining waste is classified into rock and sand heaps and slimes dams. Slimes dams contain the silt and slurry together with the chemicals – arsenic, cyanide or mercury – used to extract gold from ore. The gold ore itself typically contains uranium and significant concentrations of chromium, copper, nickel, lead and zinc.

On the Far West Rand, gold miners physically destroyed a high-quality dolomitic aquifer and also contaminated it beyond recovery by dumping radioactive mining waste into it. The ore contains high concentrations of gold making the Far West Rand the richest of all seven active goldfields of the Witwatersrand basin. It also contains the highest concentrations of uranium and, when mining started in the early 1950s, 9 of the 22 mines produced uranium as well as gold. Between 1952 and 1988, they processed the uranium into more than 11 000 tonnes of yellow cake (U_3O_8).

But miners seeking a fortune here first had to conquer the aquifer that lay above the gold reefs. The aquifer consists of caverns weathered in the alkaline dolomite by the mild natural acidity of rainwater. A number of impermeable dykes divide the aquifer into a series of 'compartments'. These dykes also ensured that pressure within the aquifer forced the water up and out through a number of springs feeding into the Wonderfonteinspruit. The water was of high quality and much prized by early black farmers and by the white farmers who displaced them. When the miners arrived and created a local market for food, the Wonderfontein Valley became a prime area for irrigation production according to mining geologist Jan Wolmarans (1984).

Early attempts to sink shafts in the area were abandoned as the shafts flooded. When real mining started in the 1950s, the corporations pumped out water into existing irrigation channels, into overland pipes or down to the Wonderfonteinspruit. They thus dewatered the aquifer. The Wonderfontein springs started drying up from 1957 and the first sinkholes – resulting from the loss of pressure in the caverns – appeared in 1960 to much public alarm. This provoked an official inquiry by DWAF[7] and, on its recommendation, government decided to sacrifice farming and the aquifer to the interests of gold mining. Ever anxious to make someone else pay for the inconvenience of the aquifer, the mining corporations bickered about who was responsible for pumping and disposing the water, so prompting the state to regulatory action. In 1963 it made dewatering compulsory for all mines in the area, confirming the sacrifice of the aquifer in the interest of peace between the mining houses. Even so, the miners do not always win against the water. In 1968, the Wes-Driefontein mine was flooded.

The dewatering led to extensive damage to farms in the area. In 1964, the Far West Rand Dolomitic Water Association was formed. Behind its bland name, it was a cat's paw for the mining companies and each had to contribute according to the amount of water it was pumping out of its mines. The association's task was to receive public complaints, buy up the farms from the complainants and then rent them out again. As a result, the association now owns large stretches of land affected by sinkholes and, as the landlord, is in a powerful position to deal with complaints. By 1984, the area had 589 sinkholes, most of them caused by dewatering and other mining activities. They seriously damaged railway lines, roads, mining infrastructure and buildings, and people's homes. Some structures just disappeared into the dolomite caverns below.

The gold miners deliberately built large numbers of slimes dams on top of sinkholes. In mining terms, sinkholes add 'stability' to slimes dams by draining away fluid and so preventing a build-up of pressure with the potential to burst the walls (Robinson and Toland 1979). Slurry thus drains straight into the caverns of the aquifer that are then made into sumps for toxic waste. In some cases, miners attempted to plug sinkholes with mining waste. Predictably, in Wolmarans's view, it didn't work. The waste simply dropped down into the water of the aquifer.

That the waste is heavily contaminated with uranium has been known to a closed circle of miners, scientists and state officials for decades. With the political transition from apartheid to majority rule, argues water researcher Anthony Turton, the mining corporations' controlling grip on this group slipped. Some began to speak out and confirmed public suspicions that the Far West Rand aquifer was contaminated with radioactive uranium. 'It is this new generation of public domain literature that has given rise to the dilemma now confronting Government, because in essence, what it has shown is that there is a massive pollution plume downstream of gold-mining activities, consisting of a cocktail of heavy metals, sulphates and radionuclides' (2008: 3).

On the Far West Rand two local farmers, the Coetzee brothers Sas and Douw, decided to clean up their farm dam on the Wonderfonteinspruit during 2007. As soon as they removed the wall and exposed the sediment, a satellite picked up the radiation from uranium that had accumulated at the bottom of the dam and alerted the National Nuclear Regulator. The National Nuclear Regulator then instructed the Coetzees to repair the wall, never to drain the dam again, not to disturb the sediment, not to allow their cattle to drink there, and not to sell any produce from their farm as it might be contaminated. The Coetzees complied because 'we were brought up to believe that it is not right to knowingly harm someone'. But they are not happy to bear the cost while those responsible for the contamination, the owners of a nearby slimes dam from which the dun-coloured slurry water traces a clear trail to their dam, face no consequences.[8] The National Nuclear Regulator has since declared that the food from the area is safe to eat. Nevertheless, its study of the catchment (NNR 2007) confirms that significant amounts of uranium are entering the Wonderfonteinspruit, that uranium is concentrated in the rivers and sediments from where it can be mobilised, and that it poses a health risk to residents. It has not explained the contradiction.[9]

The mining companies now propose to remove and consolidate all the slimes dams into two mega slimes dams situated on granite rather than on dolomite. This move is opposed by the considerable public mobilisation against the mining waste, which has given birth to a new environmental alliance, the Federation for Sustainable Environments.

Box 4.1   Radioactive waste

The status of radioactive waste has been a closely guarded secret, both because of apartheid South Africa's nuclear weapons programme and the miners' direct interest in it. However, in 2004, an unusually frank audit of radioactive waste was put together by the DME.

The report estimated that there could be 5 000 million tonnes of gold-mine tailings containing uranium, and around 1 000 million tonnes of waste rock. About 25% of the uranium in mining waste had been extracted by 2000. Vast amounts of soil were also contaminated, along with buildings and materials used in uranium plants and mines. Up to 1993, when mines first became subject to regulation by the National Nuclear Regulator, contaminated mild steel scrap – an estimated 60 000 tonnes per year – was simply sold for recycling. More than 30 mines had been identified for decontamination, to be paid for by the gold-mining industry but, by 2000, only 8 were reported to have been cleaned up.

The lax approach of the mining industry, and its regulators, can be seen in the DME report's argument for mixing contaminated materials into existing mine dumps:

> . . . there are recognised benefits of reintroducing radioactive residues from uranium and acid plant maintenance/decommissioning into the milling and gold-uranium extraction process. Apart from the financial

benefits of recovering gold and uranium, the gradual reintroduction of this material into the process has the effect of returning the radionuclide concentrations back to their original values, i.e. to the levels prevailing in the original feed material to the plant. The reprocessing of these residues therefore avoids having to dispose of them separately (a potentially risky and expensive process if they are to remain at high activity concentrations). Instead, they simply end up as being an indistinguishable part of the tailings (DME 2004: 52).

Pelindaba, the nuclear research facility near the Hartbeespoort Dam, placed its wastes in an excavated hillside called Thabana. For this waste, 'complete records are not available', as the audit politely puts it. It was foreseen that all the Thabana trenches would eventually have to be excavated. The audit anticipated that decommissioning of buildings, stores and plants (including the Safari-1 reactor) would result in 13 000 cubic metres of waste, from a total volume of 150 000 cubic metres in contaminated materials. It gave no figures but expected this to be a costly process that would last between 20 and 30 years.

At the time of the report, Vaalputs in Namaqualand contained 7 371 cubic metres of low and intermediate-level waste, which is mostly material coincidentally contaminated by radioactivity or with uranium. Vaalputs is now being considered for the burial of high-level waste. Thus far, Koeberg nuclear power station has stored its high-level waste on site. This waste is composed of spent fuel assemblies and stored in racks under water. The racks are periodically repacked to cram in more waste. According to the report, by 1999 Eskom had provided R1 164 million for the management of the spent fuel and the eventual decommissioning of Koeberg.

There is still no plan for final disposal of high-level radioactive waste.

Platinum: More precious than people

Dispossessing people of their land while trashing their environments is by no means a relic of the history of colonialism and apartheid. The lives and livelihoods of thousands of rural people in Limpopo are being trashed right now by the mining activities of the world's largest platinum producer, AngloPlatinum,[10] reports ActionAid (2008). They have lost their land, which is now being physically removed by opencast mining or covered with mining waste. They have lost access to drinking water, now polluted and unfit for human consumption. They have lost their livelihoods and have not received adequate compensation. Their ancestral graves have been removed, injuring their spiritual connection with the land. And they have been excluded from decisions about their own future, as the mining giant established front organisations – fifteen different Section 21 companies – that signed agreements on their behalf accepting relocation. Their challenges to the AngloPlatinum land grab have been met with police brutality and corporate legal action.

The villagers are traditionally almost completely dependent on farming on communal land. Jobs are scarce and social services are minimal. Their other major source of income is from government grants – old-age pensions and children's allowances. Villagers in Ga-Pila, Potgietersrus, accuse the mine of cutting off their water and electricity to force them to move. Two water reservoirs disappeared under mining waste. The municipality did not reconnect or re-establish a water supply. Even where the land is not covered by waste, villagers are not allowed to plough because it is now 'mining property'.

The villagers live – or used to live – on the richest platinum resource in the world. The Bushveld Mineral Complex hosts 88% of the world's platinum and palladium. Platinum is used in catalytic converters for vehicles to reduce levels of carbon monoxide, hydrocarbons and nitrous oxides emissions to legislated levels. These catalysts, responsible for half the demand for the platinum minerals group, are mainly produced in Britain, Germany and Italy. Platinum is also used in the electrical, electronics and chemicals industries, for glass-making and as jewellery. AngloPlatinum, which made record profits of $1.75 billion in 2007, spends less than 1% of its profits on local community development but makes extravagant claims about its positive influence. The claims are at odds with what ActionAid found on the ground.

South African law does not protect these communities from exploitation, and discriminates against communal landowners. According to ActionAid's report:

> The Mineral and Petroleum Resources Development Act of 2002, is very permissive towards mining companies . . . The law requires mining companies only to consult with the community and report back on the outcome of those consultations to the government department responsible for mining – the Department of Minerals and Energy (DME) – before a mining right is issued by the minister. The permission of the community is

not required. The DME and the minister have no obligation to consult with the community affected and usually do not do so; they depend on the report given to them by the mining company, which the community has no right to see. Once a mining right is awarded to a company, the law does not require it to obtain permission from the occupiers or the owners of the land. Rather, the law expressly authorises the company to commence laying infrastructure and undertake mining on the land. Neither does the DME require written lease agreements to be concluded between the mine and the community. The negotiation and conclusion of a lease agreement is standard practice in relation to privately owned land (land owned by white people) but is the exception in relation to communal land (land generally used by black people) (2008: 12–13).

While the guardianship of the country's mineral resources is supposed to be vested in the state, mining and prospecting rights are allocated to corporations for free. Compensation is limited by the fact that the mining corporation's offer is usually the only one on the table, reflecting 'at most the agricultural value of the land, not a proportion of the value of precious metals or minerals in the ground' (2008: 13).

Campaigners for community rights want the Mineral and Petroleum Resources Royalty Bill and the draft Mineral and Petroleum Royalty Bill to be amended to ensure:

- communities have greater rights to be fully consulted and give informed consent before mining concessions are granted;
- the consultation process is supervised by the state or an independent, non-interested party delegated by the state and strictly governed by regulations;
- environmental assessments and safeguards are retained and strengthened and remain under the control of the Department of Environmental Affairs;
- mining companies' BEE obligations include equity participation and/or community royalties for historically disadvantaged communities in mining areas (2008: 13).

The villagers have actively resisted the enclosure of their resources as is documented in Chapter 10. They have enlisted the help of environmental justice lawyer, Richard Spoor, and worked with the social movement Jubilee, groundWork and the Vaal Environmental Justice Alliance (VEJA) as well as ActionAid. Ironically, their best hope lies in the falling demand for and falling prices of platinum. As the commodity boom was reined in by the prospect of global recession, several platinum projects were cancelled or delayed. Many were revived as the price recovered in anticipation of global economic recovery.

STEELING THE FUTURE

Iscor was privatised for R3 billion in 1989 as part of the late apartheid strategy of liberalising the economy but government retained a large share through the Industrial Development Corporation (IDC). South Africa provided a low-cost base for steel production. Apart from scrap metal, all the inputs were and are cheap: energy was as cheap as it gets; labour costs were less than half the world average; and high-quality iron ore was available from Iscor's own mines. Despite this, the privatised Iscor was in trouble. It produced too many product types requiring high-cost short production runs and its gross inefficiency resulted in a high proportion of defective products. Government bailed it out with over R1.2 billion in subsidies between 1992 and 1996 on top of a 30% import tariff protection.[11]

From 1994, Iscor shut down 2.5 million tonnes of capacity, halved the number of grades produced, slashed thousands of jobs and reorganised its marketing to support exports at the cost of the domestic market. In 1996, government reduced the tariff protection to 5% in order to cut costs to downstream manufacturers, and car makers in particular, and so promote export-oriented manufacturing.

In 1995, Iscor and the IDC embarked on a joint project to build a new steel mill at Saldanha Bay – the anchor project for government's Spatial Development Initiative (SDI) – designed to produce for export. It started producing in 1998 just as the price of steel collapsed. Large steel surpluses came on to the market as the result of the International Monetary Fund (IMF) induced 'Asian crisis' and new production in China, South Korea and Brazil added to the surplus. The project bled money. It accounted for 65% of the IDC's portfolio and threatened its very existence. In panic, the IDC came up with two strategies. First, it drove a process of 'unbundling' Iscor by splitting off its iron-ore and coal-mining operations to form Kumba Resources, which is now controlled by Anglo American. Iscor opposed this move and then tried to saddle Kumba with its massive debts. It failed on both counts. Kumba would, however, supply iron ore at cost plus 3% so the deal protected Iscor's low-cost supply. Next, IDC looked for an international investor to bail it out. It found Lakshmi Mittal who was building his global empire by buying out cheap, dirty and inefficient steelmakers hit by the price collapse. His atrocious environmental record did not register as an issue with the IDC.

A fire sale doesn't quite describe it. They paid Mittal to take it away. The corporation built up its shareholding to take majority control in 2004. The unions contested the takeover. Iscor had reduced its workforce from 44 000 in 1980 to 12 200 in 2004 and they anticipated that Mittal would cut more jobs. Investors, in contrast, lauded the high profits managed by Mittal. Business journalist Ann Crotty was unconvinced. Those on the 'Iscor unemployment scrapheap' would witness a dividend pay-out that turned the previous Iscor managers into multimillionaires and gave Mittal R3 billion – which would more than cover the cost of his buying Iscor shares for R2 billion in 2001'.[12] In the meantime, Mittal achieved his ambition of building his family corporation into the biggest steel producer in the world through a takeover of Arcelor, Europe's largest steelmaker, in 2006.

ArcelorMittal leached money from the South African economy. Government had facilitated the Iscor takeover on the understanding that the benefit of dirt-cheap ore would be passed through to domestic steel users and so create a competitive advantage to local manufacturing. It did not, however, bring in measures to enforce this gentlemen's agreement. Being the dominant producer, Mittal instituted import-parity pricing, meaning that it loaded the price with the imaginary costs of transport to South Africa, handling costs at the ports, the 5% import duty, and transport inland. This added around 30% to the price of domestic steel and, between 2002 and 2005, Mittal charged domestic customers over 60% more than it charged for export steel (Roberts and Rustomjee 2009). Government has since scrapped the import duty.

In effect, Mittal used the domestic market to subsidise its export market. At a Competition Commission hearing instigated by Harmony Gold, it claimed that it no longer used import parity but instead calculates the price on an international basket of prices. This merely gives a new gloss to the local subsidy to exports. The Department of Trade and Industry (DTI) calculates that ArcelorMittal's 2009 prices were in fact higher than import parity. It happens, however, that ArcelorMittal has screwed up on the iron-ore deal in terms of which it gets the ore at about $30 a tonne – a fifth of present spot-market prices. Kumba says that ArcelorMittal neglected to update its rights and will now be charged the full market price. ArcelorMittal says it will pass the price on if it loses the fight. Government says it never passed the benefit on in the first place. Nevertheless, seeing the last chance of cheap steel for the manufacturing industry disappearing, the DTI rushed to broker a solution.[13]

Government's interest in cheap steel is not matched by its concern over pollution. Whether as Iscor or ArcelorMittal, the corporation has fought to avoid recognition of and liability for its destruction of Steel Valley. It won. Repeating the strategy of the Far West Rand mining houses, it bought out the nearly 600 smallholders in the valley and fenced it in. The municipality is now considering locating a new landfill in the valley, a sign that it is regarded as already sacrificed.

Iron and steelmaking takes place on a giant scale, consuming millions of tonnes of raw materials and very large quantities of water and energy. It is widely regarded as the most polluting industrial activity on earth. The raw materials – iron ore, scrap metal and coal – contain substantial impurities that must be removed to preserve the quality of the product and are discarded as gas through smokestacks, in liquid form or as solid wastes. The Vanderbijlpark steel plant produces 2.2 million tonnes of solid waste every year. One million tonnes of this is hazardous, containing inorganic contaminants that leak into the groundwater: manganese, aluminium, cadmium, calcium, chloride, fluorides, iron, sulphates, titanium and zinc. Various organic substances,[14] mainly derived from coal tars, pose an additional toxicity threat. Most of these materials are found in the solid-waste dumps, the evaporation dams and maturation ponds.

Impurities in iron ore include sulphur, manganese, and traces of heavy metal including cadmium, lead, zinc and mercury. Scrap is predominantly contaminated with tin, lead and copper and increasingly contaminated with plastics and paints. Some scrap metal is radioactive as described in Box 4.1. Flux materials such as limestone are used to act like 'a kind of chemical sponge' (Davis 2002: 10) to capture and remove impurities and unwanted chemicals like sulphur from the furnaces. Slag is used flux, and the scale on which it is produced is evident in the mountainous slag-heap that looms over Steel Valley.

While impurities are removed, other metals are added to the iron-carbon mixture to give the steel special properties. Nickel and tungsten add strength, chromium increases the hardness, vanadium reduces the effects of metal fatigue, and lead makes steel more pliable. Large amounts of chromium and nickel are added to make stainless steel and zinc is used to coat or galvanize steel so it does not rust. All these additives are toxic heavy metals that can and do escape from the manufacturing process into the environment.

The coke ovens are particularly toxic. Coal is purposely starved of oxygen to create coke, used in blast furnaces, and so produces carcinogenic polycyclic aromatic hydrocarbons. Water used to quench the coke catches much of this but the rest escapes as fumes and is particularly dangerous to workers. The gas created by heating the coal is led off to the coke by-products plant where ammonia and a range of volatile organic compounds (VOCs), notably benzene, xylene, toluene, phenol and naphthalene, are recovered. During recovery, the gas is sprayed with water to produce flushing liquor. 'This represents a very difficult pollution control problem,' according to steel pollution expert Frank Kemmer, 'since the liquor is very high in ammonium chloride . . . and contains such other contaminants as phenol, cyanide and thiocyanates' (1971: 10–16). In addition, dioxins are formed in coke-oven exhausts. Liquid and solid waste from the ovens includes highly toxic tars containing phenols, cresols, naphthols, acridine and pyridine.

Iscor installed its first coke ovens and by-product plant at Vanderbijlpark in the 1950s. They have operated ever since but the difficulty of handling the waste has largely been neglected. In 2004, an environmental impact assessment for Mittal reported the annual waste from the plant's coke ovens as 70 000 tonnes of crude tar, 2 400 tonnes of tar sludge, 4 000 tonnes of ammonium sulphate and 180 000 tonnes of coke 'breeze' (fine dust).[15]

Traditionally, blast furnaces – which are huge steel stacks lined with refractory brick – are used to smelt ore into liquid iron. A mixture of iron ore, coke and limestone is dropped from the top of the stack and descends through blasts of hot air to the bottom over a period of six to eight hours. Very high temperatures result. At the end of the process, the liquid iron is tapped off through one hole while the slag floats to the top and is tapped through another. While gas is caught and cleaned by special pipes, some of it is vented to the air or burnt as waste. Emissions include dioxins, sulphur dioxide, carbon dioxide, carbon monoxide and breathable iron-dust particulates.

The Vanderbijlpark plant reportedly produced 28 700 tonnes of iron dust (or particulates), 13 000 tonnes of gas-cleaning sludge, 600 000 tonnes of granulated slag and 36 000 tonnes of blast-furnace slag from its two blast furnaces in 2004. The iron dust and gas-cleaning sludge are recycled to the sinter plant and the slag is used in the cement industry and for road construction. A sinter plant prepares sinters – pellets of iron and coal dust – to feed into steelmaking furnaces.

Molten iron from the blast furnace, sinters and scrap metal are used as feed for the steelmaking furnaces of which there are two kinds: Basic oxygen furnaces (BOFs) and electric arc furnaces (EAFs). ArcelorMittal uses both at Vanderbijlpark.

In the BOFs, a lance is used to inject oxygen into the furnace at supersonic – and ear-piercing – speeds. This drives impurities off the molten steel and raises the temperature to melt the scrap metal added to the feed. Six-storey buildings are needed so that the huge oxygen lances can be manoeuvred. Fluxing materials are added to carry off impurities. Iron fumes, carbon dioxide and large amounts of carbon monoxide are released when the furnaces are charged and tapped. Water is used to scrub gases of dust and fumes.

In 2004, Vanderbijlpark's BOFs produced solid waste consisting of 12 000 tonnes of iron dust, 45 000 tonnes of desulphurisation slag and 504 000 tonnes of furnace slag, all of which was dumped. Other solid wastes – 36 000 tonnes of mud, 8 000 tonnes of grit and 36 000 tonnes of furnace slag – were re-used internally.

In the EAF, an electric arc sprung between two giant electrodes provides most of the energy to melt the scrap and iron feed. Oxygen lances are also used in this process. EAFs produce low-carbon steels and ferroalloys used in the production of ferromanganese, ferrovanadium and ferrochrome. As in the BOFs, fluxing materials carry off impurities. EAFs 'cause a rather high discharge of dust to the atmosphere' and wash water picks up very high levels of suspended solids (Kemmer 1971: 10).

In 2004, the EAFs produced 16 000 tonnes of dust that was dumped, and 100 000 tonnes of furnace slag that was reportedly re-used internally. The clouds of red dust that are regularly seen rising through roofs at the plant are from this unit.

The steel tapped from these furnaces is rolled or cast into intermediate and final forms at the hot or cold-roll mills. In the rolling mills, water picks up oils and lubricants. The steel forms are then 'pickled' – treated in acid baths with sulphuric or hydrochloric acid – to remove rust from the surface. The waste – 'spent pickle liquor' – is strongly acidic and contaminated with suspended scales. The steel forms are then galvanized at high temperatures, releasing fumes and heavy metals.

Slag-heaps are the most visible solid waste from iron and steel plants. As slag results from removing contaminants from the production process, these contaminants are again leached or blown from the heap. The scale of slag production allows other wastes to be covered up. In 2005, activists observed Mittal staff burying what appeared to be bag-house waste in the slag-heap. The bags filter particulates from the air exhaust. Altogether, a toxic brew of more than 100 chemicals is known to be emitted by steel mills. Recent research in Canada has shown that this cocktail not only affects all life forms around the mills, but goes down to the genetic level with hereditary DNA damage reported around a plant in Hamilton Harbour.[16] In addition to local health impacts, sulphur and nitrogen emissions contribute substantially to acid rain.

POWER TO ALUMINIUM

In 2001, the Australian corporation BHP merged with Billiton to create the world's largest diversified minerals corporation. Billiton was previously owned by Shell who sold it to the South African group Gencor in 1994. The deal required a major export of South African capital and Gencor sought and received an exemption from the capital controls then in place from the minister of Finance. Billiton was listed in London and it soon became evident that Gencor, the supposed parent, was in fact of subordinate interest. In an internal deal, Billiton bought Gencor's base metals assets, including the Richards Bay aluminium smelters. The deal thus preceded, and set a precedent for, the listings of other major South African corporations on the world's central stock exchanges in the late 1990s and early 2000s.

Gencor itself retained its own precious metals division but quickly unbundled, morphing into a capital holding company and selling off its last assets, a 46% holding in Impala Platinum, before closing its doors in 2003. The hollowing-out and closure of Gencor seems to have been connected with a legal claim against it by people suffering from asbestosis. The corporation bought Cape Plc's asbestos mines when the latter disinvested from South Africa in the early 1980s. Without admitting liability, Gencor made a 'full and final' settlement of R380 million to the Asbestos Relief Trust. It was then quickly liquidated, returning very substantial 'shareholder value' while terminating corporate responsibility for the ongoing ruin of the environment and of thousands of people's health. In the meantime, much of Gencor's top management had transferred to Billiton.

Billiton continued a major expansion of aluminium-smelting capacity inaugurated by Gencor. The Hillside smelter at Richards Bay, complementing the older Bayside smelter, was completed in 1996 and the Mozal smelter outside Maputo in Mozambique followed shortly after with production starting in 2000. These smelters linked with Billiton's existing bauxite mines and refineries: the Worsley mine and refinery in Australia and the mines in Suriname, in Latin America, which supplies a refinery operated by Alcoa in which Billiton has a 45% interest. The refineries produce alumina, a whitish powder, from the raw bauxite ore supplied by the mines. The process uses chemicals and heat to separate alumina from the toxic residue known as 'red mud'. Worsley appears to produce about 12 million tonnes a year of the stuff, although BHP Billiton (2006) is not exactly explicit on this point.

The southern African smelters are the primary market for Worsley's alumina – although this 'market' is obviously internal to the corporation. All three smelters were primary beneficiaries of state infrastructure investments. The original construction of Bayside, in 1971, was integral to the apartheid state's simultaneous development of the deep-water port at Richards Bay. The project required close collaboration of government departments, major state-owned corporations – primarily the IDC, Eskom and Transnet – and private interests led by Anglo

Box 4.2   Recycling red mud

The industry is busily looking for ways of getting rid of its mud – along with the costs of storing it – by touting it as a resource. In Australia, during the 1990s, Alcoa helped fund a Department of Agriculture experiment using red mud from its refinery to stabilise phosphorus run-off. The department persuaded farmers to participate in spreading it on their land, claiming that it would substantially increase yields. Instead, the farmers say, their cattle started getting sick. Spread at 20 tonnes a hectare, according to journalist Gerard Ryle, the red mud contained 'up to 30 kilograms of radioactive thorium, six kilograms of chromium, more than two kilograms of barium and up to one kilogram of uranium' together with '24 kilograms of fluoride, more than half a kilogram each of the toxic heavy metals arsenic, copper, zinc, and cobalt, as well as smaller amounts of lead, cadmium and beryllium'. The department nevertheless insisted that this had nothing to do with the cattle sickening and subsequently marketed the mud to farmers in south-west Australia as a soil dressing. Alcoa agreed that the 'product' was safe but nevertheless demanded, and got, an indemnity for any environmental damage.

Source: Gerard Ryle, 'The great red mud experiment that went radioactive', *Sydney Morning Herald*, 7 May 2002.

American. These institutional relations were, if anything, strengthened in the post-apartheid period and Billiton slipped into the seat already warmed by Gencor. Hillside was seen as an anchor project for the SDI and Industrial Development Zone intended to inaugurate another round of industrial modernisation at Richards Bay, while Mozal anchored the Maputo Corridor SDI and was accompanied by the development of a deep-water port at Maputo. Mozal also provided a vehicle for practical collaboration between the corporations at the centre of the minerals-energy complex (state and private) and the World Bank, so reinforcing local-global institutional relationships as South Africa emerged from isolation.

In contrast to ArcelorMittal's import-parity pricing, it appears that BHP Billiton uses transfer pricing to boost its profits at the cost of the South African economy. That is, it exports the aluminium to itself at below-market rates and gets an additional tax benefit for doing so. Journalist Jan de Lange reports that it has therefore refused to supply molten aluminium to downstream manufacturers in Richards Bay. Instead, they have to import solid bars and melt them. According to the manufacturers, supplying molten aluminium would cut Billiton's own electricity consumption by 924 MWh and save the manufacturers 2 640 MWh.[17]

Electric energy is the most significant input into aluminium smelting and, for Billiton, cheap electricity from Eskom was the primary reason for locating both Hillside and Mozal. Table 4.1 shows energy consumption for the three Billiton plants equating to 10% of South Africa's electricity supply and 3.6% of total final energy demand. The balance of the smelters' energy is derived from coking coal, gas and liquid fuels. Mozal, of course, is not formally included in South African energy demand or carbon emissions, but it is directly supplied by Eskom on similar terms to Hillside and Bayside. In short, it would not be there if it was not bound to South Africa's energy economy. It consumes more electricity and emits more carbon than the rest of Mozambique put together.[18]

The precise terms of the Special Pricing Agreement are secret but Billiton undoubtedly gets the cheapest electricity in the world. The normal industrial rate, at around 16 cents per kWh in 2007, was already the world's cheapest and the smelters are supplied below this price. It is known that the price of power is tied to the world price of aluminium, so this is protecting Billiton from both currency and commodity price fluctuations. In fact, much of the risk is transferred to Eskom, which lost R9.5 billion on 'embedded derivatives' when both the rand and the price of aluminium tanked in 2008/2009. Reports leaked in 2010 suggested that Billiton paid 12 cents per kWh[19] for power – about half Eskom's cost of production. During South Africa's electricity crisis in 2008, Eskom demanded a 10% reduction from the combined consumption of the three smelters and Billiton cut production at Bayside.

The smelters' high-energy consumption is largely responsible for the intensity of greenhouse gas emissions (CO_2e), contributing the equivalent of 5.7% to South Africa's emissions. This is supplemented by perfluorocarbons (PFCs), which are extremely powerful and long-lasting greenhouse gases, emitted primarily during upset conditions at the plants, according to BHP Billiton. Power outages or poor management of the smelting process therefore increase emissions and it may be estimated that 2008 was a very bad year for PFC emissions (BHP Billiton 2006).[20]

The table also shows an extraordinary intensity of sulphur dioxide emissions, with Bayside's emissions similar to that of Durban's oil refineries, and Hillside and Mozal emitting nearly three times as much. In the smelting process, alumina is saturated with fluoride to give rise to the fluoride emissions. Fluoride is toxic to a variety of plants even at very low concentrations and also accumulates in plants.

Table 4.1   Aluminium smelters: production, energy, waste (2006).

	Production tonnes	Total final energy (PJ)	Electric energy (PJ)	CO_2e million tonnes	SO_2 tonnes	Fluoride tonnes	Waste tonnes
Mozal	550 000	37	27	9.4	11 945	249	22 230
Hillside	700 000	47	45	11.6	11 161	354	48 272
Bayside	180 000	14	10	4.1	4 021	357	43 000
Total	1 430 000	98	82	25	27 127	960	113 502
South Africa[1]		2 718	816	440			

Sources: BHP Billiton 2006; DME 2006.

1. The national figures are for 2004 due to the laggard production of energy statistics. In 2010, it was confirmed that Billiton consumed about 9.3% of electricity. Billiton's consumption would have been fairly constant since its last expansion, so earlier figures would give it a higher share of national consumption.

Exposure even to low emissions thus results in fluoride concentrations accumulating over time and so entering the food chain from vegetables or grass grazed by cattle.

Aluminium is smelted in pots at very high heat. The pot-linings accumulate carbon and must periodically be renewed. Spent pot-linings form the bulk of the solid waste from smelting and the carbon is impregnated with alumina and fluoride and laced with cyanide and arsenic. It is classified as a hazardous waste. Faced with rising disposal costs, BHP Billiton entered a partnership with EnviroServ to reduce costs and 'increase the value of its waste streams into specific offset markets' (BHP Billiton 2006: 50). In other words, it was looking to sell waste with the aim, according to EnviroServ, of 'zero waste to landfill' (2007: 24). EnviroServ now 'recycles' the waste as an alternative fuel for steel and cement production and so saves 'enormous volumes of valuable landfill airspace' (25). What does not go down into the landfill, however, generally goes up into the air.

TOXICS TO CEMENT

The major cement corporations are AfriSam, Lafarge, Natal Portland Cement and Pretoria Portland Cement (PPC). AfriSam is the newest kid on the block, taking the place of transnational corporation Holcim. The latter dressed up disinvestment from South Africa as an empowerment deal that was carried through with R6 billion support from the state-owned Public Investment Corporation. These four companies are the members of the Cement and Concrete Institute whose objective is 'to increase the market share' of concrete in construction. At present, residential and commercial construction has contracted sharply and the market is being sustained by the state's infrastructure programme, starting with the 2010 stadiums and with massive demand from Eskom and Transnet's expansion programmes to follow.

The raw materials of cement production are limestone and silica and alumina from clay. They are ground to a fine powder and then fed through the kiln, where temperatures reach 1 400 to 1 500 °C, to produce 'clinker'. Kilns are traditionally fired by coal and the bottom ash is incorporated in the clinker. The clinker is then cooled and ground with various additives to the fine powder that is cement. The process is very energy-intensive and the use of coal puts cement in the same bracket as the energy sector in terms of its contribution to climate change.

Internationally, the Cement Sustainability Initiative is putting a green spin on production but, as Jane Harley comments in a report for groundWork, it 'has put out a great many documents, all of which avoid the central truth – that cement can never be sustainably produced'. Rather, the industry has focused on 'the use of . . . "alternative fuels", which translates to the use of waste as a fuel' (2006: 2). While the environmental benefits of these fuels are dubious, the economic benefits to the cement industry are evident. From 2003 to 2008, coal prices went up from around $20 to over $160 a tonne. PPC said that international demand was limiting 'the availability of the appropriate coal quality for cement manufacture' while 'spiralling' international prices were pushing up costs (PPC 2007: 24).

The industry describes burning waste as 'co-processing' or 'energy recycling'. PPC goes so far as to suggest that co-processing replaces 'fossil fuel with renewable sources' (50). Apart from twisting the notion of 'renewable' beyond recognition, the statement implies that waste will indeed be eternally renewed. The industry favours waste with a high-calorific content, much of which consists of hazardous petro-chemical wastes derived from fossil fuels. Wastes used internationally include solvents, old tyres and oil, paint and dried sewage sludge. The use of spent pot-linings from aluminium smelters has an added advantage as the alumina substitutes for alumina in the raw material fed into the kiln.

Pot-linings and dried sewage sludge are already used in some plants in South Africa with the approval of the Department of Environmental Affairs and Tourism (DEAT). It is possible that other wastes have been used without approval. Used tyres, however, would require modification of the kilns and the industry is, somewhat impatiently, 'waiting for the relevant legislation to be enacted', as PPC puts it

(2007: 50). Harley notes estimates that South Africa's used tyres could replace about 25% of the 1.2 million tonnes of coal used in kilns. The industry anticipates something better than cheap fuel. It anticipates receiving a tipping fee for disposing of tyres and has also lobbied government for an 'establishment subsidy' against the costs of modifying the kilns. A draft memorandum of agreement between DEAT and waste-tyre handlers, negotiated in 2006, looked like a very good deal for both the waste and cement industries with costs paid by the public in taxes and in the price of tyres.[21] The DEAT subsequently withdrew from the agreement without public explanation.

Meanwhile, the DEAT's proposed waste-tyre regulations, published for comment in April 2008, give priority to re-use or recycling over energy recovery, and of energy recovery over disposal. Incineration with energy recovery is thus lifted above disposal in the waste hierarchy. The regulations do not seriously address minimisation but they do impose 'extended producer responsibility' on tyre producers who must prepare integrated waste-management plans. The regulations, brought out one month after the final hearings on the Waste Bill, were published in terms of the Environmental Conservation Act but clearly anticipated the Waste Bill's enactment. It is less clear how 'recovery of energy' relates to a clause in the Waste Act requiring that any regulation pertaining to incineration be submitted to parliament or whether, in fact, early publication was designed to pre-empt that requirement. Assuming, however, that the parliamentary hurdle is either crossed or bypassed and the cement industry invests in the modification of kilns, it can be anticipated that they will provide the easiest disposal option.

Kilns fired by coal are dirty operations. Kilns fired by used tyres are even dirtier. A study cited by Harley compares the two.[22] It shows that tyre-burning emissions of hydrocarbons are lower but particulates and most gas emissions are higher while emissions of most metals are two or more times higher. Tyres, however, will not replace coal but will be burnt with coal and whatever other wastes are allowed to be added to the mix. Emissions from the combination of fuels are likely to be dirtier than the sum of emissions from each because more chemicals will be available to create more toxic compounds. Spent pot-linings, for example, would add a heavy charge of fluoride.

Waste burnt in kilns produces similar emissions to waste burnt in incinerators – sulphur dioxide, hydrogen chloride, nitric oxide, particulates and dioxins formed in the exhaust. Thompson and Anthony note that cement kiln technology has not changed much since the early 1900s and is not well adapted to 'toxic waste destruction'. Moreover, even in the European context, they are less rigorously regulated than incinerators: they are allowed to emit more and 'have poorer abatement equipment' (2005: 35). In South Africa, cement kilns have operated without any scrutiny from the authorities, even after permission was given to burn spent pot-linings at some kilns. This changed shortly after the confrontation over incineration at the parliamentary hearings on the Waste Bill. In May 2008, the DEAT announced that the Green Scorpions would do a 'blitz' on cement kilns, heralding the start of a 'clean cement' campaign. It said the cement industry was growing rapidly and might 'contribute significantly to pollution if not mitigated and managed properly'.[23] This is laudable. The suspicion remains, however, that the real intention is to head off opposition to waste incineration in cement kilns when the relevant regulations are put to parliament. In the meantime, inspection reports have yet to be made public and it is unlikely that they will reflect normal operating. The industry was given notice of the blitz and will have been on its best behaviour. High standards – for example, ensuring complete combustion – costs money. It is doubted that they are maintained outside of inspection in Europe and it seems unlikely that the local industry will be more assiduous.

Toxics generated in the kiln, including dioxins and heavy metals, have three places to go: into the air, to the dump or into the product. The kilns do not produce substantial solid-waste volumes. This is because the ash from the furnace binds with the limestone and other material inputs to form the clinker. Thus, the toxic residue in the ash is incorporated into the product. Where filters are used to reduce emissions of particulates (known as cement kiln dust), the captured waste is either sent to landfills or recycled through the kiln. The latter practice leads to a concentration of heavy metals that is ultimately incorporated into the clinker. Further, 'extenders' are added when the clinker is milled. During 2006, PPC increased its use of fly ash and limestone as extenders 'to conserve non-renewable resources' and reduce the proportion of clinker in its cement products (PPC 2006: 32). This would also reduce costs and bulk up the product to meet expanded demand. PPC does not say whether the fly ash comes from its own plant or other industries nor does it mention whether it is tested for toxic contaminants. From whatever source, however, fly ash is particulate emission and almost certainly toxic. Toxics in the clinker are thus supplemented by those in the extenders and incorporated into the product. Cement and construction workers would be most immediately exposed to any such contamination but it remains in the built environment and will be released during renovation or demolition.

LEAVING RUIN

Industrialisation massively increased the volume of resource consumption and waste. While peasants consume up to 5 tonnes of raw material per person per year, urban Europeans now use between 40 and 70 tonnes, according to Wolfgang Sachs. Most of this is for 'installations run by organizations at various levels of the system: high-rise buildings, steel plants, supermarkets, swimming baths, airports, armoured vehicles and so on' (Sachs and Santorius 2007: 36). 'Per person' is thus a little misleading. The institutions of capital and state consume more than 'consumers'. It is not just the goods on the shelf, but the shelf itself, the shopping mall, the city that sustains the mall, the machinery of manufacture and the infrastructure of energy and transport and, finally, the extravagance of arms.

It is not incidental that the financial crisis was connected to the contemporary process of urbanisation through the so called 'sub-prime' mortgage defaults, argues David Harvey (2008). Historically, grandiose urban development has repeatedly been used to absorb surplus capital when over-accumulation threatens profits. Over the last decade or so, this process has gone global reflecting the globalisation of finance capital. The urbanisation of China dwarfed everything else, but property markets boomed across the world accompanied by frenzied demolition and construction. From the towers of Dubai to the golf estates of the Western Cape, it has been marked by competitive conspicuous consumption. And, as noted in Chapter 2, this investment has been focused in enclaves to the exclusion of the poor who are driven to the urban peripheries to make way for the high value investments of 'world-class' cities.

The competition in conspicuous consumption finds direct expression in the ritualised auctioning-out of spectacularly commercialised sports festivals. Each Olympics or football World Cup competes with the last for extravagance as corporate sponsors demand yet bigger bangs for their advertising bucks.[24] When South Africa won the bid for 2010, the major cities started competing with each other for national funding of 'iconic' stadiums and transport infrastructure projects, running up debts that will settle on citizens into the future. As organisations of the poor noted, the resources mobilised for the event contrasted starkly with the repeated assertions that the state lacks capacity for 'delivery' to the poor.[25]

The industrial policy objective of expanding manufacturing production to create jobs scarcely floated even on the high tide of the boom. As Northern demand shrank following the financial meltdown, China maintained high economic growth through a massive infusion of cash into its economy and replaced the US as the largest consumer of South African exports. Primary minerals commodities –

platinum, iron ore and coal, along with ferroalloys, aluminium metal and rolled steel – accounted for three-quarters of all exports in the first half of 2010.[26] South Africa's place at the bottom of the industrial value chain, and the continued dominance of the minerals-energy complex, was thus confirmed. The economic benefits are largely taken by the corporations whose global expansion has both reduced their dependence on and increased their power over the local economy. The environmental ruin is left to the people.

5

Peak poison

THE WORLD WE LIVE IN is profoundly shaped by the abundance of oil. This was not an entirely spontaneous process. Outside of war, the problem for big oil during most of the twentieth century was that there was too much of it and oil prices and profits were constantly threatened with collapse. Big oil developed two main strategies for managing the glut: restricting supply and expanding demand. First, the big corporations established control over supplies through monopolies or cartels. Thus, the 'seven sisters'[1] used Middle Eastern countries as swing producers – opening or closing the taps to balance supply and demand – and so subordinated national to corporate interests. Their imperial manners provoked the rise of 'resource nationalism' and the formation of OPEC. In the last resort, they provoked war. 'When profits fell to what the industry called a "danger zone"', according to Retort, 'oil men turned hawkish. Each descent into the "danger zone" preceded an energy conflict, and was in turn followed by a dramatic reversal of economic fortune' (2005: 70).

War has also been significant in expanding demand. During both world wars, production was ramped up to meet military demand. Following the wars, car makers and big oil combined to expand the civilian market. In the US, they bought out and dismantled tram companies and lobbied for a massive expansion of paved roads and promoted the industrialisation of agriculture everywhere. In the 1990s, they secured the exemption of sport utility vehicles (SUVs) from fuel-economy standards applied to other family cars and, in Johannesburg as in New York, aggressive marketing and the marketing of aggression made these gas guzzlers the elite suburban vehicle of choice. More broadly, transport as a whole received massive support from the neo-liberal enforcement of open markets and export-led development. While the World Bank secured cut-rate commodities for the global market from the global South, the global restructuring of production fuelled a massive increase in trade, most of it internal to transnational corporations, accompanied by a boom in tourism. Sea and air transport grew exponentially while 'just-in-time' delivery systems created 'warehouses on wheels' to reduce storage costs. Significantly, air and sea transport are excluded from national carbon accounts. Europe could thus commit to a 20% carbon reduction by 2020 while signing an 'open skies' agreement with the US designed to increase transatlantic flights by 50% in the next five years.[2]

The age of plenty is now over. The basic assumptions of energy policy over the last century are no longer valid and, while unable to let go of these assumptions, the energy elites are nervous. The International Energy Agency's (IEA) *World Energy Outlook* (WEO) 2006 responded to a mandate from the imperial club of G8 countries to 'map a new energy future'. The stated concerns were energy security and climate change. The WEO argued that the responses to these concerns are mutually reinforcing: energy efficiency and diversification of energy sources. Its representation, however, showed them to be deeply contradictory, at least in the context of never-ending accumulation required by capitalism. Its central message was that the world must invest $20 trillion between 2006 and 2030: $11 trillion in electricity generation, transmission and distribution; $4.3 trillion in oil, mostly upstream; $3.9 trillion in gas; and $560 billion in coal. According to Claude Mandil, 'The energy future . . . is doomed to failure [because of] underinvestment in basic energy infrastructure . . . In short, we are on course for an energy system that will evolve from crisis to crisis' (quoted in Hirsch 2007). This chapter shows the growing disconnect between reality and policies founded on energy expansion and the growing intensity of pollution consequent on those policies. It starts with an overview of the energy future projected by the IEA and its response to the challenge of climate change. Box 5.1 shows why this response does not meet the challenge. It then looks at each of the sources of energy and shows why peak oil is accompanied by peak pollution.

FAILING FUTURE

The WEO of 2006 develops two energy future scenarios for the period to 2030. The 'reference scenario' is based on national policies that have been adopted in both developed and developing countries and it assumes that they will be fully implemented. The 'alternative policy scenario' takes account of additional policies aimed at enhancing energy security and/or addressing climate change. These are policies that were being considered in each country and which, in the IEA's view, they 'might reasonably be expected' to adopt (IEA 2006: 54).

The WEO of 2008 takes a slightly different line. The reference scenario is developed in the same way but, instead of the alternatives scenario, it looks at the implications for the energy system of stabilising the concentration of greenhouse gases in the atmosphere at 550 parts per million (ppm) and at 450 ppm of CO_2e. These are the targets most commonly discussed in the international climate negotiations. Following the Intergovernmental Panel on Climate Change (IPCC), the WEO of 2008 assumes that stabilisation at 550 will result in global temperatures rising by 3 °C. Stabilisation at 450 ppm is held to give a 50% chance of restricting global warming to 2 °C. These two scenarios are based on a 'hybrid' set of policies that the IEA believed negotiators might possibly adopt for the post-2012 climate regime: cap and trade for Northern countries with compulsory reduction targets; national policies and measures for Southern countries without compulsory reduction targets; and international sectoral agreements covering major industries such as iron and steel and cement making.

Table 5.1 shows global energy consumption and carbon dioxide emissions for each scenario from WEO 2006 and 2008. This excludes emissions from industrial processes and land-use change as well as other greenhouse gases. Total CO_2 emissions in 2005 were 36 billion tonnes (Gt) and total greenhouse gas emissions were equivalent to 44 Gt of CO_2 (CO_2e).

Both 2006 scenarios project massive increases in energy consumption and carbon emissions although demand rises less steeply in the alternative case as a

Table 5.1 Annual global energy consumption and CO_2 emissions (billion tonnes).

	2004 (actual)		2006 (actual)		2015		2020		2030	
	toe	CO_2	toe	CO_2	toe	CO_2	toe	CO_2	toe	CO_2
WEO 2006										
Reference	11.2	26			14.1	33	–	–	17.1	40
Alternative					13.5	32	–	–	15.4	34
WEO 2008										
Reference			11.7	28	14.1	33	–	–	17.0	41
550 scenario							14.4	32	15.5	33
450 scenario							14.3	32	14.4	26

Adapted from IEA 2006 and 2008.

Energy consumption in tonnes of oil equivalent (toe). The 11.7 billion tonnes consumed in 2006 is the equivalent of around 234 million barrels of oil a day.

result of increased energy efficiency. Assuming immediate implementation of the alternative policies, total emissions from 2005 to 2030 would be 8% less (820 Gt) compared with the reference case (890 Gt).[3] If alternative policies are implemented only in 2015, cumulative emissions for the whole period are only 2% less than in the reference case. Implementation would also be considerably more costly.

The table shows that actual energy consumption rose by 500 million tonnes of oil equivalent (toe) from 2004 to 2006. That is equivalent to over ten million barrels of oil a day. For the period to 2030, the 2006 and 2008 reference scenarios are almost identical. The implication is that the alternative policies being considered in 2006 have not in fact been adopted. The 2006 alternatives scenario and the 2008 550 scenario appear very similar. In the latter, however, emissions level off only in 2025.

Fossil fuels remain dominant in all scenarios. At present they account for 80% of total energy consumption and, in the reference scenarios, they retain this share of consumption through to 2030. The 2008 figures, however, assume higher oil prices and reduced oil demand compared with the 2006 figures. Higher coal consumption compensates for the difference. In the 550 and 450 scenarios, the fossil-fuel share decreases to 74% and 66% respectively. Consumption of fossil fuels still grows overall but coal's share is reduced and, in the 450 scenario, less coal is consumed in 2030 than in 2006. In the WEO's analysis, this is largely the result of high carbon prices at $90 a tonne in the 550 and $180 in the 450 scenario. Supposedly 'non-carbon' energies grow at a faster rate and carbon capture and storage is developed on a massive scale. Table 5.2 shows global energy sources for the WEO 2008 scenarios.

In the 450 scenario, energy emissions peak around 2020 and are then reduced to the 2004 level of 26 Gt by 2030.[4] This results in 'greenhouse gas concentration[s] initially rising above 450 CO_2e, but then declining'. The WEO argues that 'overshoot' is necessary because avoiding it would require substantially lower emissions before 2020 and 'this could be done only by scrapping very substantial amounts of existing capital across all energy-related industries'. Further, 'it is unlikely that the necessary new equipment and infrastructure could be built and deployed quickly enough to meet demand' (IEA 2008: 414). In fact, the 450 target is already overshot and, as the WEO 2006 recognised, peaking later and higher adds massively to cumulative emissions over time as well as to the costs of subsequent reductions.

While writing off investments in wells, pipelines and generators poses a barrier to early reductions, 'meeting demand' acts as an absolute imperative. All energy planning starts from the assumption of future demand growth and then organises

Table 5.2   Energy sources (million tonnes of oil equivalent per year).

Source	2004 (actual)	2006 (actual)	Scenario	2015	2020	2030	2030 share of energy
Oil	3 940	4 029	Reference	4 525		5 109	30.0%
			550		4 553	4 689	29.6%
			450		4 549	4 308	29.9%
Coal	2 773	3 053	Reference	4 023		4 908	28.8%
			550		3 694	3 575	23.2%
			450		3 639	2 381	16.0%
Gas	2 302	2 407	Reference	2 903		3 670	21.1%
			550		3 010	3 383	21.9%
			450		2 987	2 950	20.1%
Nuclear	714	728	Reference	817		901	5.2%
			550		976	1 086	7.0%
			450		987	1 364	9.9%
Hydro	242	261	Reference	321		414	2.3%
			550		389	456	2.9%
			450		391	555	3.8%
Biomass & waste	1 176	1 186	Reference	1 375		1 662	10.0%
			550		1 499	1 826	11.6%
			450		1 494	2 119	14.5%
Other renewable	57	66	Reference	158		350	2.0%
			550		237	468	3.0%
			450		235	683	4.7%

Adapted from IEA 2008.

production to meet it. If an external constraint – such as carbon reductions – prevents demand being met, the modelling tools used in planning cannot return valid results. This creates an inbuilt bias against admitting peak oil and also fixes the boundaries of carbon realism.[5] From within those boundaries, the WEO warns that the 550 target is extremely challenging and strongly implies that the 450 target is pretty much impossible. Nevertheless, this toughest of targets that the IEA can contemplate is not credible: the WEO 450 scenario will not in fact achieve stabilisation at 450 ppm; the 2 °C target does not avoid disastrous climate change; and a 50% chance of meeting it is a poor bet. Box 5.1 sets out the argument.

The collapse of the oil price from $147 to $35 in 2008 did not stimulate demand and a good deal of capital plant was then scrapped, if not simply abandoned. So it was that carbon emissions declined in the recessionary year of 2009 but, even on weak global growth, rebounded double-quick to make 2010 emissions the highest yet.[6] Meanwhile, the trillion-dollar bailouts have not lifted the economy off the rocks and, each time the markets see 'green shoots' in their economic desert, the price of oil jumps.

Box 5.1   Required CO$_2$e emissions reduction

The international consensus now defines 2 °C above pre-industrial temperatures as 'dangerous'. The 450 target follows from the conclusion of the IPCC's Fourth Assessment Report (AR4) that stabilisation in the range of 450–490 ppm CO$_2$e yields an average temperature increase in the range of 2–2.4 °C. However, this finding excludes consideration of natural feedback loops induced by global warming described in Chapter 1. These feedbacks are now being observed well ahead of the predicted dates making temperature rises of '5 or 6 °C or higher . . . plausible' (Stern 2006: 59).

Taking account of all greenhouse gases, the 2006 Stern Report said the concentrations in the atmosphere were then equivalent to 430 ppm of carbon dioxide (CO$_2$ equivalent or CO$_2$e) and rising at the rate of 2.3 ppm a year (2006: 3). This figure was in fact already dated. The IPCC's AR4, published in 2007, put the 2005 concentration at 455 ppm.[7] If, said Stern, the concentration was stabilised at 430 – in other words, carbon emissions were almost shut off – 'there is up to a one-in-five chance that the world would experience a warming in excess of 3 °C above pre-industrial [levels]' (9).

Stern underestimated the growth in emissions since 2000 on the assumption that energy and carbon intensity relative to GDP was declining as it had done throughout the twentieth century. This trend was reversed around 2000 and the pace of carbon dioxide emissions accelerated dramatically, growing faster than predicted in 'the most fossil-fuel intensive of the Intergovernmental Panel on Climate Change emissions scenarios developed in the late 1990s' (Raupach et al. 2007: 1). On the basis of actual emissions from 2000 to 2008 and taking account of climate induced feedbacks, Anderson and Bows (2008) show that stabilisation at 450 CO$_2$e will be physically impossible unless emissions peak by 2015 and global energy and industrial process CO$_2$ emissions are then reduced by 6 to 8% a year.[8] A 2020 peak could not result in stabilisation at less than 550 ppm and then only if followed by annual reductions of 9%. In contrast, the reductions proposed by the WEO 2008 450 scenario come in at just over 2%.[9]

The reasoning behind this is that cumulative emissions are more critical than final emission targets. Because CO_2 stays in the atmosphere for centuries, the total quantity pumped into the atmosphere over time determines the concentration in the atmosphere. Thus, a total emission 'budget' can be calculated relative to target concentrations. Peaking later and higher consumes much more of the budget – as WEO 2006 recognised – and so requires impossibly steep reductions following peak. Subsequent research suggests that, assuming a 2 °C target, one third of the budget for the period 2000 to 2050 was already used up by 2009.[10]

Anderson and Bows note that the economic collapse of the Soviet Union resulted in a reduction of emissions of around 5% a year. They conclude that 'it is difficult to envisage anything other than a planned economic recession being compatible with stabilisation at below 650 ppm CO_2e' (2008: 18). This implies 'an unprecedented step change in the global economic model' (15). In other words, it implies ditching capitalism.

As the AR4 was being published, new research indicated that warming of 1 °C already constitutes a '"dangerous" level of warming' (Levin and Pershing 2007: 3). Indeed, millions of people around the world, and particularly in Africa, are already faced with dangerous consequences of climate change. In testimony to the US Congress in June 2008, climate scientist James Hansen warned that "the oft-stated goal to keep global warming less than 2 °C is a recipe for global disaster, not salvation".

This conclusion was based on a paper by Hansen et al. (2008) which argues that a safe target for stabilising CO_2 (not e) concentration is 'no more than 350 ppm'[11] and may be less than that. The world had long since overshot the mark but disaster might be averted if the CO_2 concentration peaks at around 400 ppm and is then rapidly reduced. Taking account of peak oil, they argue that this is possible 'if difficult to extract oil and gas is left in the ground', coal is not used to substitute for declining oil and all coal use is phased out by 2030 unless the carbon emissions can be safely sequestered, and forest and soil sinks are restored through reforestation and changed agricultural practices. This should return 'CO_2 below 350 ppm late this century, after about 100 years above that level' (13, 14).

Even this is optimistic for two reasons. First, new research shows that the long-term carbon budget for the next 500 years only accommodates the use of two thirds of existing fossil-fuel reserves. In other words, all exploration should stop now –

there is already more than enough in the project pipeline to burn the world.[12] Second, Solomon et al. (2009) show that temperatures will not retreat along with CO_2 concentrations because of the build-up of heat in the oceans. If the temperature stops rising at the disastrous 2 °C, then that is pretty much the temperature the world is stuck with for the next thousand years. In short, scenarios in which temperature targets are 'overshot' and followed by cooling are no longer credible. This does not mean that reducing carbon concentrations is made irrelevant. To the contrary, the temperature stabilises because reduced heating from lower CO_2 concentrations is balanced by 'reduced cooling through heat loss to the oceans' (1705).

The implications of this research are shown in Figure 5.1. A 2011 peak followed by a relatively modest decline in emissions yields an 85% reduction by 2050 to keep within the carbon budget. Later peaking results in a negative carbon budget. This assumes the disastrous 2 °C target. A target of 1.5 °C, demanded by small-island states that face drowning, would require earlier peaks and steeper declines. This is not a safe target. Given the lag between emissions and temperature rise, it is probably the best that can be hoped for.

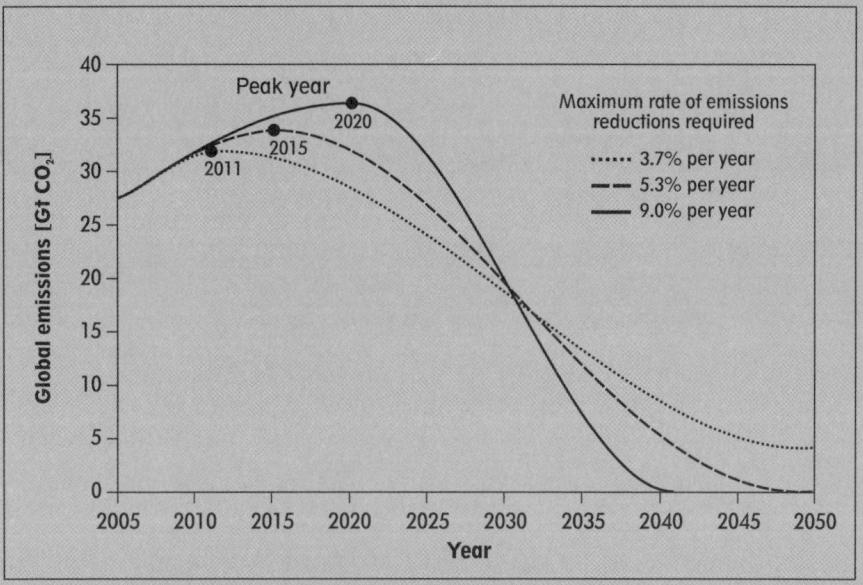

Figure 5.1 Global emissions pathways, 2010–50.
Source: German Advisory Council on Global Change (2009).

Meeting oil demand – or not

From 2000, annual demand grew by about two million barrels a day (mb/d) until high prices constricted the increase to around one million in 2005 and 2006. In these two years, demand for fuel for transport kept growing but many applications using oil, such as electricity generation and heating, switched to coal or gas. By 2007, most of the options for switching were used up while transport demand was increasing. Despite higher prices, overall consumption rose by 1.3 mb/d to 86.5 – over 31.5 billion barrels a year – according to the IEA. Production scarcely rose, however, coming in at 85.7 mb/d. The difference appears to have been made up with biofuels. The 2008 demand averaged at 86.2 but shrank fast from a high of 87.5 at the beginning of that year to 84 mb/d in mid-2009. The IEA forecasts demand at 86.9 for 2010 and over 88 mb/d for 2011.[13]

Production must not only meet demand but also compensate for declining production from existing oilfields. The WEO 2008 study showed far steeper decline rates than previously assumed and concluded that 19 mb/d of existing production would have to be replaced by 2015 and 43 mb/d would have to be replaced by 2030 just to maintain production at 2008 levels. New IEA boss Nobuo Tanaka pointed out the implications: adding in growth in demand, 'between now and 2030, we will need 64 mb/d of new oil production capacity, six times the size of Saudi Arabia's capacity today' (IEA 2008: 3).

Nevertheless, WEO 2008 asserted that new production could meet the reference scenario demand of 106 mb/d (38 billion barrels a year [bb/y]) in 2030. In contrast, Campbell put production in 2030 at 56 mb/d, down from his 2007 projection of 65 mb/d,[14] while the Energy Watch Group (Zittel and Schindler 2007) saw 2030 production falling even lower to just 39 mb/d. In claiming that the demand can be met, the WEO makes three assumptions: **technology investments** will extract more oil from existing wells; sufficient **new oil** will be discovered; production of **'unconventional oil'** will increase substantially.

Technology investments

Most oilfields yield only about 35% of the oil that is actually in the ground. A raft of new technologies for 'enhanced oil recovery' was developed in the US once its mainland production passed peak. These technologies get more oil out of old fields. In newer fields, the technology is built in or planned from the start. The potential of enhanced recovery is already accounted for and so does not add to reserves unless yet newer technologies are developed and applied. The WEO 2008 suggests that yet-to-be-developed technologies could increase yields from 35 to 50% and so 'boost world reserves by 1.2 trillion barrels – equal to the whole of today's proven reserves' (IEA 2008: 212). This is somewhat fanciful and at odds with the findings from the study of oilfields, which shows that technology investments to date have slowed the decline rate by less than a third.

Enhanced recovery is also designed to increase production rates. The result is that the newer fields have a shorter life while the final decline in all fields is much more rapid. Indeed, the attempt to maintain or increase production from old declining fields risks collapsing production altogether, leaving 'trapped oil' in the ground. In April 2006, the Saudis admitted having difficulty keeping up with demand and made an unprecedented call for energy conservation, particularly in the US. One official observed: 'When you have this kind of demand, you're forced to supply beyond the optimal rate'.[15]

Leggett concludes: 'Enhanced recovery made precious little difference to the inexorable decline of US oil production, and it will be no different globally' (2005: 68). Rather, it will sharpen the crisis of depletion when it hits because the rapidity of decline will leave little time to develop alternatives.

New oil discoveries

The WEOs assume that there are still very large fields to be discovered, particularly in the Middle East and North Africa. However, the main evidence it gives for this is that Western oil corporations have not been free to explore these areas since the 1970s. This echoes the argument, led by ExxonMobil and the US, that there is plenty of oil to be discovered but it is locked in by 'resource nationalism' and the inefficiency of state corporations. As with OPEC reserves, it is a story designed for political ends – to create pressure in support of big oil's bid to regain effective control of the world's largest reserves, irrespective of whether there is more oil to be found, while leaving OPEC with the blame for any shortfall. War and sanctions have left Iraq with the largest untapped reserves of 'easy oil' and, following deals signed in January 2010, the supermajors have taken control of them.[16]

The scale of discovery has long since diminished. The biggest discoveries were made in the 1950s and 1960s. Since then, many more wells have been found but, despite increasingly sophisticated exploration technologies, there has been a consistent downward trend in the amount of new oil discovered. The last time discovery matched production was in the early 1980s and the gap has been widening since. Moreover, new discoveries are smaller in size, contain lower-quality oil and are more often located in extreme environments than the earlier fields. Thus, Chad's oil, brought on stream in 2004, still makes it into the category of 'easy oil' but is of such low quality that only two or three refineries in the world can process it.[17] The implication is higher costs and dirtier production at both wells and refineries while the energy return on energy invested (EROEI) is lowered all along the production line. According to Heinberg, imported oil in the US had an EROEI of 8.4 in 1996 compared with 30 in the 1970s and 100 for US-produced oil before 1950 (2005: 138).

Several offshore finds in very deep water have been announced to much fanfare in recent years. In April 2010, BP's Deepwater Horizon drill rig was blown out of the water by the force of oil and gas released under enormous pressure from the great depth of the well. The rig sank with the loss of eleven lives and, over the next five months, some 4.9 million barrels of oil gushed into the Gulf of Mexico from the well-head 1.5 kilometres below the surface. Efforts to cap the well failed because of the difficulty of working at such depth and it was finally 'killed' only after a relief well was drilled. The blowout occurred just weeks after the US administration said it would lift a moratorium on offshore exploration. The announcement 'reflected the widely-held belief that offshore oil operations, once perceived as dirty and dangerous, were now so safe and technologically advanced that the risks of a major disaster were infinitesimal'.[18] That view was propounded by the oil industry to secure lax regulation from the US state, enabling BP and its contracted service companies – Halliburton and Transocean – to cut costs on the blowout prevention system. But the issue is not just about regulation. Big oil is pushing for rights to explore ever deeper and in ever more remote and extreme environments including the Arctic.

Unconventional oil

The WEO 2008 sees production from unconventional oil growing from 2 to 8% of global production in 2030. This in itself is an indication of the peaking of regular oil and of the poor state of reserves under control of Western oil corporations. In contrast, the 1970s oil shock resulted in interest in unconventional resources, but very little production.

The oil deposits in Canada's tar-sands and Venezuela's extra-heavy oils are truly enormous. The problem is getting it out. Canada's tar-sands have seen huge investments since 2002 to produce 1.2 mb/d in 2007. Only 60% of this is a low-quality 'syncrude', the rest being bitumen. The WEO 2008 sees this rising to 5.9 mb/d in 2030, a million barrels more than the 2006 forecast. In David Strahan's view, this was always unlikely but necessary to balance the IEA's projected demand.[19] New tar-sands projects need $85 a barrel to get a return and, with the price in free-fall, expansions totalling 1.7 mb/d were deferred or cancelled in 2009.

To date, most of the extraction has been done by opencast mining of tar-sands, creating pits some 80 metres deep and 7 kilometres wide, which is then 'washed' to separate the oil. Getting to the deeper reserves requires the injection of super-heated steam to make the tar more liquid, separate it from the sand and then pump it up. Both processes are very energy-intensive, yielding an EROEI as low as 1.5 in Heinberg's calculation (2005: 128).

The energy used for the process comes from Canada's rapidly declining gas reserves which are thus diverted from the future supply to heat houses and fuel electric generators. The Canadian Centre for Policy Alternatives (CCPA) notes that households in Alberta, where the tar-sands are located, are already turning to dirtier coal for heating. In principle, the natural gas can be replaced by gasification of the tar-sands themselves, but this further reduces the EROEI, is more expensive and more polluting. Substituting coal or nuclear power for gas is also being considered, according to Hirsch, Bezdek and Wendling (2005), but will similarly increase the energy costs.

As it is, the pollution is intense. Carbon emissions of around 125 million tonnes have broken Canada's Kyoto commitments and are accompanied by severe sulphur emissions. Energy intensity is matched by the intensity of water use and waste. The water is drawn from an environment that is already drying out under the influence of climate change. The effluent and tailings ponds,[20] up to 15 square kilometres and 50 metres deep, litter the landscape and 'no one knows where [the toxics] go after that' (CCPA 2006: 31). Scaling up production by five times does not present a pretty prospect.

The documentary film *H₂Oil* (Walsh 2009) shows that at least some of the toxins leach into the Athabasca River. They have been found downstream at Fort Chipewyan, home to a largely Native American community who traditionally fish the river to provide a substantial part of their diet. A local study showed high levels of 'arsenic, cadmium, polycyclic aromatic hydrocarbons and resin acids in the [river] sediment, as well as high levels of mercury in tested fish'. Not surprisingly, the local doctor has recorded increasingly high levels of cancer along with other ailments associated with toxic pollution.

The tar-sand industry's response is familiar to fenceline communities around the world: they didn't do it – the river was naturally polluted by the tar-sands prior to development. The Canadian establishment has rallied in industry's defence. Politicians routinely puff the industry as a wonder of the world, the official environmental and water agencies avoid relevant investigations, and the Fort Chipewyan doctor 'was summarily silenced by Health Canada and reprimanded by the College of Physicians and Surgeons in Alberta for causing "undue alarm"'.[21] Although cleared of the charge, he is effectively on notice that he'll be taken down at the first opportunity.

> **Box 5.2   Third-Worlding Alberta**
>
> CCPA sees Canada being turned into a peripheral energy province of the US. Both oil and gas are pumped south in ever greater quantities while North American Free Trade Agreement rules subordinate domestic energy security and conservation to the demands of the larger US market. More specifically, the US sees Canada as a secure source of fuel for its imperial war machine. At the same time, the aggressive neo-liberalism of Alberta's provincial government seems calculated to make a Third World enclave in the tar-sands area: the oil corporations have been given major subsidies in tax and royalty breaks while labour unions are purposely subverted, environmental regulation trashed and local government is starved for funds for social services or even to maintain the infrastructure in the oil boomtown of Fort McMurray.

Venezuela's sulphurous extra-heavy oils are only marginally easier to extract. Existing projects, hitherto managed by big-oil corporations, have recovered only 5 to 10% of the resource and production rates are slow. Technology developments could improve this but gains would be offset as the more accessible reserves are extracted. The WEO 2006 saw little development by 2030, with production rising from 100 to 400 thousand barrels a day (IEA 2006: 93). This view was clearly informed by Venezuela's inversion of the usual relationship between oil and social investment: in place of society subsidising oil production, oil is made to subsidise social investment. The IEA's view was no doubt reinforced in February 2007 when President Hugo Chávez decreed that Petróleos de Venezuela SA, or PDVSA, the state petroleum corporation, would take a majority stake in the heavy-oil projects. In May of that same year, PDVSA took operational control as well. The WEO 2008 is more optimistic, forecasting 1 mb/d from the Carabobo field alone. However, state auctions of oil blocks have been delayed several times as prices dropped and oil corporations held off.

Slow development retards both social and environmental impacts but is not intentional. In 2008, Venezuela secured OPEC recognition of heavy-oil reserves. This boosted its reserve figures from about 99 to over 170 billion tonnes and lifted it from fifth place in the OPEC reserve rankings, based on its 'regular' oil reserves, to second behind Saudi Arabia. Heavy oils thus secured a larger OPEC quota although they will contribute little to filling it.

> **Box 5.3 Challenging empire**
>
> In stark contrast to Canada, Venezuela has articulated an explicitly anti-imperialist agenda and is loosening the ties of dependency. The US is still its largest market but it is cutting back exports to the US while expanding exports to China. In Chávez's view: 'The United States as a power is on the way down, but China is on the way up. China is the market of the future'.[22] Venezuela is also pursuing Latin American integration under the banner of the Bolivian Alternatives for the Americas as a direct challenge to the failing US agenda for an all-American free trade zone.[23]
>
> At the same time, Venezuela is loosening big oil's grip on its petroleum industry. ExxonMobil, BP, Total, Chevron and ConocoPhillips are now minority partners in existing Orinoco projects while PDVSA is entering new partnerships with state-owned oil corporations, mostly from other Southern countries. Partnership deals with the China National Petroleum Corporation, Brazil's Petrobras, Cuba's Cupet, Iran's Petropars and Russia's Lukoil variously cover the certification of Orinoco reserves as well as exploration and extraction.
>
> These South-South partnerships may yet upset the IEA's estimate of what is technologically possible in the absence of the traditional big-oil corporations. Indeed, they reflect a deeper shift in the meaning of what constitutes big oil. Ten of the biggest fifteen oil and gas corporations by production are now state-owned according to WEO 2008. Saudi Aramco tops the list and ExxonMobil, the leading supermajor, comes in fourth. If ranked by revenues or profits, however, the supermajors take the top six places.

Gas

Gas is closely linked to oil. As noted above, it has replaced oil in applications such as heating and electricity generation and thus moderated oil demand in 2005 and 2006. It is also found in the same places and 'associated' gas is a by-product of oil production. Further, natural gas liquids, otherwise known as condensates, are produced from what McKillop (2006) calls hot greasy gas. In the US, such gas represents the tail end of production from otherwise depleted oil wells. The natural gas liquids share of crude-oil production is rising rapidly. It now contributes 13% to oil production and, on current trends, WEO 2008 sees natural gas liquids production doubling to account for 20% in 2030. This is 20 mb/d and a massive increase on the 2006 projection of 15 mb/d by 2030.

Several big oil corporations are also constructing gas-to-liquid (GTL) plants using ordinary natural gas and WEO 2008 sees production rising from just 50 thousand barrels a day to 650 in 2030. However, 'much of the gas used by GTL plants is for the conversion process, which is extremely energy intensive' (IEA 2006: 113). The EROEI, in other words, is dismal. In February 2007, escalating development costs prompted ExxonMobil and Qatar Petroleum to abandon a joint GTL project. The partnership will instead supply gas to the domestic Qatar market.

Overall, WEO 2008 forecasts gas demand (excluding natural gas liquids) rising from about 49 mb/d of oil equivalent to over 75 mb/d[24] in 2030, with electricity generation accounting for the largest part of the growth. North America and Europe remain the largest consumers, with the US by far the largest single consumer. Reserves are increasingly concentrated in Russia and other Central Asian countries and in the Middle East. Sour gas – with a high content of hydrogen sulphide and/ or carbon dioxide – constitutes over 40% of reserves. Most of it will be transported by lengthening pipelines as supplies close to the main markets are depleted. As the network expands across Europe and Asia, China will increasingly compete with Europe for Russian gas.

Piped gas is increasingly supplemented by liquefied natural gas that can be shipped – although at a major cost in energy as liquefied natural gas must be refrigerated to minus 176 °C. Although WEO 2008 sees North American production holding up through to 2030, this is not enough to meet future demand. Being isolated from the expanding pipe network in Eurasia, WEO 2006 assumed that US imports of liquefied natural gas would 'make good . . . the shortfall' (IEA 2006: 120). However, 'some productive activities have stopped or been shifted overseas, where gas prices and overall production costs are lower. The US chemicals industry, which relies heavily on natural gas feedstock, has contracted sharply in recent years' (293). This seems something short of 'making good the shortfall'. It also indicates an accelerated movement of energy-intensive dirty industries to locations where the energy is available, mostly in the global South.

Since 2006, the US has massively expanded production of non-conventional 'tight' gas from sands, coal-beds and, most spectacularly, from shale. This has more than compensated for rapidly declining conventional production. Developing tight gas is typically costly and energy-intensive and the wells are quickly depleted. In 2008, constant drilling – with 33 000 new wells each year – barely kept pace with demand according to investment touts *Energy and Capital*. When demand and prices dropped in late 2008, drilling was abruptly halted on most shale fields.[25]

According to the IEA, the growing global demand for gas can be met by existing reserves, not including new discoveries, for 40 years. It also remarks that gas is preferred on environmental grounds because it has a lower carbon intensity than oil.

Both these conclusions are contested by McKillop (2006). First, gas reserves are overstated in the same way that oil reserves are. Russia, supposed to have the largest reserves, is having difficulty maintaining production. Second, peak oil is provoking a sharp increase in gas-based oil production. Third, 'the loss rate is increasing much faster than production'. As long as oil is more valuable than gas, 'associated' gas is treated as waste and vented, flared or re-injected into the well unless there is an infrastructure to gather it. Gas is also a by-product of natural gas liquids production and is similarly vented or flared. In both cases, the proportion of gas to oil increases as the oil reserve is depleted. On the World Bank's estimate, gas equivalent to 943 million barrels of oil is flared every year. Finally, natural gas is prone to leaking and the scale of losses will grow faster than the infrastructure. Larger storage at the market end will leak more as will the lengthening pipelines tapping ever smaller reserves in increasingly harsh environments. Meanwhile, the costs of maintaining production and infrastructure will spiral.

McKillop concludes that gas depletion is happening much faster than assumed and those who hope that gas will provide a 'bridge' to a clean energy future will find the bridge collapsing. And while gas burns cleaner than oil, the scale of losses undermines the environmental claims. Flaring releases millions of tonnes of carbon dioxide while venting and leaks release methane. Getting at the shale gas, meanwhile, is both energy- and carbon-intensive. It requires horizontal drilling combined with 'hydraulic fracturing', which involves injecting a combination of water, sand and toxic chemicals into the well at high pressure to force the gas from the shale. Evidence has already emerged of large-scale poisoning of groundwater.

Coal

Coal consumption has increased faster than either oil or gas since 2000. The WEO 2008 puts global coal demand in 2006 at 4.4 billion tonnes, equivalent to about 63 mb/d of oil, and rising to over 7 billion tonnes, equivalent to 100 mb/d, in 2030.

As with gas, coal prices are dragged up by the high oil price. 'Steam' coal competes with both oil and gas for electric power generation and for industrial process heat. In addition, coal-to-liquid (CTL) and coal gasification for chemical production or for use as gas becomes more competitive at higher oil and gas prices.

As with unconventional oils, the 1970s oil shocks provoked interest in CTL but projects collapsed as crude-oil prices crashed in the 1980s. The exception was Sasol's construction of the Secunda CTL plants, which was driven by apartheid South Africa's increasing international isolation.

The WEO 2006 saw minor growth in CTL and this is little changed in WEO 2008. Rising coal prices offset rising oil prices and the plant cannot be viable unless it sits on top of a very large cheap coal reserve. Capital costs are exorbitant, estimated at over $5 billion in 2006, compared with $2 billion for GTL, for a plant producing 80 000 barrels a day. The process is even more energy-intensive than GTL and carbon emissions are astronomical. The WEOs do not comment on the intensity of pollution from other emissions such as sulphur dioxide or the intensity of water use and pollution.

Farrell and Brandt (2006) believe the IEA underestimates the likely use of all synfuels including CTL, GTL and syncrude from oil sands and extra-heavy oils. Hirsch, Bezdek and Wendling (2005) positively advocate a major CTL building programme in the US in anticipation of peak oil. In 2007, Sunita Dubey of groundWork reported that at least nine CTL plants were being planned in the US and the industry, including Sasol, was lobbying hard for subsidies to build capacity of around 2.6 mb/d by 2025.[26] Sasol's plans are furthest advanced in China. In 2007, two 80 000 barrels a day (b/d) plants were on the cards but, with the collapse of markets, this has been reduced to one. It is also investigating CTL production in India and Indonesia (see Chapter 6).

Electricity generation is the biggest consumer of coal. Compared with earlier projections, WEO 2008 sees 'slower economic growth' (IEA 2008: 141) and therefore reduced additional demand for electricity: instead of doubling, consumption increases by 80% and coal increases its share of production from 41 to 44%. In the 550 scenario, electricity consumption rises less steeply as energy efficiency is promoted although more electricity is used for transport. Switching to other energy sources also reduces coal's share but there is still an absolute increase in the amount of coal burnt for electricity.

Despite massive global reserves, WEO 2008 remarks that 'the rapid increase in demand in recent years has seen the reserves-to-production ratio fall sharply, from 188 years in 2002 to 144 years in 2005' (128). It attributes the decline to a 'lack of incentives' rather than of available resources. Nevertheless, supply lines have been stretched and new coal is harder to mine. The US holds the largest reserves but started importing because of increasingly high mining and transport costs. In 2007, China overtook the US as the biggest producer and consumer. It was also a major exporter but is now importing more than its exports. India, the third-largest consumer, similarly holds major reserves but is already importing large amounts of coal. The WEO 2008 projects world trade rising from 613 to 979 million tonnes in 2030, implying very large infrastructure construction. Major exporters, including South Africa, have been expanding railways and ports to handle bulk exports.

Coal is the dirtiest of the fossil fuels and has the highest carbon density. 'Clean coal technologies' are now being promoted to justify the continuation of the industry in the context of climate change. Carbon capture and storage (CCS) is the main hope. This involves separating carbon dioxide from the emissions stream – leaving other pollutants to go their way unless separately scrubbed – and injecting it in liquid form into deep geological strata or the ocean. The scale required for meaningful CCS makes it improbable. Technology and environment academic Vaclav Smil comments: '... [T]o sequester just 25% of CO_2 emitted in 2005 by large stationary sources ... we would have to create a system whose annual throughput (by volume) would be slightly more than twice that of the world's crude-oil industry...'[27]

The ocean has already absorbed an overload of carbon dioxide and is consequently becoming more acidic. This is already affecting the reproduction of krill, the foundation of the ocean food chain, and so threatens to collapse fisheries. Risking accelerated acidification through ocean sequestration thus seems like a really bad idea. As a liquid, it may also spread across the ocean floor creating dead zones.

That carbon dioxide, injected on the scale required, will stay where it's put in geological strata is also uncertain and is possible only in particular geological formations. Such formations do not necessarily coincide with the location of power plants and other big industrial emitters. South Africa, for example, has recently mapped its CCS potential and the best prospects are offshore and remote from the carbon-intensive power and CTL plants. Many other industrial regions would need to construct 300-kilometre pipelines to take the carbon dioxide to suitable locations. A peculiarity of the CTL process is that it already separates out a portion of carbon dioxide and so makes capture relatively easy. Adopting CCS, whether or not it actually works, therefore requires the additional costs of compressing and injecting it. Power stations would, in addition, have to separate the carbon dioxide, which is very costly and consumes around 30% of the energy produced by the power station – so producing even more carbon to be sequestered.

CCS was not recognised under Kyoto so carbon credits cannot be claimed but the pressure is on to change this. CCS has long been pushed by the US, the World Bank and corporations to avoid cutting fossil-fuel use. Europe has joined the clamour in order to meet its own unilateral target to cut carbon emissions by 20% by 2020.

A European Union directive, issued in April 2009, indicates that new power plants should be 'CCS-ready' and power corporations anticipate that they will be required to implement CCS. This puts the coal industry in a quandary. One the one hand, it has promoted CCS as a response to climate change. On the other, it is concerned that the cost will wipe out coal's price advantage and generators will turn to nuclear instead.[28]

Nukes

The WEO 2008 sees a modest expansion of nuclear-power generation in the reference scenario and much greater expansions in the 550 and 450 scenarios. It argues that this will enhance energy security and reduce carbon emissions. The claim of carbon savings is widely disputed, however. Nuclear energy does not emit carbon from the generating plant but the full cycle of production is both energy- and carbon-intensive. Heinberg (2005) argues that, in energy terms, nuclear has been subsidised by cheap oil just as it has been subsidised economically by governments for (usually unacknowledged) military reasons.

The IEA says uranium deposits are plentiful and widely distributed. At current usage, this may be the case. A worldwide turn to nuclear would, however, soon test the limits of supply and production. Again, this is not just about whether or not there is uranium in the ground, but how fast it can be extracted and processed to supply a greatly expanded industry as the high-grade 'easy' uranium is mined out. At present, the world's 443 nuclear power stations consume 68 000 tonnes. Only 40 000 tonnes comes from mining. The rest is supplied from decommissioned Russian warheads that will be used up by 2013. Like oil, uranium prices are volatile now and a recent sharp fall in prices is putting investments, and hence future supplies, in jeopardy.

The mining industry has been prone to disaster. In October 2006, the Cigar Lake mine in Canada flooded with groundwater. This is a new mine still under construction by Cameco, the world's leading uranium producer. It was advertised as the world's largest undeveloped uranium deposit and expected to supply 10% of world demand from 2008. Following the flood, Cameco said it would bring the mine into production in 2010. A second flood interrupted remediation work in 2008 and the corporation now says it will bring the mine on line in 2013. The scale of groundwater contamination is unknown but remediation plans involve pumping it out to the surface.[29] Short of disaster, miners are routinely exposed to radiation while mine tailings leave a radioactive legacy for tens of thousands of years. Niger supplies most of the uranium for France's nuclear power stations from mines operated by French nuclear corporation Areva. Radiation levels on the streets of local towns are up to 500 times higher than normal and drinking water in some areas is also contaminated, according to a report by Greenpeace. The mines are also depleting water sources and threatening to wipe out the local pastoral economy.[30]

In addition to fuel production, nuclear construction is enormously costly in energy, carbon and money – with a history of over-running large budgets by three times or more[31] – and again in the disposal of waste and in the final decommissioning. Taking account of the full nuclear cycle therefore substantially lowers the EROEI of nuclear and destroys its carbon claims. The Eco-Institute in Darmstadt, Germany, calculates that a 1 250 MW nuclear power station in Germany emits 33 grams of CO_2e per kWh, amounting to 250 000 tonnes per year. Carbon emissions are higher for lower grades of uranium ore: for grades between 0.1 and 1%, CO_2e emissions are 120 grams/kWh.[32]

The last step, decommissioning and disposing of high-level nuclear wastes, has a particular significance. First, no satisfactory solution has been found for either. Second, in a post-peak oil context, decommissioning will compete with other resource demands and may simply be beyond the capacity of a declining energy system. Nuclear power will then leave an irredeemable toxic legacy to future generations. Economist David Fleming calculates that, by 2020, it will take more energy to clean up nuclear sites and deal with their wastes than the whole nuclear industry will be able to generate from the remaining uranium ore (cited in Heinberg 2007: 7).

Nuclear power claims an above average safety record because it is tightly regulated. This is partly achieved simply by secrecy. Many incidents at nuclear plants have come to light years after the fact. Even if it were true, the claim does not address the real issue that a single incident can be catastrophic. The 1986 meltdown of the reactor at Chernobyl in Ukraine spread radioactive fallout across Europe. Recently published research puts the death toll at close to a million people.[33] The area surrounding the plant is effectively sacrificed forever. The multiplication of plants around the world clearly increases the risks of catastrophic failures.

Finally, the proliferation of nuclear power cannot be dissociated from the proliferation of weapons. The Non-Proliferation Treaty has been discredited and now appears as a tool for maintaining the military advantage of the great powers and their allies. The US has abrogated its own obligations under the treaty, supported Israel's nuclear capacity in defiance of the treaty, and used the treaty as a diplomatic weapon against Iran.[34] In this context, it proposed a Global Nuclear Energy Partnership that is little more than a move to take control of the world's nuclear supply chain.

Big hydro

Large dams currently supply about 16% of the world's electricity, according to WEO 2008. Despite a doubling of capacity, this share of supply drops to about 14% by 2030 in the reference scenario. The share increases in the 550 and more so in the 450 scenario as more dams are built. The IEA claims that less than one third of potential hydro power has been exploited with most of the potential in developing countries. It shows the largest potential expansion in Africa. It also repeats the common assumption that big hydro has low-carbon emissions: 'In Brazil . . . where more than 80% of electricity is hydropower, the power sector accounts for just 10% of the country's CO_2 emissions, four times less than the world average' (IEA 2006: 142).

This claim omits the very large carbon emissions associated with dam construction and even larger emissions of methane from rotting submerged vegetation. This is just the beginning of the social and environmental impacts. As noted in the introduction, the 45 000 big dams already built have had a major impact on earth's freshwater hydrology. They have also forced the removal of over 100 million people worldwide and submerged their most productive river valley fields. Not surprisingly, they have provoked massive resistance and are routinely accompanied by heavy state repression.

Big dams are the 'economic hit man's' project of choice. The economic benefits are invariably overstated while the costs understated – even when the cost to those dispossessed is ignored. Thus, the crisis of surplus petrodollars in the 1970s led to a massive round of dam building to the benefit of corrupt Northern banks, construction corporations and Southern elites while the debt burden was mostly imposed on ordinary citizens through such instruments as structural adjustment programmes.

The World Bank says the Congo River has the potential for 100 000 MW of hydropower. This is no doubt a salesman's figure to provoke investor interest in the Grand Inga project, which the Bank is touting in partnership with the World Energy Council. The potential capacity of the project is advertised at 40 000 MW – equal to South Africa's total power production and twice the size of the Three Gorges Dam in China. Visiting the site, Bank president Robert Zoellick said that what 'brings the biggest change to people's lives [is] bringing electricity to rural communities. It transforms the lives of women most of all because they get labour saving devices, they get lights so they can study at night, and it helps the kids with school'. But he also 'urged African governments to design more "bankable" infrastructure projects . . .'.[35] That means getting the product to market to pay the debts. In the case of Grand Inga, the intended markets are Europe and energy-hungry industries in South Africa – not poor rural African women. The cost is estimated at $80 billion and this will certainly rise dramatically if the project goes ahead. The debt will accrue to the already indebted Democratic Republic of Congo (DRC) and provide a pretty income for finance capital with the World Bank as enforcer.

There are already two dams, built in the 1970s and 1980s, at the Inga Rapids. They have produced little power and drained 'the country's finances for decades'[36] but are now being refurbished with World Bank support. A third project, the 5 000 MW Inga 3, has been the subject of lengthy negotiation between southern African state-owned utilities. It was originally driven by South African interests and primarily intended to transmit power to South Africa. The DRC government has apparently pulled the plug on this and is in talks with BHP Billiton to both fund the project and to build an aluminium smelter or two as its primary market. The people displaced by the existing Inga 1 and 2 dams have not been compensated and live in wretched conditions. Most do not get electricity. And all are excluded from negotiations on the planned projects and from information on the terms of the deals already done.[37]

Many of the east and southern African countries now produce most of their electricity from hydro. The supply, however, has proved erratic as the regular droughts in the region cut the flow of water. Climate change will exacerbate this vulnerability. Tanzania relies on hydro for nearly 90% of its electricity. In 2005, drought cut capacity from this source from 559 MW to 120 MW[38] and resulted in widespread outages. In Uganda, the priority given to power production at the Nalubaale Dam[39] resulted in over-use of water from Lake Victoria and lowered the lake's level. The World Bank nevertheless approved a $360-million loan package for the construction of the Bujagali Dam downstream of Nalubaale. Hydrologist Daniel Kull comments that the Bank's studies ignored the 'true damage done to Lake Victoria by the existing dams and follows with a selective and optimistic view of current lake levels and possible climate change impacts' (2006).[40]

Hydropower projects and competition for water intersect in many parts of the world. All the countries east and south of the Himalayas, home to half the world's people, have big dam-building and water-transfer ambitions both to compensate for depleted aquifers and to generate power. China controls the Himalayan headwaters of most major rivers and it has the money, skills and resources to carry out grandiose projects. The biggest is a 40 000 MW hydropower scheme at the 'great bend' of the Yalong Zangbo River in Tibet. The Yalong Zangbo is a major tributary of the Brahmaputra which, together with the Ganges, feeds the great delta that defines Bangladesh and West Bengal in India. Both countries suspect that the great bend scheme is also intended to divert water to the Yellow River to supply China's dry north. India itself, however, has plans to divert water from both the upper Brahmaputra and the Ganges (Pomeranz 2009).

Biofuels

There are two basic forms of biofuel: ethanol is an alcohol produced from just about any plant matter to blend with, or substitute for, petrol; biodiesel is produced from vegetable oils. Biofuels are heavily advertised as a renewable fuel source and carbon-neutral because the carbon emitted when they are burnt is supposed to equal the carbon absorbed during the plant's growth. This may be so where they are produced from recycled cooking oils or, on a small scale, from organic agriculture. It is certainly not so where the agriculture is itself energy- and carbon-intensive, besides being a major polluter, and the scale of production threatens food security.

Production more than doubled between 2003 and 2007, largely driven by high oil prices and concerns over energy security but justified also by the supposed climate benefits. Nevertheless, biofuels accounted for less than 1.5% of liquid fuels for transport (equivalent to 600 000 b/d of oil) in 2006. By 2030, this rises to 5% (3.2 mb/d) in the WEO 2008 reference scenario. The volume of biofuel production is substantially increased in the 550 scenario and doubled in the 450 scenario.

Together, the US and Brazil produce 80% of biofuels, prompting the US to propose, in February 2007, a biofuel partnership to promote the industry. Brazil has been the industry leader, developing ethanol production from sugar following the 1970s oil shocks. The subsequent fall in oil prices squeezed the industry but it was maintained through compulsory blending of fuels. Production reached new highs in 2007 with ethanol supplying around 14% of domestic fuel. Brazil is the world's leading exporter but this market slumped along with the oil price in the second half of 2008. Massive expansion of ethanol production from maize in the US, motivated primarily by national energy security, accounts for most of the growth in world production. In 2005, the US overtook Brazil as the largest consumer and producer of biofuels, consuming around 15% of the maize crop but producing only 1% of US liquid fuel demand. Production was set to double by 2008 but, even if the whole maize crop were used, biofuels could provide only 7% of US demand, according to Pimentel, Patzek and Cecil (2007). The industry is made viable only by heavy subsidies to ethanol production, on top of the extravagant subsidy of US industrial agriculture, supplemented by tariff protection. Despite this support, many of the corporations that rode the boom went bust in 2009, according to the *Wall Street Journal*.[41]

Biodiesel accounts for only about 15% of biofuels. Europe is the main centre of production and consumption. It has more than tripled output since 2000 and also become the largest importer as it chases a European Union (EU) target of 20% biofuels in the liquid-fuel mix by 2020. The expansion is ostensibly motivated by climate concerns but the timing, as well as the dubious nature of environmental claims, indicates that energy security and agricultural policy are the real drivers. For Britain to meet the EU's target, however, 'would consume almost all our cropland' (Monbiot 2006: 158). The whole of Europe will not do much better so the policy implies a heavy reliance on imports from Southern countries where land and labour are cheap.

The claim that biofuels reduce carbon emissions assumes a positive EROEI: fossil-fuel inputs in agriculture and in the production process must be less than the energy content of the biofuel. A study cited by WEO 2006 supports this claim, but Pimentel, Patzek and Cecil (2007) show that this study[42] does not take account of the full range of energy inputs. Taking all farming, ethanol production and marketing energy inputs into account, they find that US production of maize ethanol requires '43% more fossil energy than the energy produced as ethanol'. If wastes can be turned into by-products, this would be reduced to 28%. Tropical sugar has the best EROEI but it is still negative in their view. Biodiesels, according to Pimentel and Patzek (2005), also show negative energy returns.

The carbon equation does not end with the energy equation. Soil is a major carbon 'sink' – that is, it absorbs carbon from the atmosphere. Industrial agriculture destroys this function as heavy machines compact soils while agricultural chemicals kill the microbes that give structure and life to soil. The effect will be to reinforce one of the feedback loops created by global warming. Higher temperatures are expected to convert soil from a major sink to a major source of carbon dioxide. Journalist George Monbiot notes that this reversal 'was not supposed to happen for several decades but in 2005 British scientists reported that soils in England and Wales had already become carbon sources' (2006: 10). Meanwhile, the conversion of land to industrial agriculture results in a massive loss of carbon to the atmosphere. European demand for biodiesel has driven a rush into palm oil production in Malaysia and Indonesia. Natural forests are being cleared and peat bogs drained on a very large scale to make way for industrial palm plantations that scarcely begin to compensate for the carbon losses caused by the clearance.[43]

This conversion of land use is associated with dispossession. The Brazilian land movement, Movimento Sem Terra, notes that it gives new intensity to established patterns of rural dispossession, gross exploitation of labour and environmental destruction associated with sugar throughout its history. In a joint statement with social movements from other Latin American countries, titled 'Full tanks at the cost of empty stomachs', they denounced the US-Brazil 'biofuels partnership' as part of a US geopolitical strategy to counter Venezuela's influence in the region. The partnership was also intended to support the interests of Northern transnational gene and agribusiness corporations. Biofuels thus created the basis for novel partnerships between agribusiness, big oil and motor corporations and now represented 'an important source for the accumulation of capital'.[44] Brazil's role 'would be to provide cheap energy to rich countries which would represent a new phase of colonisation'.[45]

In May 2007, a group of African NGOs responded to Britain's proposed Renewable Transport Fuel Obligation that sets out biofuel targets.[46] They noted that their response was uninvited because the consultation process was restricted to Britain but that meeting the targets implies large-scale land conversion in Africa. Already, the Ugandan government planned to give over 7 000 hectares of the 30 000 hectare Mabira Forest, a reserve area, to sugar corporations interested in ethanol. The forest is part of a people's commons, contributing to the livelihoods of over a million people who draw on it for water, firewood, honey, mushrooms and materials for making baskets. Ironically, it also conserves the Lake Victoria catchment adjacent to the Nalubaale and the proposed Bujagali dams, and its preservation was agreed as necessary to optimising the new dam's performance. Opposition to the giveaway was intense and demonstrations in Uganda during April 2007 were accompanied by a police crackdown and rioting. In October, government backed down, scrapped the sugar deal and said it was committed to conserving Mabira.[47] Elsewhere in Uganda, palm-oil plantations are displacing forests while Benin is planning a major expansion of palm oil in peat bogs.

More broadly in Africa, the potential use of cassava for ethanol poses 'an especially grave threat to the food security of the world's poor,' according to agricultural economists C. Ford Runge and Benjamin Senauer (2007). It is the staple 'for over 200 million of Africa's poorest people . . . the food people turn to when they cannot afford anything else' and a reserve against the failure of other crops. Higher cassava prices will certainly be advertised as benefiting peasant producers but, the authors note, 'the history of industrial demand for agricultural crops . . . suggests that large producers will be the main beneficiaries'. An African NGO group is similarly concerned that dispossession and the privatisation of common lands, along with environmental degradation, will follow from large-scale biofuel mono-cropping. They argue that biofuels will be sustainable only within diverse farming systems controlled by local people and 'produced for household, local or domestic use, in order to meet the energy needs of the poor' rather than the demands of export markets constructed 'as a quick-fix replacement to fossil fuels'.

Environmentalist Lester Brown calculates that '[t]he grain required to fill a 25-gallon SUV gas tank with ethanol will feed one person for a year'. The expansion of US biofuels in 2006 doubled the price of maize, which in turn dragged up wheat and rice prices and pushed up feed costs to raise meat, dairy and egg prices. Because the US dominates world grain trade by virtue of farm subsidies – which have precisely that intention – US prices set world prices. In Mexico, the price of tortillas went up 60% and millions faced the prospect of empty stomachs. In February 2007, 75 000 workers and peasants took to the streets of Mexico City in protest and extracted a promise of price controls on maize products from the government.[48]

Journalist John Ross notes the deeper roots of Mexico's food crisis. The North American Free Trade Agreement signed in 1994, subjected Mexican peasant farmers to direct competition with US corporate agriculture that receives 'up to $21 000 an acre in subsidies from the US government, enabling them to dump their corn over the border at 80 percent of cost'. In consequence, six million peasant families have been forced from their land and joined the stream of migrant labour. The North American Free Trade Agreement also enabled US corporates, in partnership with Mexico's dominant firm, to take control of distribution and retailing. They are now aiming for control of the seed market and used the food crisis to attack a ban on genetically modified seed, claiming that 'bio-tech is the only solution to growing more corn and keeping the tortilla affordable'. The corporations are in fact focusing on genetic modifications to enhance biofuel production.[49]

High food prices were not only driven by biofuels. Australia is the second biggest wheat exporter after the US. The longest and worst drought on record collapsed production in the Murray River basin. Independent of climate change, industrial agriculture is undermining its own resource base, resulting in the global loss of 5 to 7 million hectares every year from land degradation and another 1.5 million hectares from water-logging and salination, according to the Food and Agriculture Organisation (FAO).[50]

As with oil and other commodities, speculative capital in flight from the crashing equity markets also crowded into food taking short-term profits, effectively traded for people's lives, until the commodities bubble burst in 2008. Prices then eased somewhat but were again rising sharply in 2010. Meanwhile, the crisis provoked a

Third World land-grab. It is driven by two distinct sets of interest. First, several cash-rich countries faced the prospect of food shortages and responded by looking for cheap land in other countries. Second, farm land is now seen as a long-term safe haven investment by finance capital as well as agribusiness transnational corporations.[51]

Renewables

'Other' renewables – wind, solar, ocean, geothermal – are the fastest-growing energy sector, according to the IEA, but off a very low base. In 2006, they produced around 0.6% of global energy supplies and by 2030 will produce only 2% in the reference scenario. In the 550 and 450 scenarios, they produce 3 and 4.7% respectively. Renewables have a greater share of electricity generation: 2% in 2006 and 8.5% (reference), 13% (550) or 19.6% (450) in 2030. These figures are buffed up by the inclusion of 'modern' biomass. Growth is fastest in the global North but, in the 450 scenario, there is very substantial growth of renewables in the South as well. The IEA gives more credence to renewables in WEO 2008 than in previous years. They nevertheless remain something of a niche market except in the 450 scenario.

This perhaps indicates a turning point in the energy establishment's traditional hostility to renewables. While fossil fuels benefit from immense subsidies from the World Bank and national states, renewables have generally been discriminated against. Thus, former World Bank president James Wolfensohn thought them an interesting option but 'we also have to remain realistic: renewable energy is expensive' (quoted in Simms, Oram and Kjell 2004: 20). The New Economics Foundation responded that this view 'reflects the interests of the Bank's major donors' fossil-fuel industries'. Further, oil, coal and gas are used to catch poor countries 'in a nexus of dependency relationships with other nations, multilateral donors, and foreign companies' (2004: 23). Renewables are dangerous to this establishment because they offer poor countries and, more particularly, poor people a potentially autonomous energy supply and the possibility of throwing off the shackles of dependency.

It does not follow from the New Economics Foundation's argument that renewables can replace fossil fuels to maintain profligate consumption by industry and the world's rich. The opinion of environmentalists is sharply divided on this issue. Leggett argues that renewables can produce enough energy to meet the global 'demands of 10 billion people wasting energy at the level your average wasteful European does today' including energy for transport (2005: 201). The more immediate problem is the transition from fossils to renewables. Building a renewable systems and infrastructure will take time and consume vast amounts of energy that is not presently available from renewable sources. In other words, a renewable future needs an energy subsidy from fossil fuels to get started and, past peak, that subsidy will not be easily available. Leggett therefore sees a period of chaos followed by the explosive growth of renewables. A massive programme of energy conservation and efficiency will ease the transition. In the meantime, people should mobilise to persuade governments and corporations to start the switch now. Every new investment in fossil energy is not only a commitment to future carbon emissions but also a misdirection of resources.

Many environmentalists, mostly but not exclusively Northern, share a basic assumption with Leggett: preventing runaway climate change must be achieved within the present order of power – that is, within the context of capitalism and economic growth. Time is now so short that the powers must be persuaded to a heroic international effort, equivalent to war-time mobilisation and combining the resources of nation states and corporations. They must be persuaded that renewables can keep the world economy powered-up for growth. Against this, many environmentalists and most peak oil analysts do not believe renewables can come close to replacing the flow of energy from fossil fuels. While renewables can be and must be expanded, along with a massive drive for energy conservation, the overall energy supply will contract and it will not be possible to sustain economies based on growth.

Renewables include a very wide range of technologies and energy sources and the definition of what is or isn't renewable is contested. From Britain to South Africa, countries that have adopted renewable energy targets include biofuels, landfill gas and big hydro in their definitions of renewable. Paul Mobbs (2005) distinguishes between low-carbon energy, including biofuels, biomass and biogas, and renewables that use natural flows of energy ultimately derived from the sun or from gravity, including solar, wind, hydro, wave, ocean current and tidal energy.

For practical purposes, low-carbon technologies are renewable if they are founded on sustainable production systems. In industrial contexts, most are not. As noted above, industrial production of biofuels scarcely qualifies even as low carbon. Similarly, biogas from waste dumps is heavily promoted by governments and the World Bank. Landfill gas is produced from rotting organic matter but, being contaminated by other matter in the dumps, is toxic. In contrast, biodigesters capture the gas from sewage and organic matter before it becomes a pollution problem and also produce compost (energy in a different form). However, they would require a thorough transformation of waste management and scarcely register in the official energy future.[52] Firewood is the primary fuel for the rural poor in the global South and the carbon account is balanced if new trees are grown to replace what is burnt.

Renewable energy resources are abundant and free but, in contrast to fossil energy which is very dense, renewables are diffuse or 'thin'. This means that only a small fraction of the potential can be used in practice. Some, but not all, renewables are also limited because they are intermittent sources of energy and they therefore need backup from more constant supplies or from storage. Renewables may produce heat energy – as with solar water heaters – or be converted into another usable energy, usually electricity. Given the monetary subsidy to fossil fuels, renewables have competed on unequal terms but, with rising fuel prices, they become more competitive. The costs of a number of technologies are also falling as production is scaled up. Wind energy is now more or less competitive with conventional power in many countries and new innovations are reducing the very high costs of solar photovoltaic (PV) power. In contrast to fossil fuels and nuclear energy, both of which will face rising fuel costs in the future, the costs of production from many renewable systems falls over time because the energy source is free. Nevertheless, if renewables must provide the energy 'to make and operate' the renewable system itself – including mining and manufacturing – the EROEI declines and 'in some [not all] cases is negative', according to renewable energy researcher Ross McCluney (2005: 161).

There are also environmental limits to renewables. They do incur environmental costs, not only in mining, manufacture and construction but also in their operation. Solar PV panels contain metal toxins that will require sophisticated waste management starting with the design of panels to enable safe recovery and re-use. For most other renewables, the issue is one of scale. Small-scale and dispersed systems have a negligible impact but, if they are scaled up to replace a major portion of fossil energy, the impacts will be substantial because they draw energy from the natural system. Thus, a large tidal dam has had a severe impact on an estuary in northern France, large wave systems will cause coastal erosion and may affect marine life sensitive to noise, large ocean-current systems will slow and may divert the current, very large wind farms have been shown to affect local climates although this impact does not begin to compare with fossil-fuel impacts.[53]

In short, concentrated large-scale production from most systems will likely become as controversial as big hydro. Yet it is precisely such systems that tend to be favoured by national states and big corporations. The Tyndall Centre, a British climate-change think tank, observes that the 'existing regulatory system for electricity distribution operates within the paradigm of centralised generation and one-way flow of electricity from large power plants to users. The "passive" user has co-evolved with such a supply system' (2005: 73).

The implication is not only that national grids are designed on the assumption of a few big power sources but, more importantly, the system embodies the concentration of political and economic power. The proposed Desertec project, initiated by a consortium of twelve major European energy, engineering and banking corporations, fits the bill. It involves massive solar installations in the Sahara Desert to produce 15% of Europe's electricity by 2050. The New Economics Foundation's understanding of renewables linked to people's autonomy is expunged here. Instead, renewables are fixed within the 'nexus of dependency'.

Centralised power is also extremely wasteful. Heat losses from big plants combined with losses from long-distance transmission through the grid means that less than 40% of the primary energy used to fuel the generator ends up as useful energy. A growing body of environmentalists therefore advocate decentralised energy systems based on numerous micro-generators producing electricity for localised mini-grids and heat where it will be used. In this way, districts and even households would produce a surplus and the surplus from the mini-grid would feed into the national grid. In the Tyndall Centre's view, it could also 'stimulate new user/consumer identities as awareness of energy per se, and of sustainable energy in particular, rises' (74).

Energy decentralisation relates to a wider set of demands. In the North, a variety of social movements call for 'localisation' and an end to dependency on the plunder from the South. In the South, movements of resistance to corporate plunder call for local food and energy sovereignty. In the peak oil perspective, localisation is likely to be forced by declining oil production. Renewable energy is then a matter of survival and there can be no assumption of economic growth. Mobbs observes that 'when you cut your energy consumption by 75% renewable energy options become far simpler . . .' Moreover, the term 'energy-poor' will lose its meaning in a future where none are energy rich because 'the amounts of energy will not be as relevant as the extent to which people can control and operate their own energy systems' (2005: 172, 173).

The use of small-scale and dispersed renewables is thus linked to local and democratic control of production because, unlike fossil and nuclear fuels, they do not require centralised corporate empires to manage them. We will return to this in Chapter 10. In the present, however, the democratic potential associated with decentralised renewables does not mean that they cannot or will not be managed by corporate empires. Indeed, several of the oil supermajors, including BP and Shell, are already established as leaders in the renewable energy field. They are widely criticised for co-opting renewables for corporate greenwash because their investments in renewables are dwarfed by their fossil-fuel investments. Nevertheless, their deep pockets give them a leading role in defining renewable technology options and business models to centralise profits even from decentralised plant. Nor does local necessarily mean more democratic. Corporate control of local generation may well be accompanied by local despotism in order to enforce the returns on investment.

Efficiency and the ends of energy

Whatever the energy source, efficiency is held to be the easiest and cheapest means both to energy security and carbon reductions. In a market system, however, energy efficiency leads to an overall increase in energy use. This is known as the 'Jevons paradox'. For capitalism, increased energy efficiency is another form of increased productivity. It increases the work done by energy but the benefit is taken in profit and economic growth rather than a reduction of overall energy use. Put differently, the priority is the efficiency of capital, not energy, and the additional returns to capital must then be reinvested in further economic activity that requires more energy.

Moreover, efficiency generally assumes 'grandfathering'. That is, existing technology systems are assumed and efficiency is advocated within that system. Thus individual units such as cars are made more efficient but the transport system and the interests that promoted the car are not. For most of its history, the oil industry has been concerned to manage a glut of supply and has promoted expanded consumption. Thus, big oil purposely sabotaged public transport in the US to promote the use of cars and so created a system-wide reduction of efficiency. Beyond this, Tadit Anderson notes that supposed efficiencies associated with economies of scale in manufacturing relied on a profligate supply of fossil energy to drive out local industries and concentrate power in markets constructed over ever larger regions.[54] The technologies thus embed relations of power and it is really these power relations that are grandfathered in the discourse of energy efficiency as well as in the Kyoto Protocol's carbon reduction commitments (see Chapter 8).

A limit on that expansion is not compatible with economic growth. If the quantity of energy is fixed then growing use for some can only be had at a loss to others and this equation grows more acute in the context of peak oil. Ultimately, Mobbs observes that 'energy efficiency is meaningless in the face of actual shortages' – efficient or not, the car will not go without fuel (2005: 143). In a context of declining energy supplies, the choice is what – or whose – energy use to cut.

6

The chains of petro production

THE OIL INDUSTRY IS generally talked of in terms of upstream and downstream production. Upstream production is about exploration and extraction – finding the oil and getting it out – while downstream is about refining and marketing. The previous chapter showed that peak oil brings on peak poison as the 'easy' oil goes first leaving heavier and dirtier oil increasingly supplemented by unconventional sources including coal-to-liquid (CTL). But the 'easy' oil was never clean. This chapter opens with an all too brief account of the subordination of upstream producers in Africa to the interests of big oil and of the blood on the pipelines. A fuller account is given in *The groundWork Report 2005*. The main focus here is the petrochemical value chain in South Africa, starting with the refineries and CTL plants and looking further downstream at plastics, one of the sectors identified for expansion in industrial policy. The petro corporations extract value from every point in the production chain and have a strategic view of it that is not confined by locality or nation. They also provoke resistance all along the line but this invariably starts locally – in each place where people's lives are disrupted by the incursion of the industry. Connecting people in different localities thus becomes an important strategy of resistance and, in 2005, people living on the refinery fencelines in South Africa visited the Niger Delta to build solidarity in resistance upstream and downstream. There they witnessed the intensity of the upstream war against the people. The village of Odioma had recently been razed to the ground by the Nigerian army while everywhere the gas flares roared and spilt oil saturated the ground and slicked over the waters of the delta.

AFRICA'S OIL RUSH
The US invasion of Iraq added impetus to Africa's oil rush. The Gulf of Guinea off west and central Africa was already 'viewed by the oil industry as the world's

165

premier "hotspot" ' and the biggest discoveries worldwide in 2001 were made there (Gary and Karl 2003: 9). Security and the cost of crude supplies were top of the agenda for consuming countries. The US in particular stepped up diplomatic and military activity in the region, edging in on the regional hegemonies of the former colonial powers of Britain and France. Also reflecting the changing pattern of power relations in the globalised world order, several Third World countries developed active interests in African crude supplies, including China, Malaysia, India and South Africa. While corporate and state interests are not necessarily identical, they are closely aligned and move pretty much in lock-step into the oil regions. The dominance of particular corporations in each country thus tends to reflect international relations of power at the time that oil was discovered.

The international financial institutions – the International Monetary Fund (IMF) and the World Bank – are key actors in support of the Northern agenda. Vallette and Kretzmann (2004) show that the World Bank first invested in the oil sector at the behest of the US in the late 1970s with the aim of opening production in new countries and so reducing OPEC control of prices. It was also in this period that the neo-liberal Washington Consensus began to emerge as the US started to make more direct use of the IMF and World Bank to extend its control over the economic policies of Third World countries. Almost all Bank loans for oil projects have been to the benefit of 'Northern fossil fuel corporations, especially those based in the United States' (Vallette and Kretzmann 2004: 7). Further, 82% of Bank oil projects were designed to export the oil to the major Northern markets and a good many of them are located in countries ruled by despots and warlords. The Bank itself makes substantial profits from resource extraction but its broader role is to provide a political guarantee to oil and finance corporations that they will get their profits out from projects in unstable countries. Contrary to its stated mission of alleviating poverty, it thus appears that the Bank's real mission is to secure the flow of resources to 'the market'.

Box 6.1   The Extractive Industries Review

The World Bank commissioned the Extractive Industries Review (EIR) in 2000 in response to mounting criticism from civil society organisations that lending to oil, gas and mining projects contradicted its stated mission of alleviating poverty. The review, published in December 2003, found that the Bank's 'project funding in the extractive industries has not had poverty reduction as its main goal or outcome' (Vol. I: 18). Indeed, '[o]ver the course of two years of examination, the World Bank . . . was unable to provide an example of a single instance where an oil project alleviated poverty. Many examples were provided of oil projects that exacerbated poverty' (Kretzmann and Nooruddin, 2005: 13).

The central recommendation of the EIR was that the World Bank 'should phase out investments in oil production by 2008' and focus on sustainable energy (Vol. I: 64, 65). It also said there should be no support for oil projects in a context of human rights abuse or corruption. The Bank rejected the phase-out – repeating its discredited claim that such projects were necessary for poverty reduction and the delivery of energy to the poor – but said it was adopting most of the other EIR recommendations. Dr Emil Salim, who headed the EIR, commented that in fact the Bank's response intended to justify business as usual and made 'few commitments to addressing these recommendations fully or to implementing them' (quoted in Stockman and Muttitt 2005: 16).

Producing countries, and would-be producers, are no less enthusiastic. Their economic interest is primarily in oil revenues as well as balance of payments. Politically, they benefit from the international recognition that comes with sitting on top of a strategic resource and oil discoveries may reinforce their grip on power – provided they play the game. Thus, Equatorial Guinea moved from being a virtual outcast nation to being courted by the great powers with the US re-opening its embassy, closed in 1988. Their relationship with corporations tends to be as close as that between corporations and their 'home' countries but, given the superior economic clout of big oil corporations, it is an unequal relationship and marked by duplicity and corruption.

As oil prices rose from a low of $10 a barrel in 1999, everyone in the extraction business did well except ordinary people in oil-producing countries. While the fabulous wealth of oil was paraded before them, they were driven ever deeper into poverty. The very common association of oil wealth with the impoverishment of people and the failure of national economies has given rise to the notion of the 'resource curse'. Conventional accounts of the resource curse emphasise the effects on currency values, the devaluation of other sectors of the economy and the consequent dependency on oil both for state revenues and the economy as a whole but generally leave out the devastation of environments. At the same time, these accounts narrow the frame of reference to conceal the actual politics of resource extraction. They treat each oil-producing country as a separate economy and leave out the broader context of global corruption and the purposeful subordination of Southern countries documented in *The groundWork Report 2005*.

Africans doing it to themselves

The New Partnership for Africa's Development (NEPAD) embodies the vision of Africa's governmental leaders 'to enable the continent to catch up with developed parts of the world' (NEPAD 2001: para. 65).[1] The 'partnership' of the title is between Africa and the major powers who are asked to help finance NEPAD as a 'Marshall Plan' for Africa. From its inception, NEPAD has been presented to successive meetings of the G8 rich country club. Consultation with African civil society – primarily labour and faith-based organisations – appeared as something of an afterthought following intense criticism.

NEPAD opens by criticising the role of colonialism in impoverishing Africa and acknowledging 'poor leadership, corruption and poor governance in many countries' (para. 21) in the post-colonial period. It commits African leaders to democracy, respect for human rights and good governance and to the pursuit of the United Nations Millennium Development Goals which set targets for addressing poverty. This sounds good but is already given away in the core vision of emulating Northern development.

NEPAD commits Africa to capitalist development. Its key goals are to attract capital investments – from donors as well as private-sector investment – and to gain 'market access' through increased productivity and expanding exports. By this means, it reckons to achieve 7% economic growth per year for fifteen years. This is not credible. The central problem is that Northern development was substantially financed by plundering the Third World and there is no other Third World left to plunder unless African leaders plunder, once again, their own people. This is more or less what they do anyway. As Manuel Castells argues, the Northern powers and African elites have a common interest in Africa's fragmented integration into global capitalism. Europe and the US benefit from the extraction of valuable assets and 'what is a human tragedy for most Africans continues to represent a source of wealth and privilege for the elites' (2000b: 127). In this light, the 'partnership' in NEPAD is hardly encouraging. Money will be made as the poor are made poorer.

Developing regional infrastructure corridors for transport and energy is a key focus for NEPAD. The West African Gas Pipeline – to take Nigerian gas to Benin,

Togo and Ghana – fits the bill. Originally proposed by Chevron in the early 1990s, it is now NEPAD-approved. NEPAD argues that this breaks with the colonial infrastructure that connected African countries only to the colonial power. So it appears. Yet the money flows are as colonial as ever. The project is subsidised by 'generous exemptions from taxes, rates and customs duties' while the $400 million invested will return to the Northern countries which provide the engineering resources (ERA and Oilwatch 2000: 13). The profits will follow in the same direction. A similar project supplies gas from Côte d'Ivoire direct to AngloGold Ashanti's Ghanaian gold mines. It is an infrastructure developed for the benefit of capital, not people who are unlikely to be able to afford the energy at the end of the pipeline.

The West African Gas Pipeline also reflects NEPAD's bias towards capital-intensive mega-projects. Such projects are favoured by financial institutions, particularly the World Bank, partly because they are easier to administer than numerous small initiatives but, more substantially, because they reflect the interests of global capital. Mega-projects produce 'enclave economies' divorced from local needs and dependent on transnational corporations and expatriate resources. They give concrete form to Africa's fragmented integration into global capitalism.

A regional power

The South African government's contribution to NEPAD was critical and, allowing for differences in context, the document reflects its development thinking. This role reflects the pan-African sentiments of its leaders but also its growing economic interest as the leading economy in the region – producing 44% of sub-Saharan Africa's GDP. As Deputy Minister of Foreign Affairs Aziz Pahad put it in 2005, 'economic diplomacy is a central pivot around which to anchor all our efforts to address underdevelopment and poverty. In this regard, there are many new economic opportunities in Africa, the Middle East and Asia which should be investigated and exploited by the South African private sector'.[2]

South Africa's corporations, public and private, have certainly followed the diplomats into Africa, investing in a wide range of sectors from cell phones and supermarkets to resource extraction and accounting for nearly half of all foreign direct investment in the Southern African Development Community. Mining and metal processing have been prominent, as indicated by the acquisition of Ashanti by AngloGold. The country's interest in crude extraction is relatively new although it has the best developed refining sector in the region. PetroSA, the state oil corporation, has two relatively small fields off South Africa itself. It has been moving into Africa since 2003, acquiring exploration and production licenses in Gabon, Nigeria and Equatorial Guinea. In 2005, following on the heels of South Africa's peace-keeping troops in Darfur, PetroSA signed an exploration agreement with Sudan's state-owned corporation. It has also acquired exploration rights in Egypt's Gulf of Suez.

Sasol has expanded rapidly into upstream oil and gas exploration and production with operations concentrated in Africa. Gas fields in Mozambique were brought into production in 2004 to supply its South African production plants through a dedicated pipeline. In Gabon, it has shares in several oilfields and is the operating partner in an exploration project. It is entering into deep-sea exploration in Nigeria and is reviewing operations in Equatorial Guinea and South Africa. Sasol is also developing a gas-to-liquid (GTL) project with ChevronTexaco at Escravos in the Niger Delta. Clearly, it is not deterred by the idea of dealing with regimes that abuse human rights although it says it is determined 'to bring world-class environmental standards to all new and planned future projects, irrespective of location and project type' (2004: 39).

Corporate South Africa's march into Africa is backed by state-directed funding through the Industrial Development Corporation (IDC). The IDC has typically backed large-scale capital-intensive projects, reflecting an historical bias towards mega-projects that is reinforced by the development of institutional relations with the World Bank. It has a specific interest in the petroleum sector and provided substantial backing to Sasol's Mozambique gas development. Not surprisingly, the IDC, Sasol and PetroSA are all keen to identify their African projects with NEPAD.

Frontline Nigeria

On 10 November 1995, Ken Saro-Wiwa and eight other Ogoni activists were executed on the order of a rigged military court. In his closing statement to the court, Saro-Wiwa wrote:

> I and my colleagues are not the only ones on trial. Shell is here on trial . . . the ecological war that the Company has waged in the Delta will be called to question sooner than later and the crimes of that war will be duly punished (quoted in Doyle 2002: 174).

The ecological war in the delta starts with enclosure. Nigerian law gives control of land and oil to the state but the practical effect is that oil corporations can and do take what they want from the people within their areas of operation. The corporations themselves call this the 'land take'. The pollution has poisoned a much wider area, filling the air, spreading over the waters and saturating the land.[3] For people who depend on fishing and farming, said Saro-Wiwa, 'that's . . . saying we don't have any right to live' (quoted in Doyle 2002: 161). Christiana Mene from the Escravos Women's Coalition, involved in shutting down ChevronTexaco's export terminal in 2002, echoed the point: 'Our farms are all gone . . . we cannot catch fishes and crayfish' (quoted in Turner and Brownhill 2004: 67).

Niger Delta communities have a long history of resisting the enclosure of their land. The Movement for the Survival of the Ogoni People (MOSOP) became the best-known organisation of resistance and an inspiration for communities across the delta. In 1993, it organised mass protests throughout Ogoniland and forced Shell to close down its Ogoni production wells although active pipelines still cross the territory. Resistance was met with brutal repression. It started with security force attacks thinly disguised as inter-ethnic violence. Then, in 1994, four 'moderate' Ogoni chiefs were murdered at Giokoo. The circumstances indicate that they were killed by security operatives acting under cover. Saro-Wiwa and his fellow MOSOP leaders were immediately accused of the murders and arrested without even the pretence of an investigation. According to Owens Wiwa, Shell's managing director told him that he could secure his brother's release 'but would only do so if MOSOP called off its international campaign against his company' (Okonta and Douglas 2003: 58). Saro-Wiwa refused. Shell denies making this offer.

The Giokoo murders provided the pretext for the military occupation of Ogoni. A special task force closed off media access and launched a terror campaign marked by arbitrary detention, torture, rape, murder and military assaults on towns. Colonel Paul Okuntima, who led the operation, later claimed that Shell helped finance it. Shell denied it and Okuntima subsequently retracted. Human Rights Watch established, however, 'that all through the Ogoni crisis Shell Nigeria representatives met regularly with the commander . . .' (Okonta and Douglas 2003: 135).

The use of brutal security force violence did not begin or end in Ogoni. From the early 1990s protests across the delta became more organised and numerous ethnic groups adopted charters loosely modelled on the Ogoni Bill of Rights. They also looked for a broader unity that would give expression to Saro-Wiwa's vision of a pan-delta solidarity based on people's common experiences. The savagery of the security force response also intensified throughout the decade. Ijaw youth greeted the new year of 1999 by mobilising in support of the Ijaw Youth Council's Kaiama Declaration. In response, security forces killed over 100 people and burned down ten or twenty homes. In many similar incidents around the delta, corporate helicopters and boats were seen carrying security forces. At Odi, the army razed the entire town to the ground and killed several hundred people according to Human Rights Watch (2002: 21ff.). On the other side, people occupied oil facilities and forced repeated shutdowns across the delta.

In this period, gun trafficking in the delta escalated and armed youth groups, sometimes known as 'cults', emerged. Mostly, it appears that they were armed by politicians to intimidate opposition party supporters, by local elites to secure their control over oil sub-contracts and pay-offs against rival factions, or through 'illegal bunkering' networks responsible for the wholesale theft of oil. Cult leaders have also been used to infiltrate and subvert resistance movements but, in the ambiguity of the delta, that works the other way too as cults turn to resistance against their erstwhile sponsors. From this, a more direct insurgency was spectacularly announced by a direct attack on Port Harcourt, Nigeria's oil capital, by Dokubo Asari's Niger Delta People's Volunteer Force. Human Rights Watch (2005) sees this as part of a battle with rival gangs for control of illegal bunkering. Nigerian scholar Ike Okonta observes to the contrary that Asari was an insurgent leader whose actions were 'symptomatic of a larger and quickly spreading national crisis' (2005).

This was emphatically confirmed with the emergence of the Movement for the Emancipation of the Niger Delta (MEND). In February 2006, government sent helicopters and gunships to attack the small village of Okerenkoko on the Escravos River, claiming it to be a centre of oil bunkering. 'It was this bloody incident that triggered the birth of MEND,' says Okonta (2006). MEND captured expatriate oil workers, destroyed flow stations and pipelines across the delta and even attacked offshore platforms 'to remove any notion' that deepwater production is beyond its reach.[4] It has shut in something like half of Nigeria's production and reduced the flow of oil from the delta itself to a trickle. Yet MEND is more an idea than an organisation, according to Okonta, and composed of shifting alliances of resistance groups and 'cults'.

The oil corporations portray MEND and the cults as merely criminal. They represent themselves as the victims of illegal bunkering, sabotage and the loss of state authority. They are, however, directly complicit in creating Nigeria's outlaw economy. Ridding the delta of the transnationals is a key MEND objective. It routinely warns them to leave and several service companies have done so. Shell and the other majors hang on but have 'lost their [social] license to operate' observes Watts (2009). That licence was always a fiction, a cover for coercive force. Nigeria demonstrates the scale of rebellion needed to withdraw it. Further, '[w]hat is on offer in the name of petro-development is the terrifying and catastrophic failure of secular nationalist development . . . From the vantage point of the Niger Delta . . . development and oil wealth is a cruel joke . . . The government's presence, Okonta notes, "is only felt in the form of the machine gun and jackboots"' (Watts 2008).

The oil regime is bereft of all legitimacy. Getting the oil out, whether it is done by the corporations or the bunkerers, is dependent on gangs of armed men whether or not they are uniformed. This situation in the Delta is replicated in numerous producing countries. And at the global scale, the invasion of Iraq, the plunder of its treasury and the attempt to rewrite its legal framework in the interests of US corporations, show that all is now the product of protection rackets.

REFINING ENVIRONMENTAL INJUSTICE

South Africa and Nigeria are the main centres of refining in sub-Saharan Africa. Nigeria's three refineries, operated by the state-owned Nigerian National Petroleum Company, are badly managed and run at less than 50% efficiency. They do not meet the demand for petrol even in Nigeria, let alone the rest of West Africa, so much of the region's demand is supplied from US refineries. In contrast, South Africa's refineries massively expanded production in the 1990s to supply growing domestic demand and to export refined products to southern and eastern Africa. South Africa's own demand has since overtaken this expanded supply and several new refineries are now proposed both in South Africa and in the region.

Petrochemicals in South Africa

The chemicals sector makes up a major slice of South African industry, producing 24% of the value of all manufacturing. This includes liquid fuels production that dominates chemicals, producing close to 33% of value within the sector and creating the feedstock for chemicals production. Liquid fuels are produced from imported crude oil, coal and gas. Table 6.1 shows the location, ownership, fuel source and capacity of the refineries. Sasol Chemical Industries, located primarily in Sasolburg, uses the same technology as its Secunda CTL 'synfuel' plant to produce basic chemicals. The process is particularly polluting and consumes about 41 million tonnes of coal each year.

South Africa's four crude-oil refineries are all complex, using 'catalytic cracking' to produce a higher proportion of high-value products such as petrol but at a major cost to the environment. They are Sapref and Engen in Durban, Caltex in Cape Town and Natref at Sasolburg. Natref is the only inland crude refinery and is supplied by pipeline from Durban.

Table 6.1   Refineries and ownership in South Africa.

Refinery	Location	Owned by	Fuel source
Calref	Cape Town	ChevronTexaco (Caltex)	Crude oil
Engen	Durban	Petronas 80% & Worldwide Africa Investment Holdings 20%	Crude oil
Sapref	Durban	Shell 50% & BP 50%	Crude oil
Natref	Sasolburg	Sasol 64% & Total 36%	Crude oil
Secunda	Secunda	Sasol	Coal
Mossgas	Mossel Bay	PetroSA	Gas

In 1995, over two thirds of the crude was sourced from Iran. Saudi Arabia is now the largest supplier followed by Iran. These crudes have a high-sulphur content with major implications for pollution. Since 2000, Nigeria has become an increasingly significant supplier while South Africa itself has developed three small oilfields and is the fourth largest source at less than 3%. Oil is South Africa's biggest import item, with around 21 million tonnes imported annually, most of it through the port of Durban.

Structured for profit

Although privatised in 1979, Sasol remained intimately linked with the state both before and after the political transition. With sanctions lifted, Sasol repositioned itself as a transnational corporation in its own right. It has listed on the New York Stock Exchange and has major investments in Europe, the US, China, the Middle East and Africa. This expansion has been made possible by a massive accumulation of subsidies at public expense, not to mention the additional subsidy of being allowed to pollute.

Petroleum is dominated by transnational corporations – Shell, BP, Caltex and Total – which were also complicit with apartheid and sanctions-busting. In return for this co-operation, the state guaranteed corporate profits by regulating the price of fuel in relation to the supposed costs of importing oil.[5] As part of the deal, the transnationals were required to buy Sasol's synfuel to blend with their refined crude-oil products while Sasol was restricted to a few symbolic pumps and could not develop a significant retail market. Industry regulation has thus centred on pricing and the use of Sasol's synfuels. The pricing mechanism is still in place.

When the price of crude oil is low, as it was for most of the 1980s and 1990s, Sasol's synfuel is hopelessly uncompetitive. From 1989 to 2000, it enjoyed nearly

R8 billion in subsidies paid out of the 'fuel equalisation fund'. Sasol was paid from the fund when the oil price fell below a benchmark figure ($23 a barrel in 1995), and was supposed to pay back into the fund if it rose above a second benchmark ($28 in 1995). In 1996, government announced that this subsidy mechanism would be phased out. The last subsidy payment was made in 1999 when the price hit bottom at $10. The escalation of prices since then reversed the competitive relationship between synfuels and oil products because Sasol controls its own supply of cheap coal and is insulated from rising global energy prices.[6] It thus enjoyed windfall profits guaranteed by the oil-based pricing mechanism. Sasol then argued that the equalisation mechanism had lapsed so it did not have to repay the subsidy.

Government did not quite share this view. It initiated a review of the equalisation mechanism in 2000. It was not made public but apparently recommended that the mechanism be retained – in other words, Sasol should pay back the subsidy. This was revealed in the report of a second investigation, announced by finance minister Trevor Manuel in March 2006, into whether Sasol should be slapped with an additional tax on windfall profits. According to this report, the equalisation mechanism was in fact based on a gentleman's agreement. 'When in 2003 Sasol believed that it no longer required tariff protection it refused to reintroduce such a "gentleman's agreement"'.[7]

The report also made clear that Sasol Synfuels, Natref and the oil refineries more generally, secured numerous other hidden subsidies besides the equalisation and pricing mechanisms: Natref did not pay for piping crude oil from Durban for seventeen years; Natref also received oil from the strategic reserve at Ogies at cut rates; Mossgas received about R1.5 billion from the equalisation fund; the benefits to the oil majors from a deal cutting them into coal exports are not known; the state over-invested in pipelines in the 1960s and 1970s and the costs have not been recovered.

Sasol's response suggested that more subsidies might be appropriate. It argued that the international trend was to provide incentives for 'alternative fuels' and that its subsidy paled beside those given to defence industries – R200 billion, mostly paid before 1994 – and the motor industry – R90 billion paid out through the Motor Industry Development Programme. In all, it said, the state had paid out some R334 billion to industry between 1989 and 2000.

Deregulation[8] of liquid fuels under the 1998 policy banner of competition and industrial restructuring envisaged a three-phase 'managed transition' to 'allowing market forces to set prices' in Phase 2, with government monitoring and measures to correct market failures in Phase 3.

Phase 1 centred on terminating the requirement that the oil majors purchase Sasol's product and allowing Sasol independent access to the market. The most immediate effect was a merger deal between Sasol and Petronas, owner of Engen, together with their respective BEE partners, to form a new company called Uhambo. The other oil majors – BP, Shell and Caltex – opposed the deal at the Competition Tribunal and, in February 2006, the Tribunal refused to allow it. It found that Uhambo would dominate the market, control most inland refining capacity and also the existing pipelines from Durban, and entrench import-parity pricing that enables Sasol to reap the windfall profits from the difference between its costs and high-priced oil imports.

When, and whether, Phases 2 and 3 of the restructuring will take place is not known. The policy is formally unchanged but the official view is that South Africa is not 'ready' for deregulation.[9] At the Uhambo hearings, Department of Minerals and Energy (DME) officials indicated that the pricing mechanism would be maintained well beyond 2010 to guarantee petroleum profits so that BEE partners would be able to pay for their shares. *Business Report*'s Ann Crotty calculates that this implies that 'consumers are paying about R594 million a year towards the cost of empowerment . . . a cost that is generally carried by the shareholders of companies'.[10] It does not follow that deregulation would benefit consumers. More likely, regulated prices would be replaced by cartel prices.

Government, meanwhile, is building up another state-owned petroleum enterprise within the Central Energy Fund (CEF) group of companies. PetroSA was established in 2002 from a merger of the state's exploration and refining businesses. It owns the Mossel Bay GTL refinery and has relatively small gas and oilfields off South Africa itself. Other CEF subsidiaries are the Strategic Fuel Fund, responsible for securing supplies and ensuring reserves of oil, the Petroleum Agency SA and IGas which respectively promote oil and gas exploration, and the Energy Development Corporation, tasked with facilitating the development of 'commercially viable' renewable energy projects.

For the industry, BEE has been critical to securing a 'social licence' to operate. It creates a black interest group ready to defend industry interests and to take the political offensive to legitimize profits. This is particularly necessary for the refineries because local organisations have questioned the legitimacy of profits bought at the cost of people's health. The petroleum industry pioneered the concept of sector empowerment charters with the Liquid Fuels Charter. It includes the target of 25% black ownership of the industry and all the corporations operating in South Africa have moved rapidly to meet it.

The industry also makes much of its contribution to the South African economy. Thus, the annual reports of individual corporations and of the South African Petroleum Industry Association (SAPIA) give prominence to 'value-added statements'. SAPIA claims R44.6 billion as the combined value added for the industry in 2006, up from R24.8 billion in 2001. They made R6.7 billion after tax, down on 2005 but close to double the profits of the early 2000s. State taxes, duties and levies amounted to about R31 billion. Reflecting the capital-intensive nature of the industry, salaries and wages – including the exorbitant remuneration of top directors – account for less than 10% of this in most years.[11]

Box 6.2   Oil stain

South African corruption does yet not match Nigeria's but it has grown very rapidly. At local level, corruption has been named as a core grievance in many of the so-called 'service-delivery' protests. Arms and oil have been at the centre of higher-level corruption. A series of reports by the *Mail and Guardian* newspaper showed that the ruling ANC received election funds through a front company which diverted money meant to pay for oil consignments.

The story starts in 2001 when the company, Imvume Management, received support in negotiating with Iraq for oil allocations from senior ANC office-bearers, officials of the DME and a director of the Strategic Fuel Fund, the state agency responsible for crude-oil reserves. The deal implied the exchange of diplomatic support to Iraq for oil-supply deals, with the profits contributing to ANC funding. This was followed by Imvume being awarded a contract for the delivery of Iraqi oil by the Strategic Fuel Fund. The newspaper details a series of irregularities in the tender process and concludes that the award was rigged. However, the US invasion of Iraq put paid to these deals and the expected profits were not realised.

Following its establishment in 2002, PetroSA immediately awarded Imvume a contract to supply oil condensate to the corporation's Mossel Bay GTL refinery. Imvume acted as a go-between in these deals, buying the oil from resource trader Glencore. Several cargoes were delivered in terms of this arrangement. In December 2003, Imvume requested, and was granted, an advance payment of R15 million from PetroSA for the next cargo. Five days later, it donated R11 million to the ANC, which was then short of funds for its 2004 election campaign. But

Imvume failed to pay Glencore for the cargo. Under threat of having the next cargo withheld, PetroSA then paid Glencore directly. The *Mail and Guardian* reporters conclude: 'The effect of the entire transaction was that PetroSA, and ultimately the taxpayer, subsidised the ruling party's election campaign: a blatant abuse of public resources.'

The ANC, Imvume, SSF, DME and PetroSA all denied wrongdoing and the ANC and Imvume each served notice of court actions against the newspaper. This seems to have been a bully tactic as they have not followed through with the actions.

Source: Reports by Stefaans Brummer, Sam Sole and Wisani wa ka Ngobeni in *Mail and Guardian*, 20–26 May 2005; 15–21 July 2005; 22–28 July 2005.

Expansions

The formidable process of refinery expansion is shown in Table 6.2. Despite this, SAPIA warned in 2005 that the local refineries will not be able to meet rising demand for petrol and diesel in the countries of the Southern African Customs Union – South Africa, Namibia, Botswana, Lesotho and Swaziland – by 2010. The deadline arrived early starting with diesel imports in 2005. By 2008, South Africa was importing about 1.4 billion litres of petrol, diesel and kerosene. Demand for diesel was stoked up first by the Western Cape electricity crisis and then by the national crisis as Eskom used its Open Cycle Gas Turbine (OCGT) generators, normally reserved for peak demand, as a substitute for base-load while private corporations brought in diesel backup generators.[12]

In late 2005, and coinciding with the start of the Cape Town power blackouts, fuel shortages hit several areas of the country. The causes were largely technical, relating to a switch-over to tighter fuel standards – 'Cleaner Fuels Phase 1' – and failures of refinery planning. Government was duly alarmed at the prospect of future shortages. A DME study estimated that a complete collapse of supply would cost the country close to a billion rand a day and the department then developed its *Energy Security Master Plan: Liquid Fuels* (DME 2007).

In the plan, the priority for growth is absolute. The document uses 'affordability' to link poverty alleviation to growth, emphasising that energy is a 'strategic input to a resource-intensive South African economy' (14). In contrast to most economic planning documents, the master plan repeatedly refers to environmental management and climate change. However:

Table 6.2   Refinery expansions 1990 to 2004.

	Capacity in thousand barrels per day (or equivalent)								
	1990	1992	1993	1994	1995	1998	2001	2002	2003
Calref	50				90			100	
Engen	67	85		105			125		150*
Sapref	120				165	180			
Natref	78		86					108*	
Secunda	150								
Mossgas	45								
Total	510	528	536	556	641	656	676	708	733

Compiled from industry sources

* Engen currently runs below capacity at a nominal 135 thousand barrels per day while Natref's
  production was cut to 92 thousand barrels per day consequent on the 2009 power shortages.

> In the short-term, South Africa cannot sacrifice its development at the
> altar of the environment but in the long-term, unless South Africa,
> along with the rest of the world, does something about global warming,
> its own economy is threatened by climate change (31).

It also sounds a warning on oil depletion: 'a transport strategy that is over 90%
dependent on oil is guaranteed to land South Africa in serious trouble in a few
years' time. No form of planning will find South Africa oil, when it has all been
mined or acquired by those with more might or insight' (8). As with global warming,
however, peak oil is effectively treated as a long-term issue and deferred to the
future.

The plan proposes a range of interventions, most of which were already being
pursued. Fuel transport and storage infrastructure is to be massively expanded. A
proposal for a privately operated pipeline to import refined product through Maputo
to Gauteng was approved in 2007 while a high-capacity 'multi-product pipeline'
from Durban to Gauteng is a major component of Transnet's infrastructure
programme. Rail-transport capacity and port-handling capacity is also under
construction. Finally, the plan calls for an expansion of refining capacity.

Sasol is obliging with a 20% increase – to 180 thousand barrels per day – in
synfuel production at Secunda. This is part of a broader expansion, including
chemical production, dubbed Project Turbo. The pace of expansion was retarded
first by the 'instability' of new plant and then by cuts in capital spending as Sasol
reacted to the collapse in oil prices in 2008. This delays the scale-up in production

but does not reduce it. In 2007 Sasol was also talking up the need for a new inland refinery and initiated feasibility studies for Project Mafutha, an all new 80 000 barrel-a-day CTL plant that would require a whole new Sasolburg. The Waterberg area in Limpopo province, where Eskom is already polluting the air, is the favoured site. Being a dry area, CTL's exorbitant water demands can only be met through massive cross-watershed transfers as will be discussed in Chapter 8.

Irrespective of the oil price, PetroSA has pushed its Mthombo crude-oil refinery as hard as it can. In 2007, it announced its ambition to build a new 200 000 barrel-a-day refinery at an estimated cost of R39 billion. A month or so later, it said that the Coega Industrial Development Zone (IDZ) had 'won' selection as the preferred location in competition with four other sites, an outcome that appeared pre-determined. The scale of the project has since expanded to a 400 000 barrel-a-day behemoth initially estimated to cost $11 billion. This estimate was then reduced to $10 billion as steel and other input prices declined. Given the volatility of both the exchange rate and input prices, these figures imply a cost of somewhere between R70 and R120 billion.[13] That is assuming PetroSA got its figures right in the first place and there are no cost escalations along the way. In 2009, PetroSA lined up with every other state-owned corporation to ask the Treasury for funding support. It is also looking for partners to invest up to 63% in the refinery but suggests that government may, for strategic reasons, prefer PetroSA to retain a controlling interest of over 50%. It also wants control of crude inputs and has signed a joint venture with Venezuela's PDVSA and held talks with Angola's Sonangol amongst others. BP and Shell oppose the project, arguing that there is now a global surplus of refining capacity and it would be cheaper to import than build a new refinery.

PetroSA says Mthombo will be designed to refine low-quality crude to high-quality liquid fuel specifications while yielding a high proportion – up to 90% by volume of crude – of light high-value product. The large quantity of sulphur and other pollutants in heavy sour crude must find a destination other than the fuel product – the atmosphere, solid waste, the already saturated sulphur market or bunker fuel for ships. The Environmental Impact Assessment (EIA) for the project was expected to start in 2009 but appears to be delayed. Halliburton subsidiary KBR,[14] fingered for corruption in Nigeria, has been appointed as the lead engineering consultant for the project.

High-tech pin-up

Sasol is the poster boy of South Africa's industrial strategy. Set up by the apartheid government in the 1950s, it developed the only commercial CTL plant in the world together with a string of heavy chemicals plants. It is now positioned as a global technology leader, active in 35 countries and linked into global production networks through partnerships with a range of leading transnational corporations, including ChevronTexaco, and state-owned corporations such as Qatar Petroleum. Its exports are founded on high-value chemical-design services as much as on the export of commodities – primarily coal and heavy chemicals.

The corporation made hay in the global sunshine of high oil prices. In partnership with the Shenhua Ningxia Coal Group, it is planning to build an 80 000 barrel-a-day CTL plant in China. The project is now in the feasibility stage. It has entered a similar partnership in India with the Tata group. In March 2009, the partnership secured a massive coal block in Orissa state, so opening the way for a pre-feasibility study for another 80 000 barrel-a-day CTL plant. In Indonesia it has initiated preliminary studies for a CTL plant while in the US it has joined with local corporations to lobby for government handouts for new synfuel plants. In South Africa, government waived the windfall tax and agreed that it would facilitate expansion while Sasol would undertake the feasibility studies for the Mafutha CTL plant. Mafutha is advertised as a partnership with the IDC. Sasol is also in a leading position on the development of gas-to-liquid (GTL) plants. Its new Oryx plant in Qatar, a joint venture with state-owned Qatar Petroleum, is the largest in the world though soon to be overtaken by Shell's neighbouring Pearl plant. It is also building a GTL plant in the Niger Delta in partnership with Chevron and has recently signed a joint venture with Uzbekneftegaz and Petronas to look at building one in Uzbekistan. It plans another GTL plant in Canada where it has joined the rush for shale gas through a partnership with Talisman Energy. It then hopes to bring 'fracking' home to South Africa to produce the feedstock for yet another plant.

Rebranding itself as an environmental leader is perhaps Sasol's greatest innovation. Within the discourse of ecological modernisation – the World Bank's version of sustainable development – it has indeed made significant improvements, but off an appalling base. The essential problem for Sasol is that its processes are inherently energy-, carbon- and pollution-intensive.

For GTL, production is more energy-intensive than oil refining but Sasol claims that the superior performance of GTL fuels offsets higher carbon emissions at the plant. Over the life cycle of production and consumption, 'total [greenhouse gas] emissions of the GTL system may vary between 12% less and 11% more than the refinery system, depending on assumptions about the nature of the operating conditions' (Sasol SDR 2006: 21). Sasol claims more significant reductions in sulphur dioxide, nitrogen oxide and volatile organic compound emissions from GTL.

'The case for promoting coal,' according to Sasol, 'is strengthened by the development of clean-coal technologies and the need for energy security' (Sasol SDR 2006: 20). For CTL, 'clean coal' comes down to carbon capture and storage (CCS) and a choice of integrated gasification combined cycle (IGCC) or gas to power the plant itself. Regrettably, neither CCS nor coal-fired IGCC are proven technologies. Be that as it may, Sasol compares three variations of CTL plants – coal-fired plants, IGCC-fired plants and gas-fired plants – with a standard refinery. It then runs the comparison again using CCS with each type of CTL plant. In Table 6.3, the numbers are relative to a refinery set at 1.0.[15]

Table 6.3   CTL CO_2e emissions compared with a conventional refinery = 1.0.

CTL	Coal-fired	IGCC-fired	Gas-fired
Without CCS	2.5	1.5	2.0
With CCS	1.5	0.8	1.0

The results of its studies show that a conventional CTL emits 2.5 times as much CO_2e as a conventional refinery. The credibility of this claim seems doubtful. In 2004,[16] Sasol's Secunda plant emitted 52 million tonnes of CO_2 while the larger Sapref refinery emitted one million tonnes. With CCS, it claims that CO_2e emissions drop to 1.5 times those from a refinery. The combination of IGCC and CCS yields the lowest greenhouse gas emissions, a little below the level of a conventional refinery.

In short, on Sasol's own information, replacing conventional oil with GTL or CTL, however modified, presents no climate change advantage over the conventional oil system that got us into the climate crisis in the first place. These results should also give pause to those who have advocated, however reluctantly, CCS as a last-ditch solution. Even assuming the carbon stays where it is put, CCS does not necessarily reduce greenhouse gas emissions to anywhere near zero.

Pollution

The petrochemical plants produce massive wastes to air, water and land. All boast continuing environmental improvements. These improvements are off a very poor base. Under apartheid, they had a virtually unlimited licence to pollute. The walls of secrecy started to crumble in the 1990s and national and local environmental justice organisations began putting real pressure on the corporations and on government. Much more is now known about the extent of pollution but this is very uneven. Information still relies heavily on what industry chooses to reveal or is forced to reveal. This means that most is known about those areas where resistance is most active.

Sasol's Sustainable Development Reports (SDRs) give aggregate figures on worker safety and environmental wastes for its global operations but do not break the figures down for specific sites, plants or component businesses. Since it claims real reductions of some pollutants at some plants, it is curious that it will not release site-specific figures.[17] Its reporting is thus of limited use to local communities but responds to the corporate social responsibility framing of global institutions and business organisations.[18] The corporation's Sasolburg Health, Safety and Environment brief quantifies solid wastes produced by Sasol Chemical Industries and Natref but not air emissions. It has not been updated since 2006. Sasol has not produced a similar brief for Secunda although this plant produces 90% or more of its global wastes and is the largest single source emitter of carbon dioxide in the world.

Sapref and Engen now publish information on source emissions and other wastes from the Durban refineries. Their reporting is clearly designed to respond to – and often to rebut – local criticisms. The refinery wastes appear small by comparison with Sasol. The latter's coal-based process is undoubtedly the filthiest way of producing either fuel or chemicals. Nevertheless, Sasol's figures largely represent the integration of extraction and production whereas the refinery figures exclude the appalling costs of oil extraction in other countries. Engen claims that it produced no high hazard waste (H:H) in 2006 but does not report 'low hazard' (H:h) waste. Sapref avoids specifying what proportion of its waste is hazardous.

Chevron gives no account of its environmental wastes and has consistently reneged on promises made to local environmental activists to reduce emissions. PetroSA is a state-owned corporation but its annual reports show no sense of public accountability in respect of the environment and say nothing about its wastes.

Information is also contested. Industry has on occasion been forced to admit under-reporting – as when Sapref admitted in 2000 that it had under-reported sulphur dioxide emissions by 12 tonnes a day for the previous five years. Government, however, has not developed its own capacity to verify or dispute industry claims on pollution from source. Unless it does so, the credibility of basic data will remain suspect.

Table 6.4 is constructed from industry reports or direct communication with the refineries. It shows some improvements in source emissions over previous years. The Durban refineries made environmental improvements conditional on expansions. They have reduced sulphur dioxide emissions in absolute terms but improvements in the rate of emission of some substances, such as nitrogen oxides, have been offset by expanded production. Sasol is converting from coal to gas, extracted from Mozambique, to fire its synfuel plants and as the feedstock for chemicals production in Sasolburg. This mitigates the astronomical pollution from coal-based processes but emissions remain high even by the standards of the oil industry.

Sasol's recent SDRs leave out the mountains of ash produced each year, previously reported at over 10 million tonnes (mt) in Secunda and 1.8 mt in

Table 6.4 Wastes from selected petrochemicals in 2006 (tonnes).

Pollutant	Sasol Global	Sasol Synfuel (Secunda)	Sasol Chemical Industries	Sasol Total Natref	Chevron Calref	Sapref	Engen
Air	**2006**	**2004**	**2004**	**2004**	**2004**	**2006**	**2006**
Carbon dioxide	60 009 000	52 164 000	8 872 000	829 000	746 479	978 000	930 385
Sulphur dioxide	223 000	189 923	30 989	1 333	4 940	4 015	4 668
Nitrogen oxide	162 000	148 300	25 824	686	953	1 301	1 935
Hydrogen sulphide	78 000	85 682	16 496				
Particulates	7 560	6 128 (flyash)	1 218 (flyash)	583	85		255
VOCs	461 000	409 783	17 663		2 189	3 529	971
Solid Waste	**2006**		**2006**	**2006**		**2006**	**2006**
Hazardous General Unspecified	270 000 1 126 000		5 755 11 557	880		6 335	2 391

Notes:
Figures are for direct emissions and do not include emissions associated with electricity consumption.
Engen and Sapref report average daily emissions from which the annual figure is calculated.
Sasol now reports methane under greenhouse gases rather than volatile organic compounds (VOCs) but they are here returned to the VOC line and make close to half the total.

Sasolburg. This is because Sasol is now 'recycling' its ash by selling it off to brickmakers. The ash contains various toxic residues, including heavy metals such as mercury, that leach from the dumps and into water, so recycling this ash should result in local environmental benefits. Nevertheless, the toxic residues will remain in the bricks and, as with PPC's cement, are dispersed into the built environment and will be released over time. Sasol is now also selling spent catalyst and waste waxes to clay brickmakers. This contributed to a substantial reduction in hazardous waste from its Sasol 1 site from 14 851 tonnes in 2005 to 4 257 tonnes in 2006. The corporation makes no comment on the final fate of the toxic material. As with ash bricks, however, this is effectively a strategy of dispersing toxic waste in space and time.

Catalysts are used to create chemical reactions in both the refining and chemicals industries. In Sasol's process, catalysts react with 'syngas' to produce synfuels and a variety of chemicals. The catalyst is designed for specific processes but is generally composed of grains of metal oxide coated with other metals. Commonly used minerals include iron, aluminium, nickel, cobalt, vanadium and potassium. In production, the catalyst is contaminated and is constantly regenerated until it is degraded beyond use. Waste catalyst is choked with heavy metals and is highly toxic.

Sasol's Natref refinery is now disposing its waste catalyst to recently established waste-recycling companies for export 'to companies abroad for metal recovery and final treatment' (SH&E 2005: 7). 'Abroad' is in fact China. This has reduced the refinery's toxic waste from 4 000 to 880 tonnes a year. In so far as the metals are recovered, this is likely to result from the commodity boom creating high metal prices. A sharp drop in metal markets, as in 2009, might collapse such enterprises and so return the problem to Sasol – or leave it at sea.[19] It should also be recalled that toxic-metal recovery has a poor record as was demonstrated at Thor Chemicals. The problem of toxic waste may therefore be transferred to the workers in the metal-recovery factories. Given the documented experience of IT recycling in China, it is possible that a proportion of the waste is simply dumped on arrival depending of the recoverable value.

Engen has increased its consumption of catalyst but claims zero H:H waste. This is achieved by 'de-listing' H:H to H:h waste which is then dumped at Bulbul Drive landfill. It is not clear how the metals in catalyst waste can be treated to warrant the de-listing. Replying to queries from groundWork, Engen said that catalyst waste batches are variable and it may not always be possible to de-list. In this case, it would be disposed to an H:H site. Engen is also exploring the possibility of recycling catalyst wastes that have 'a lot of monetary value due to their constituents'. It justifies this also in terms of the requirement for waste reduction.[20]

Water pollution from these processes is intense. All plants produce effluent and say the quality of their effluent is within their permit conditions. These permits are in many cases 'exemptions from the general standard' – meaning that they are permitted to meet lower requirements than in the national regulations.

The figures in Table 6.4 show normal emissions from normal operations. Abnormal incidents are pretty normal too. Fires, explosions, gas leaks spills and excessive flaring occur with appalling regularity at the petrochemical plants. Sasol's Polymer plant in south Durban leaked clouds of chlorine gas over the neighbourhood on five successive occasions in 1999 and 2000. In south Durban 2001 was another bad year with four major fires, five major flaring incidents, five oil spills and nine chemical spills. In the worst spills, Sapref lost 26 tonnes of tetra ethyl lead – which is as toxic as it sounds – from a badly maintained tank. Later that year, it spilled between one and two million litres of fuel from a pipeline buried under a residential street. The spill forced the evacuation of local people and marked the beginning of a lengthy struggle to make Sapref replace its 40-year-old pipes rather than just patch them. These were two of twenty-six spills from the Engen and Sapref refineries recorded by the South Durban Community Environmental Alliance (SDCEA) from 2001 to the end of 2004, not including spills from road tankers.

Toxic air releases, usually through the flares, are as common as spills. Such incidents are interspersed by earth-shaking explosions followed by fires. In south Durban 2007 was a year of fire. At the Island View chemical storage on Durban docks a series of explosions ripped through eight tanks that burnt through the night of 18 September. The air was thick with chemical smoke and fish turned up dead in the water a few days later. Three major fires at Engen spread fumes and soot across the neighbourhood. In July, a fire in the alkylation unit was caused when a corroded metal flange failed. In November, a storage tank was, according to management, struck by lightning and burned for three days. Just a week later, a leak at the lubricants plant caused an explosion and a fierce, if brief, fire. At Sapref, in November, a fire broke out in the catalytic cracker unit. The fires have not stopped. In November 2008 another major fire at Engen's crude-oil feed shut down the entire refinery for weeks.

The Chevron Refinery's neighbours in Cape Town fare no better. Apart from the gas clouds, they have had crude oil raining down on them. It got so bad that the Green Scorpions broke the official silence on incidents in 2006. They invited groundWork and the Table View Residents Association to a meeting to discuss measures for dealing with the refinery and revealed a stack of incident reports from the past three years.

Sasol's inland plants are, if anything, even more dangerous. A massive fire closed the Natref refinery in Sasolburg for several months in 2001. Sasol's record in 2004 was particularly dismal. At least fourteen people were killed in a series of incidents, four of which were particularly serious: In June, a gas liquor-storage tank exploded at the Phenosolvan facility in Secunda. In July, Sasol Collieries was rocked by an explosion. In August, a Sasol gas pipeline ruptured at the Gateway Industrial Park and the flames shot more than 30 metres into the air. The most serious incident was in September. Ten people died and more than one hundred were injured in an explosion at the Sasol Polymers' ethylene plant in Secunda. Sasol then initiated a major safety programme. Nevertheless, from a low of fifteen in 2006 the number of reported 'fires, explosions and releases' has increased steadily to thirty-six in 2009. Following the ethylene plant explosion, Labour Minister Membathisi Mdladlana said that if Sasol killed any more people he would shut them down. They have. He hasn't.

The SDCEA put the spotlight on excessive flaring from 2003. Flares are necessary safety valves in case of a build-up of explosive gases but, in the US, the Environmental Protection Agency found that they are frequently used to evade limits on emissions, for example, by burning off unwanted sulphur at night. The South African regulators would not have the capacity to detect such practices if they occur in South Africa.

Following tighter regulation, the Engen refinery reported 109 flaring incidents in 2003. Sapref reported just one because it regards all flaring as 'normal' unless caused by an external event. People in Durban remember 21 April 2004 as 'black Wednesday'. A power failure resulted in Sapref shutting down and a dense plume of smoke from the flare spread over the city. Sapref said there were no ill-effects beyond irritation from smoke and odour. No one believed them.

Box 6.3 Disputed production

The *Comparison of Refineries in Denmark and South Durban in an Environmental and Societal Context: A 2002 Snapshot* was produced by Danmarks Naturfredningsforening (DN) and SDCEA. It includes a technical comparison of the refineries based on information supplied by the refineries. At the launch, the Durban refineries attempted to block publication and have subsequently sought to discredit the report claiming technical inaccuracies.

The central point of contention related to the refinery technologies. In a paper to the World Congress on Environmental Health in February 2004, Engen's Sustainable Business manager Alan Munn, complained that:

> The study compares the performance of two simple hydro-skimming refineries in Denmark, that process low sulphur North Sea crude oil, with two large complex refineries in South Durban processing higher sulphur crude oil from elsewhere. This is like comparing a single engine aircraft with a jumbo jet and trying to draw meaningful conclusions. For example, for the authors to suggest that catalytic crackers, which are the heart of any modern refinery, are obsolete is ridiculous (2004).

The study is, in fact, at pains to point out the difference (DN and SDCEA 2003: 21) and does not suggest that the catalytic crackers are obsolete. It says:

> The quality of crude oil has wider implications because it determines the limits on technology options . . . The additional production units necessary to fully exploit the lower quality crude – particularly the cat cracker – are major sources of particulate and sulphur dioxide pollution (45).

It goes on to encourage the use of high-quality crude inputs and suggests that the cat crackers could then be sacrificed in the interests of reduced pollution. It clearly indicates that this would also involve a sacrifice in the proportion of high-value products – particularly petrol – produced from each barrel of crude and hence also of refinery profits.

There would certainly be broader implications to such a decision. First, South Africa's vehicle fleet would need to shift from a bias to petrol to more efficient diesel. Second, applied globally, it would imply a radical reduction in usable oil reserves and so push the transformation of energy and transport systems. This transformation is essential if catastrophic climate change is to be avoided.

At another level, the dispute brings into focus the question of who makes society's technology choices. At present these choices are largely made by corporations. For the refinery managers this is only natural and their response to the report perhaps indicates some shock that anyone should have the temerity to challenge this prerogative. For civil society, the deeper question must be how to democratise production and what technologies are compatible with such democratisation.

Burn-up

The petrochemical corporations sell around 30 billion litres of refined product annually. At the end of the fuel chain, most of it goes into vehicles. They burn the largest part of hydrocarbons and contribute proportionately to air pollution. In 2006, government introduced cleaner fuel regulations to eliminate lead in petrol, and thus also enable the use of catalytic converters that reduce a range of other pollutants from vehicle exhausts, and to reduce the permitted sulphur content in diesel from 3 000 to 500 parts per million (ppm). In a second phase, this will be further reduced to 50 ppm. These improvements are driven as much by new engine technologies as by environmental concerns and are linked to South Africa's strategies in manufacturing cars and catalytic converters. The regulations should mitigate pollution particularly in congested urban areas although this will be partly offset by expanding demand. As with the refinery improvements, however, they are off a low base. European regulations now have an upper limit of 50 ppm sulphur content and require the introduction of 10 ppm in both diesel and petrol.

The reduced sulphur content implies the removal of more sulphur at the refineries. Refineries can no longer legally vent the difference in the form of increased sulphur dioxide emissions. Globally, intensified regulation has created a glut on the sulphur market since the production of sulphur is now driven by the demand for petroleum rather than for sulphur.[21] If this additional sulphur cannot be sold, it becomes a waste product. The main market for sulphur is agricultural chemicals and the sulphur industry is pushing it aggressively. Stripping the sulphur out of fuel thus relates to the continued practice of toxic farming. No doubt following the lead of Northern refiners of low-quality crudes, South Africa's coastal refineries have found a way out by concentrating surplus sulphur into bunker fuel. It will thus convert into intensified sulphur dioxide emissions from ships and be dispersed over the sea – out of sight, out of mind, beyond regulatory authority. Natref described this as an unfair competitive advantage in order to argue that it should be subject to less stringent fuel standards. It is rather better described as environmental hooliganism.

CHEMICALS AND PLASTIC

Apart from liquid fuels and tars, petrochemicals are the source of all carbon-based or 'organic' chemicals. They are used in the manufacture of an extraordinary array of products including agricultural fertilizers, herbicides and pesticides, all plastics and most rubbers, synthetic textiles, explosives, medical products, cosmetics, detergents, paints, varnishes, waxes, glues and solvents. There are two other primary sources for chemicals: those derived from plants are largely used in the pharmaceuticals and cosmetics industries and 'inorganic' chemicals derived from minerals are used to produce chlorine, caustic soda, acids and fertilizers.[22]

Chemicals from different sources are mixed in production. The endless manipulation of molecules, the basic building blocks of chemistry, results in some 2 000 new products coming on to the global market each year, many of which are toxic. Chemicals are also pervasively used in production processes, including the production of other chemicals. In the process, they are contaminated and so become unusable and often end up as toxic wastes from production.

Sasol dominates production of primary chemicals. It has six distinct chemicals businesses with production plants in the US, Europe and East Asia in addition to the Sasolburg and Secunda plants. The products are marketed globally, mostly supplying chemicals for industrial use. Here we focus on the plastics production chain.

Box 6.4 Making plastic

Plastics are produced from polymers which, in turn, are produced from monomers. Monomers are composed of simple chemical molecules. Catalysts and energy are used to produce long-chain molecules that make up polymers. Thus, ethylene is a common monomer and the basic molecules can be joined up to create the polymer polyethylene. However, not all polyethylenes are the same: the longer the chain composing the molecule, the higher the density of the polyethylene. High-density polyethylene is used to make thicker and more rigid plastics while low-density polyethylene is generally used to make flexible light products such as film-wrap.

The most common monomers are ethylene, vinyl, styrene and propylene. Where a polymer is made from two or more monomers it is called a copolymer. Polymers are also combined with other chemicals such as chloride used in polyvinyl chloride (PVC). Finally, various other chemicals can be added in the process of producing polymers or plastics: plasticisers such as phthalates are used to add flexibility, pigments are added for colour and flame retardants are added to products subject to heat. The mix is called a resin and is sold in the form of liquids, solid rods or pellets, as the raw material for plastic fabricators or 'converters'.

According to the Plastics Federation of South Africa (PFSA),[23] the South African plastic industry consumes over 1.1 million tonnes of polymers a year of which 800 000 tonnes are produced from Sasol's monomers. Sasol is the only producer of monomers and the largest producer of polymers. As part of Project Turbo, Sasol Polymers has nearly doubled its capacity. Like monomer production, polymer production is capital- and energy-intensive and there are just three other producers: Safripol (formerly Dow Plastics) located in Sasolburg, SANS Fibres in Cape Town and Hosaf Fibres in south Durban. Converters are considerably less capital-intensive and there are some 850 firms ranging in size from small local firms to transnationals.

Power within the industry lies upstream, primarily with Sasol but also with Safripol. While state regulation of the petrol price awards import-parity pricing to Sasol's fuel business, both Sasol and Safripol impose import-parity pricing on polymers that are not regulated. Sasol is the monopoly producer of low-density polyethylene and PVC while Sasol and Safripol share the market for high-density polyethylene and polypropylene. Industry analyst Ralitza Dobreva, writing shortly before Dow's sale to Safripol, observes that the behaviour of Sasol and Dow is 'implicitly coordinated' as 'their prices are consistently in line . . .' (2006: 9). In short, they operate as if they were a monopoly and, like ArcelorMittal, use import-parity pricing to appropriate added profit equivalent to the transport, handling and tariff costs of polymer imports. SANS and Hosaf both produce polyethylene terephthalate, used to make soft drink and water bottles, and must either import the ethylene monomer or buy it from Sasol.

While profits are concentrated upstream in the industry, labour is concentrated downstream. According to Dobreva, the plastics industry employs 35 000 people with 30 000 employed downstream. At both Sasol and Dow, new investment has been associated with labour shedding or with dramatically increased output per worker. Sasol Polymers reports a 26% increase in production per employee from 2006 to 2007 following the investment in Project Turbo. Dobreva concludes that policy should aim for the expansion of the downstream industry in the interests of job creation and this recommendation is reflected in the Department of Trade and Industry's (DTI) *Industrial Policy Action Plan* (2007). The longer-term benefits are doubtful, however. Expansion would certainly be accompanied by mergers and acquisitions predicated on expanding economies of scale and increased labour productivity. It thus appears as a short-term response that will reproduce job-shedding growth over the longer term.

Plastic and packaging

Packaging consumes 52% of plastics by value. For their part, plastics make up 70% of the R29 billion packaging market and are rapidly expanding production and market share. In its submission to the parliamentary portfolio committee hearings on the Waste Bill, the packaging industry claimed it had 'achieved impressive results' in reducing, re-using and recycling packaging in accordance with the waste hierarchy.[24]

Reduction is claimed because the weight of such items as beverage cans and glass and polyethylene teraphthalate bottles has been reduced over time. Reduced weight reduces the embodied energy, and hence production wastes, in such items as well as the energy required for transport. The growth of the industry, however, means that the number of items and the total volume of packaging material are rising rapidly. As a whole, the industry now consumes two billion tonnes of raw material a year. Plastic is held to be particularly virtuous for its lightness and the substitution of plastics for other materials is claimed as environmental progress. What is not said is that reduced weight is not associated with reduced embodied energy in the substitution. Corporate Accountability International comments caustically that a water bottle embodies energy equivalent to filling a quarter of the bottle with oil[25] while the Berkeley Plastics Task Force notes that producing one polyethylene teraphthalate bottle results in 'more than 100 times the toxic emissions to air and water than making the same size bottle out of glass' (Stover, Evans and Pickett 1996: 11).

The plastics industry claims to recycle 33% of plastic packaging, implying that 66% is destined for dumping. The bulk of what is recycled would appear to be factory waste – off-cuts and trimmings from the plastics production floor – and industrial packaging. Consumer waste is rather less easy to recycle but is nevertheless the focus of industry PR aimed to justify plastic in environmental terms. The underlying strategy, however, continues the core business of expanding the market.

Understanding this requires a step back in time to see how, as Heather Rogers puts it, 'today's polymer-laden reality is not simply the inevitable outcome of some natural process; it is the direct result of an industry that was nurtured by massive public spending, unrelenting lobbying, and sophisticated public relations' (2005b). In the two decades after the Second World War, the industry discovered the virtues of packaging designed for dumping. Returnable glass bottles, for example, were re-used up to 40 times. Single-use plastic, glass and can throwaways thus made for a massive expansion in the market and in profits. Emphasising that the industry should aim for 'low cost, big volume, practicability, and expendability', one farsighted participant told his colleagues at a plastics conference in 1956: 'Your future lies in the garbage wagon!' (quoted in Rogers 2005a: 121). To make this future, they had to persuade people who were used to mending and re-using things that throwing them away was the natural thing to do. They had to make people think of themselves as consumers.

Throwaways, together with the massive expansion of long-distance transport infrastructure, also enabled market concentration and centralisation. Returnable bottles were generally tied to local markets within easy transport range of bottling plants where they were refilled. Throwaways thus became a weapon in the hands of large corporations as they centralised production and used their financial clout to undercut and bypass local bottlers. In the two decades following the war, Coke and Pepsi established dominance in the soft-drinks market while the number of brewers in the US dropped from 400 to 100. A study for the US Environmental Protection Agency confirmed that this trend to monopolisation was 'encouraged and permitted by the introduction of nonreturnable bottles' (Rogers 2005a: 137).

The US packaging industry responded very quickly to environmental campaigns against the ever-growing torrent of waste. It launched Keep America Beautiful in the early 1970s, ran advertising and 'education' campaigns on the virtues of plastic and funded lobbyists to prevent legislative restrictions. Finally, it initiated industry-driven recycling and re-advertised itself as a green champion. The strategies honed in the US have been repeated around the world and the Plastics Federation of South Africa (PFSA) has taken up the US slogan: 'Plastics don't litter, people do!' The problem is thus individualised and confined to the domain of consumption in order to deflect questions about production and the structuring of markets.

The PFSA was established in 1997, in time to participate in the final round of lobbying on the National Environmental Management Bill as well as in the drawn-out waste-policy process. Mimicking the message of the US industry, it advertises its recycling initiatives as reducing waste in support of the waste hierarchy. Yet the intention of expanding waste is evident in its explicit promotion of incineration. According to its website: 'We need to recover as much as we can for recycling or energy recover' [*sic*]. The Department of Environmental Affairs and Tourism (DEAT) added 'energy recovery' to the waste hierarchy in the Waste Bill brought to parliament in 2007 and so elevated waste-to-energy incinerators above disposal. Incinerators, however, demand the waste that feeds them, particularly if they produce energy, and so subvert the waste hierarchy's first priority to avoid creating waste.

In the energy sector, plastic and paper are known as 'non-energy' because they are produced from energy resources. Plastic has a much higher energy content than paper. In 2004, eThekwini waste managers argued that the South African waste stream does not have a high enough proportion of plastic to make energy production from waste incineration viable. The reason given is that most South Africans are poor.[26] If poor people do not throw enough plastic into the bin, rich people certainly do and it ends up in dumps located in the neighbourhoods of the poor. Incinerators will be similarly located. In promoting incineration, both government and industry are promoting a particular meaning of 'development': that it produces more waste and more energy-intensive plastic waste in particular. Development, as the idea of 'a better life for all', is thus made to serve the active construction of the market in throwaway packaging. Such development, however, is not only unsustainable on environmental grounds. It also produces, rather than alleviates, poverty and inequality.

Behind this better life lies never ending brutality. In September 2005, activists from across Africa met in opposition to the Eighteenth World Petroleum Congress in Sandton, Johannesburg, whose theme of 'shaping the energy future' sounded more ominous than reassuring. They responded:

> As the bosses of big oil gather in South Africa for the 18th World Petroleum Conference, concerned citizens and activists around the world unite to condemn the oil industry. At every point in the fossil fuel production chain, where the bosses 'add value' and make profit, ordinary people, workers and their environments are assaulted and impoverished. Where oil is drilled, pumped, processed and used, in Africa as elsewhere, ecological systems have been trashed, peoples' livelihoods have been destroyed and their democratic aspirations and their rights and cultures trampled.

7

Power trip

CAPE TOWN'S POWER is supplied through the long transmission lines from Mpumalanga and from the Koeberg nuclear power station near the city. In November 2005, transmission line failures caused a series of electricity blackouts. Then, on Christmas Day, a loose bolt ripped the huge turbine rotor at one of the two Koeberg nuclear generators. The reactor closed down, removing nearly a fifth of the supply and plunging Cape Town and the Western Cape into a prolonged electricity supply crisis marked by repeated blackouts and load-shedding over the next eight months. The crisis was declared over when Koeberg was restored to full power in August 2006.

The blackouts caused chaos. Trains stopped and workers were stuck on – or between – platforms and arrived late for work, if at all. Businesses closed whether or not their workers arrived. Food rotted in fridges in supermarkets, restaurants and people's homes. Computers crashed and the city was abruptly taken offline. Cape Town's garment industry, already under pressure from cheap Chinese imports, stopped in mid-stitch. Fruit and grape harvests halted. Water pumps stopped and sewage works overflowed into Cape Town's already polluted streams. And, as the streets went dark at night, they were felt to be unsafe.

While the Western Cape suffered the most severe power crisis, blackouts were increasingly experienced across the country. These local blackouts were generally attributed to failing municipal distribution systems but the national grid was also strained. Eskom warned that the margin of supply over peak demand was dangerously narrow and would remain so until it brought the first unit of its new and very big coal-fired power stations into operation in 2012. In January 2008, the lights went down across the country. At the height of the crisis, Eskom abruptly shed over 4 000 MW from an already reduced supply and 'begged people to "turn it off", as the country hit the brink of disaster'.[1] The chaos of the Western Cape

195

was repeated on a national scale. National government led the country to panic stations with little evident effect. Municipal distributors desperately rotated blackouts from area to area. The mines shut down on 25 January for two days. Eskom first begged and then demanded a 10% reduction in industrial consumption. In May it declared the immediate crisis over and 'suspended' load-shedding but reiterated its warning of tight supply margins through to 2012.

This chapter starts by locating the crisis in the context of Eskom's history. It then focuses in on the crisis in the Western Cape to explore how things played out in a specific local context. Finally, it draws some conclusions about the broader implications of an overall decline in the energy system following peak oil.

Box 7.1 The power supply system

Base load is the backbone of the electricity supply system. South Africa's system is primarily designed to supply cheap power to energy-intensive mining and industry. It has therefore developed a high base-load capacity founded on the largest coal-fired plant available at the time of building. Prior to the 'new build', it had ten coal plants located atop major coalfields with eight clustered on the Mpumalanga highveld. Their combined capacity is 33 878 MW. The Koeberg nuclear plant in Cape Town provides an additional 1 900 MW base-load capacity.

Base load is supplemented by peaking power plants that kick in with additional power during peak-demand periods. In the daily cycle, demand peaks on weekday mornings and is higher in the evenings when middle-class people return from work. Working-class people contribute comparatively little to this peak because most of the cooking is done by unemployed women during the day.[2] In the annual cycle, winter demand is highest for both base and peak load. The electricity supply has to be managed to match demand as either an under-supply or an over-supply can blow out distribution systems or sections of the national grid. The 'disaster' feared by Eskom in 2008 was a total collapse of the grid. In that case, getting the system going again would take several days as the big power plants would have shut down. Starting them requires a major charge of electric power so the whole process would need careful sequencing.

Two kinds of peaking plant are used in South Africa. Open Cycle Gas Turbines (OCGTs) are something like jet engines and fuelled by diesel. They are designed for short runs and are thirsty and expensive to operate, but get to full power very quickly and so can respond to demand spikes. Continuous operation can damage them. Pumped storage systems consist of two dams, one uphill of the next. Water is pumped uphill in off-peak periods when there is a surplus of power, and released to generate electricity in response to peak demand. Pumped storage systems help balance the grid but, overall, they use more energy than they generate. They can only respond to an emergency if the water is already in the top dam.

CENTRALISING POWER

The crisis has long roots in Eskom's history, starting with the original mandate to deliver cheap and abundant power for industry. The development of the grid in the 1970s enabled Eskom to rationalise power generation and centralise administration, planning and information systems. It also determined policy, effectively becoming the apartheid government's energy arm and more or less running the power section in the Department of Minerals and Energy (DME). Its own inclination for secrecy was reinforced and protected by security legislation and its monopoly on strategic information prevented any serious challenge to its decisions.

This institutional and technological regime enabled changes to the labour regime. Both black and white workers had periodically demonstrated their power to disrupt production in the first half of the twentieth century and had been brutally suppressed by force of arms. Eskom had long aimed to minimise and isolate its labour force. Its new power stations were always capital-intensive and, with the grid in place, they could be built in remote areas while still subject to centralised management. White workers, bought off with the apartheid privileges that they had demanded, collaborated in the despotic management of the workplace while black migrant workers on contract were made to feel their vulnerability to dismissal and were housed in tightly controlled compounds distant from the urban centres of working-class agitation.

In the 1970s, electricity demand soared on the back of the commodity boom associated with the oil shocks. As international capital punted cheap loans to lay off surplus petrodollars, Eskom borrowed heavily to build seven new giant stations between 1979 and 1992. It was then caught in the debt trap induced by the Reagan-Thatcher counter-revolution. The price of gold and commodities collapsed from record highs, and the costs of the loans multiplied as interest rates rocketed and the value of the rand dived. The economy went into recession and Eskom's projection of future demand proved wrong. By the end of the decade, generating capacity exceeded peak demand by 62% (Eberhard and Van Horen 1995: 49) and Eskom had to mothball power plants while desperately trying to boost demand.

Government was stuck with the debt. It declared a moratorium on repayments in 1985 but also initiated the neo-liberal policies that would mature over the period of the political transition. The privatisation of Eskom was mooted, although not implemented, and the founding requirement that it operate without profit was revoked, marking a significant turn towards commodifying public services (Gentle 2009). Eskom raised tariffs, provoking industry and the mines to call for tighter government control to force it to operate on 'business principles'. If this sounded contradictory, Eskom then raised the alarm about 'politicians in the engine room' even as it maintained its occupation of the DME. Its corporate sense that it was a law unto itself was even more sharply revealed as the political transition got underway. According to its then boss, Ian McRae, staff feared that the new ANC government would 'nationalise' the corporation (McRae 2006: 78).

In fact, the ANC government's 1998 White Paper on Energy proposed privatisation on the assumption that 'the market' would lead the action to create economic growth and jobs. It predicted that new power plants would be needed by 2007 and said that building them should be left to the private investors. Eskom then found itself defending against proposals to break up its generating monopoly into supposedly competitive bundles to be sold off to the private sector and to hand the grid over to a separate state entity. It was supported by the real heart of the energy policy – the long-term commitment to cheap energy for industry as the foundation of international competitiveness. While government barred Eskom from planning new plants, private investors were not interested so long as there was no price escalation in prospect.

The privatisation policy was suspended in 2004 as government adopted the rhetoric of the developmental state. Amid alarms that economic growth was now overtaking the capacity to deliver power, Alec Erwin, then minister of Public Enterprises, announced that 'R107 billion will be needed between 2005 and 2009 to meet the country's growing energy needs. Eskom will invest R84 billion over the next five years. The balance of R23 billion is reserved for independent power producer (IPP) entrants'.[3] New generator plants were supposed to be up and running in 2008. By 2010, the IPPs had built nothing because the electricity price was too low to yield a profit while Medupi, Eskom's first new base-load plant, was expected to start generating only in 2012. The five-year capital spend had meanwhile increased six-fold.

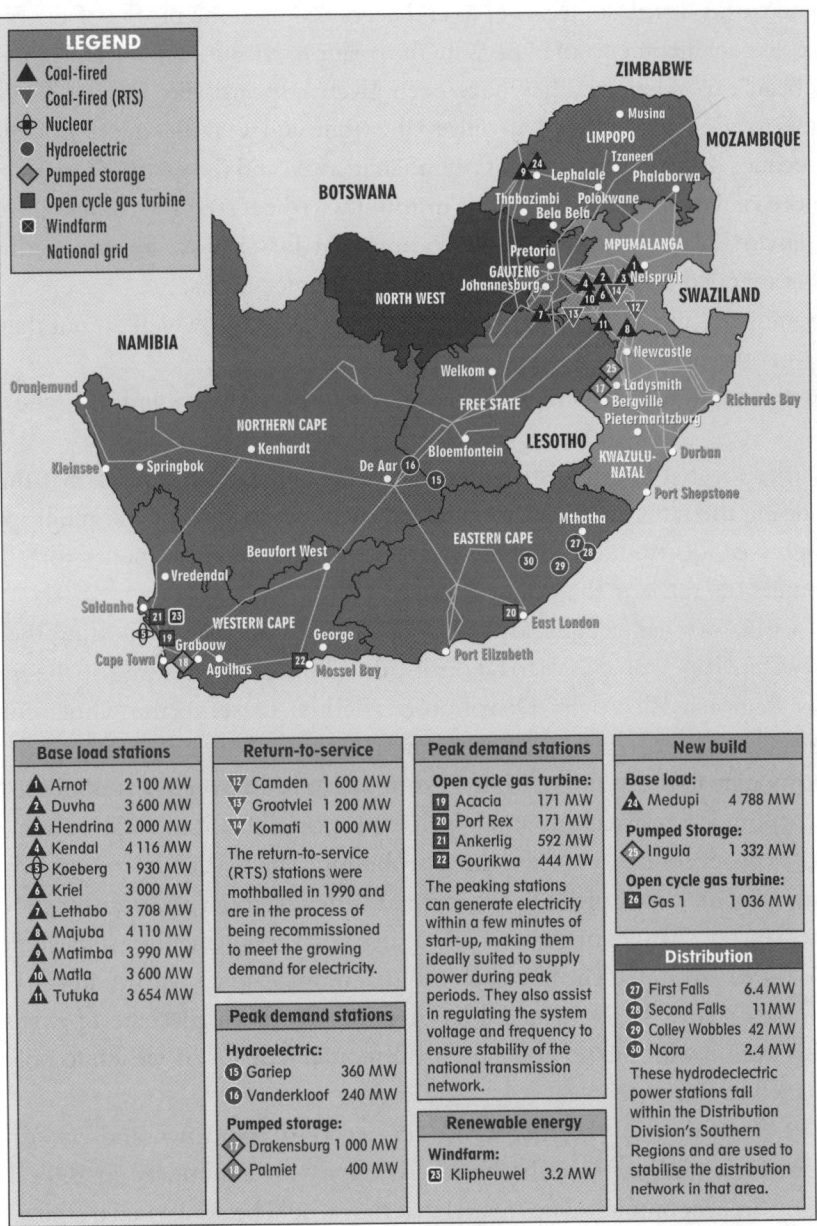

Figure 7.1   Eskom's power stations in 2008.

Adapted from www.Eskom.co.za.

Distributing power

Historic tensions between Eskom and the municipalities have revolved around distribution. White municipalities bought bulk electricity from Eskom but controlled local networks distributing power to end users and used the profits of distribution to supplement income from rates. With the political transition, municipal boundaries have been expanded and they have been given responsibility for the delivery of services to all citizens rather than only to the white and wealthier minority. Demands on revenues have thus expanded faster than income and the profits from electricity are more vital than ever. Eskom itself distributes to the rest of the country including:

- many black townships and commercial farms now located within the expanded borders of previously white municipalities;
- most municipalities located in the old homelands – which do not therefore get the benefit of a supplement to the rates;
- most, but not all, energy-intensive users – big industry and the mines.

Industries located in city distribution networks generally complain that they are subsidising the rates and there is some rivalry between Eskom and municipalities for rights of distribution to these users. Overall, Eskom accounts for 40% of customers but 60% of the value of sales.

This history has created a patchwork distribution map. Following the 1998 policy White Paper, proposals to rationalise distribution have centred on the creation of six Regional Electricity Distributors (REDs) to serve the whole country. Municipalities were to pool their distribution assets and obtain shares in each RED proportionate to those assets, while Eskom was to hand over its distribution assets. The corporation resisted this and it is now proposed that it too will take shares in all REDs proportionate to its existing distribution. The DME's *Electricity Distribution Industry Blueprint Report* of 2001 recommended that national government should be able to 'restrict changes in ownership' for five years following the establishment of REDs (DME 2001: 8). The implication is that municipalities will eventually be able to sell their shares. REDs would then pave the way to the piecemeal privatisation of distribution with the potential for a substantial transfer of wealth to politically connected business people.

The formation of the first RED, centred on Cape Town and covering the Western Cape and parts of the Northern Cape, was announced in 2005. It was stillborn. Smaller municipalities feared that they would be swallowed by Cape Town and refused to buy in. Cape Town feared that the RED would develop distinct institutional interests to the detriment of its notional shareholders and return diminished profits to municipalities. By 2010, government was pushing through legislation, including a constitutional amendment, to force municipalities into the still-to-be-formed REDs. Their argument for doing so was reinforced by the deteriorating state of municipal infrastructure. 'We dare not allow a situation where collapse of distribution networks would plunge us back into darkness,' according to Energy Minister Buyelwa Sonjica.[4] Exactly how REDs would save the light was not clear.

For the anti-apartheid movement, the discriminatory distribution of services was a key issue and providing access to energy, and specifically to electricity, became a political imperative during the transition period. The well-lit, high-consumption suburbs of white South Africa contrasted starkly with the dark and grimy black townships. In 1992, in the majority of formal townships 'more than 80% of households did not have access to electricity', while hardly anyone living in informal settlements, African rural areas or farmworker households had access (Greenberg 2006: 28).

Eskom started an electrification programme in 1991, both to reposition itself politically and to use up some of its excess generating capacity. In 1993, a National Electrification Forum – including the power industry, the political actors in the transition, trade unions and civics – was established and agreed to an accelerated programme. The ANC's Reconstruction and Development Programme (RDP) set a target for connecting 2.5 million homes between 1994 and 1999 and this was one of the few RDP targets that was met and exceeded. By 2004, access to electricity had doubled to 70%, prompting energy analyst Anton Eberhard to comment that 'doubling access to electricity . . . in a matter of years is probably without precedent' (2005: 6). Five years later, according to the president's fifteen-year review, 80% had access but for lights only, 67% had sufficient access to cook with and 59% had enough for heating as well (Presidency 2008: 21).

For poor people, access to the electricity supply is compromised by the costs of making use of it. The suspension of privatisation was not accompanied by a suspension of cost recovery on services to the poor and electrification has not fulfilled the promise of affordable energy services to poor households. For Eskom, electrification certainly brought political dividends. The anticipated economic returns did not materialise, however, as newly electrified households consumed less than expected and so did not generate the scale of returns to cover infrastructure costs. Eskom funded the electrification programme until 2000. When government subjected it to taxation it announced that it would no longer 'subsidise' the programme. This implied that the programme was exchanged for its tax-exempt status and allowed Eskom to escape what was arguably a miscalculated investment rather than a public-interest initiative. Subsequently, the DME has funded the programme through subsidies to Eskom and municipalities.

Since 2001, when Eskom was 'corporatised' and restructured as a tax-paying company, its profits rose steadily through to 2008: 'a comparison with the Fortune 500 top global companies shows that Eskom's after-tax-profits-to-revenue ratio is nearly twice the median produced by 23 electricity utilities listed' (Greenberg 2006: 39). The strategy to minimise employment has also met with success as the corporation shed over half its workforce, reducing the number of workers from over 65 000 in 1985 to 30 000 in 2004.

Thus far, Eskom has retained its strategic grip on the industry. It will remain the dominant generator and, assuming that its distribution assets are in fact turned over to the REDs, it will be a major shareholder in each of them.[5] Throughout the transition period it has defended its monopoly on strategic information and planning capacity in the power sector, dominated the DME and held fast to its centralised and secretive corporate culture. This was weakened, but not broken, with the establishment of the National Energy Regulator of South Africa (NERSA), which has developed some independent capacity. The national power crisis, and public outrage at its exorbitant tariff demands, has shaken Eskom's hold more profoundly. Most immediately, Minister of Energy Dipuo Peters says that its systems operator division – in control of the grid and responsible for balancing supply and demand – will be hived off into a separate state-owned entity in order to facilitate access to the grid for IPPs. This was a key element of the privatisation model put forward following the 1998 White Paper and, if carried through, will mark the end of Eskom's corporate model of vertical integration. Nevertheless, Eskom still exercises disproportionate power in policy.

DARKEST CAPE

Cape Town was energy-poor from the start. The more spectacular fynbos species contributed to early Dutch botanic collections, but the settlers complained of the lack of firewood and started growing European species such as oak. The British, who ousted the Dutch in 1806, later planted Australian acacias – notably the invasive Port Jackson Willow – for firewood and to stabilise the sandy soils. As the energy infrastructure was expanded, the Port Jackson was left to the poor, who used it both for firewood and building materials. By the late twentieth century, Port Jacksons were identified as a threat to the fynbos and to the Western Cape's tourism economy and were infected with a gall that is now killing off whole stands of the trees.

Following the discovery of gold, the mother city was thrust aside from the centre of the southern African economy, being geographically remote from the minerals-energy complex that emerged at the heart of South Africa's economy. The Western Cape has no coal deposits, no mining economy and, apart from the Chevron oil refinery, no energy-intensive industries. Cape Town's pattern of energy consumption is therefore not typical of South Africa: First, liquid fuel supplies about 60% of final energy – most of it for transport – as compared with 32% nationally, making the region particularly vulnerable to rising oil prices. Second, industry consumes a lesser proportion of energy than is the norm in South Africa. Commerce and industry consume 29% of total energy and 59% of electricity – as compared with 49% of total energy and 73% of electricity nationally. Cape households consume 15% of total energy and 38% of electricity.[6]

The city was connected to the coalfields of the Eastern Highveld, 1 600 kilometres inland, when Eskom completed the national grid in the 1970s. It was more intimately linked with the minerals-energy complex when the Koeberg nuclear power station opened in 1982. Koeberg was supplied by uranium mined in the former Transvaal and enriched at Pelindaba, and was a central link in setting up the nuclear supply chain needed to service apartheid South Africa's military nuclear ambitions (Fig 2005). In the 1990s, the supply chain was dismantled as the last apartheid president, F.W. de Klerk, renounced such ambitions before relinquishing power. Koeberg's fuel has subsequently been supplied by France.

Cape Town's economy has grown strongly in the 'new' South Africa. Manufacturing, particularly the textile industry, has declined precipitately but the services sector has grown significantly. Tourism, financial services and real estate have boomed along with a film industry able to make capital out of the dramatic landscape and relatively cheap labour costs. The Western Cape's wine and fruit farms constitute the most 'dynamic' sector of South African agriculture, having been integrated into global production chains supplying European supermarkets. Investment in these sectors has not created jobs. Unemployment is rising while the 'distribution of economic activity in the city has been highly skewed towards those with greatest skills and access to resources, with a large majority of the city's population precluded from meaningful participation in the economy'.[7]

Rich city, poor city

Apartheid Cape Town was designed to preserve both the economic advantage of white people and their sense of superior Western identity within what McDonald and Smith describe as a 'mixed economy [with] racial welfarism' (2004: 1461).

Black people were needed for their labour but were physically removed from the urban centre to the remote and bleak edges of the Cape Flats and excluded from the high-level services afforded to whites. This was a racist variation on what urban specialist Mark Swilling calls the 'consumption city', planned and built around the needs of the rich and catering to

> . . . the need within capitalist economies to create a mass of consumers that provide the markets for the suppliers of the basket of urban goods that are now defined as the basic elements of urban living . . . The basic building block of the 'consumption city' is the 'consuming neighbourhood' that, in particular, needs to buy in the necessities for daily living from the outside (often from very distant locales) – energy, water, waste removal services, building materials, food, vehicles, etc. The city's urban infrastructures had to be planned and managed to make sure these goods and services can be supplied, transported, removed, financed, and extended (Swilling 2006: 5).

Consumption is highly unequal. While the richest 16% of households used nearly 60% of all domestic water, 20% of all Capetonians had no piped water supply in 2000. Profligate consumption by the rich – for gardens, swimming pools, deep baths, etcetera – is expected to exhaust Cape Town's limited water supplies in 2025. These richest households also produce over half of Cape Town's 895 000 tonnes of residential waste every year. Half of this is organic (food and garden) waste, and so produces methane gas as it rots, but only 6.5% of all waste is recycled. This is one of the highest rates of domestic waste production and one of the lowest rates of recycling in the world. The waste is taken to dumps located in poor neighbourhoods on the Cape Flats where it pollutes both the air and the shallow water aquifers.

Electricity use is similarly unequal. According to The State of Energy Report for Cape Town, 95% of households are now electrified but the poor remain 'very dependent' on paraffin. The poor spend up to 25% of their income on energy while medium and high-income households spend only up to 5% and 'use electricity almost exclusively'. These richer households – about 39% of the population – consume four or five times as much energy as the poor and 'emit an immense 737 kg' of carbon each every month (SEA 2003: Section 4:3). These figures exclude transport where consumption is dominated by private vehicles.

The disparities in consumption are in fact higher than suggested here, first because there are more people in poor households than in rich ones, and second because electricity and water to a large proportion of poor households is routinely cut off because they cannot pay for it. Even after the introduction of the free 'lifeline' provision of water and electricity, 'service disconnections and household evictions continue in the city on a daily basis, supplemented by aggressive efforts to introduce prepaid water and electricity meters in an attempt to deal with non-payment of services' (McDonald and Smith 2004: 1475).

While apartheid discriminated on racist grounds, the neo-liberal city aggressively asserts the order of the market. McDonald and Smith show that, as an ideology, neo-liberalism is embraced by all political parties and most city managers and planners and is, indeed, represented as the means of addressing apartheid inequalities. In part, they are responding to fiscal constraints imposed by central government which slashed financial transfers to local government by 85% between 1991 and 1997, and by a further 55% between 1997 and 2000. This financial squeeze was accompanied by the expansion of municipal mandates to deliver services to all citizens rather than just the white minority. Private-public partnerships, punted by the World Bank and national government, then appeared as an efficient and cost-effective means of serving 'unfunded mandates' while the notion of extending public service delivery was systematically downgraded. Local departments, including Cape Town's Electricity Department, still responsible for delivering services are meanwhile corporatised – meaning that they are fenced off from the rest of local government so that they can be run like businesses.

Since 2004, the cuts in transfers to local government have been reversed. National government is also investing substantial sums through the Municipal Infrastructure Grant programme supplemented by additional funding for the 2010 football World Cup. This adds up to big money, although dwarfed by Eskom's billions, but it does not reverse the neo-liberal assumptions that now frame local institutional relations and planning. The City authorities have stated their dual priority for economic growth driven by global competitiveness on the one hand and pro-poor development on the other – a dual development strategy that mirrors ASGISA. For its first priority, Cape Town adopted the 'world-class city' slogan, which sounds like a good thing but in fact expresses a commitment to keeping the city within the circuits of global capital. That means creating and servicing the high-value locations and infrastructure to attract corporate investors and enable top managers to link with high-value locations in other global cities. Both public and private investment is consequently concentrated in wealthy areas. The pro-poor strategy is meanwhile failing. Urbanist Charlotte Lemanski comments that 'the two goals appear mutually incompatible'.[8]

Swilling argues that planning assumptions favouring the consumption neighbourhood are deeply rooted in the discipline of urban planning and linked to the priority given to growth. As Cape Town reaches the limits of its water, energy and other environmental resources, it becomes ever-more evident that satisfying the demands of capital is possible only at the cost of the poor:

> . . . it is difficult to see how poverty eradication in Cape Town is a realistic goal if scarce financial resources and free services from nature . . . are wasted on maintaining an ecologically unsustainable system that works in financial terms for the middle and high income communities . . . but tends to be too costly for those poor households that are lucky enough to be serviced (Swilling 2007: 38).

Energy analyst Tristen Taylor takes a more caustic view, concluding that 'electricity is granted to non-elites in accordance with their ability to serve the elite class' (2007: 6). The same might be said of other amenities and the war on poverty more often seems like a war on the poor as the city 'cleans up' to present itself to investors. This has provoked a growing movement of resistance as people struggling to keep their homes have acted in solidarity with each other. In the late 1990s, the Western Cape Anti-Eviction Campaign linked local Cape Town groups, and these groups are now linking more widely with people in other South African cities and specifically with Abahlali baseMjondolo.

Cape Town's flagship N2 Gateway Housing Project shows what world class means. It was intended to clean up the shack settlements that are highly visible to foreign visitors on the N2 'gateway' from the international airport to the city centre and the shack-dwellers were supposed to be the first 'beneficiaries'. Cost overruns resulted in unaffordable rentals and the first houses completed remained empty for some time. Those who moved in found shoddy building work and believed they were being made to pay the costs of corrupt profiteering by construction companies linked with the ruling party. Protesting outside parliament in July 2007, tenants demanded that rent be suspended until construction defects were made good. They were rebuffed by Minister of Housing Lindiwe Sisulu who 'reportedly advised them to "give a month's notice, pack their bags and make way for people who are willing to pay"'.[9]

The people of the Joe Slovo shack settlement know that they will not be able to pay. They were promised housing in the Gateway project in 2005 but faced removal to Delft on the city periphery to clear the way for construction of Phase

2 of the scheme. They were particularly aggrieved that government excluded them from decisions about their own future and appealed to the Constitutional Court. The court allowed their eviction but required 'meaningful engagement' by the authorities with the residents, that alternative accommodation should meet a minimum standard and that 70% of the new houses should be allocated to Joe Slovo residents. This, according to Kate Tissington (2009) of the Centre for Applied Legal Studies, is merely a palliative. At best, only 1 050 of the 3 000 Joe Slovo households will eventually get the new housing. Meanwhile, the Delft sites are already full, the housing mostly consists of 'government shacks', and the supposedly temporary settlements are turning into permanent camps that are less about housing than holding the poor.

Blackouts

When the bolt hit the rotor at Koeberg Unit 1 on 25 December 2005 it removed 900 MW capacity from an already stressed system. Two major blackouts followed in February 2006 when the failure of the national grid supply had the further effect of shutting down Koeberg Unit 2. Each event resulted in extended unplanned blackouts of several days duration for large parts of the Western Cape. Unit 1 was brought back on line in May when a replacement rotor obtained from France's nuclear utility was installed. Unit 2 then had to be closed for delayed maintenance and refuelling. Full power was finally restored in August. During the entire period from December 2005 through to August 2006, scheduled 'load-shedding' – a euphemism for planned rolling blackouts – was used to prevent overload on the available electricity supply and hence more general blackouts.

Explaining the unplanned blackouts, Eskom emphasised external technical causes, including fires under the lines, high pollution levels and misty conditions, causing 'flash-overs' that tripped the transmission lines. NERSA, however, subsequently found that Eskom's maintenance and commissioning procedures were inadequate and concluded that Eskom had 'transgressed its licensing conditions and was negligent'.[10]

The Minister of Public Enterprises, Alec Erwin, looked for a more political, outside cause when he tried to blame the bolt in the rotor on sabotage. Somewhat lamely, he later denied saying what he said. Anti-nuclear activists saw this in the context of earlier government pronouncements threatening to silence them in the name of public order. The allegation carried the threat of redefining dissent as terrorism. Opposition political parties saw it as a ploy to divert attention from government mismanagement ahead of local government elections, indicating their sense that political legitimacy is now reduced to technical competence. Eskom has subsequently noted that it was prevented from making critical investments following 1998 when the Energy White Paper pronounced in favour of privatisation, a policy strongly supported by the political opposition.

This was a crisis foretold, says Leila Mahomed of Sustainable Energy Africa. In early 2005, Sustainable Energy Africa wrote to Eskom warning that demand was rising and the transmission lines appeared vulnerable. It concluded that 'all eggs are in the one Koeberg basket'. Eskom responded that Koeberg was a very safe basket. This response was not merely complacent: it entirely missed the point that highly centralised energy systems reliant on large-scale generators are inherently vulnerable to a major loss of power. It is a point that Eskom keeps on missing because localised small-scale production is not compatible with its corporate culture or its interests.

Box 7.2   The French connection

A replacement for the 200-tonne rotor, supplied by Electricité de France, arrived on a South African warship in Cape Town on 5 April. At a press conference on board the ship, the French ambassador revealed that the replacement had been the subject of high-level negotiation, including discussion between Presidents Mbeki and Chirac. The pomp and ceremony is a theatre of obligation and dependency. It has been continued with Presidents Sarkozy and Zuma who met in South Africa prior to the Copenhagen climate conference and again on the way to the conference. Behind the glitz, the South African and French nuclear establishments are closely connected.

Like Eskom, Electricité de France is a state-owned electricity monopoly. It produces 70% of its electricity from nuclear power and is a central player in the global nuclear industry's current push for expansion. It is closely tied to Areva, France's state-owned nuclear construction and supply corporation. Areva executive Anne Lauvergeon represented the energy sector on President Mbeki's International Investment Committee and 'gives her utmost attention to South African projects in the nuclear energy field'. Areva was formed from a merger between Framatome and Cogema. Framatome built Koeberg, completed in defiance of the anti-apartheid campaign in 1982. From 1976, Cogema bought uranium from South Africa's Nuclear Fuel Corporation and financed the uranium concentration factory at Randfontein.

The French connection is now tighter than ever. The Areva University trains the staff of the South African National Nuclear Regulator and advises the Nuclear Energy Corporation of South Africa on options for deep geological storage of radioactive waste. Areva plays a role in all stages of the nuclear fuel cycle in South Africa, generating revenues of 30–50 million euros in South Africa each year. It holds contracts with Eskom for maintenance, services, technical assistance and fuel supply and it seconds 50 to 60 technicians to Koeberg during maintenance shutdowns.[11] Not surprisingly, Areva is positioning itself as a frontrunner in the bids to build South Africa's next conventional nuclear power station.

The pomp and ceremony that attends the French connection does not hide the fragility of the nuclear chain as plants age and breakdowns increase and as qualified nuclear engineers also age and are not replaced. When it comes to decommissioning, there may not be enough skills left.

Managing participation

The public reaction to the crisis was intense and highly critical of Eskom and government, creating a forceful dynamic for some form of participation in decision making. Given the centralised power of Eskom, there was no readily available institutionalised forum for participation and the demand found its own outlet. In February and March 2006, a series of meetings of the bureaucratically named Energy Risk Management Committee (ERMC) were convened. The ERMC effectively emerged with the crisis and quickly evolved from a discrete stakeholders' forum to a more or less open forum with increasing numbers of people at each successive meeting. The original stakeholders were Eskom, the provincial government, the City of Cape Town, business and labour, with labour calling in environmental and other civil society interest groups.

Eskom apparently thought it enough to assert that it was in control. The February blackouts shredded its credibility – clearly it had no plan beyond load-shedding while fixing Koeberg. Government's credibility was also on the line but provincial and city governments also had no plan – beyond relying on Eskom. Not being 'in the engine room' suddenly seemed hazardous as business loudly announced the threat to economic growth. The politics of energy was suddenly centre stage and open to question. The questioning, however, was effectively contained within two parameters: holding Eskom accountable but not otherwise challenging its institutional power; and managing the crisis until Koeberg was back on line. Thus, the provincial Member of the Executive Committee (MEC) responsible for energy, Tasneem Essop, demanded a 90-day plan from Eskom but prevented civil society representatives from presenting a set of proposals to the ERMC.

She also proposed that the ERMC appoint a more orderly and restricted stakeholder body, which came to call itself the Provincial Monitoring Team (PMT). Chaired by a representative of business and including representation from civil society, it defined its primary role as 'to monitor the Western Cape Integrated Recovery Plan on behalf of the EMRC' and it did indeed act to hold the central actors to account and enable a freer flow of information. However, its closing report, written as the dust of crisis settled in August, was addressed to the provincial MEC and not to the ERMC. This chaotic but more open forum had simply evaporated.

Outside of this quasi-official process, the Congress of South African Trade Union's (COSATU's) Western Cape region attempted to broaden public participation through a public meeting early in March. This meeting was predominantly working class and left, with a strong showing from the unions, social movements and environmental activists, but also included some business representatives. Eskom and the Western Cape premier were invited but did not attend. The meeting issued a declaration with a stinging critique of both Eskom and government. It identified the cause of the crisis in government's earlier commitment to privatisation and the consequent 'under-funding of the generation and transmission capacity', in the 'age and servicing levels' of Koeberg, and in the labour regime of outsourced and casualised work. It denounced 'the entire electricity restructuring process' as inadequate or 'downright dangerous'. Inequity was built into the system as domestic consumers subsidised energy-intensive industries. Inequity was also evident in the response to the crisis as 'load shedding has been applied unevenly with some areas being preferred above others in the absence of any clear [and] agreed to objective criteria'.[12]

Noting the hazards of nuclear power, the declaration proposed that government's power-sector strategy should focus on renewable energy, starting with solar water heaters and wind turbines, and that the relevant manufacturing capacity should be supported through the industrial strategy. In the short term, it called for the subsidised use of gas for cooking and supported official proposals for energy-efficient lights and the insulation of geysers.

This meeting thus began the work of creating a new politics of energy, providing a trenchant criticism of the current order and proposing practical responses that went beyond the immediacy of crisis. The forcefulness of the declaration showed the strength of networked relations in Cape Town's civil society. The follow-up showed civil society's weakness: the meeting proposed action – but nothing happened. Labour and civil society representatives duly took their seats on the PMT. Some believe that they were able to play an important defensive role there but the opening created by the crisis for a more radical engagement was quickly closed down and the voice of the working class was muted after this single expression. PMT proceedings were dominated by Eskom staff through sheer weight of numbers while Eskom's fiercest critics were silenced. According to Maya Aberman, who represented environmental group Earthlife Africa on the PMT, 'There was an explicit understanding that while we were part of the team we would not criticise it from the inside. I now think that we should have been on the outside, criticising and mobilising people so that the flaws of the current energy system would become more apparent'.[13]

'Sharing the load'

On 31 March Eskom produced the 'recovery plan for the winter of 2006' demanded by Essop and the ERMC. It anticipated peak demand exceeding the available supply and envisaged a multi-stakeholder process accompanied by an intense communications strategy aimed at:

- restoring public confidence in 'the industry', that is, in Eskom;
- managing load-shedding combined with disseminating information to enable consumers to anticipate cut-offs;
- minimising load-shedding by 'demand management' aimed to reduce peak loads by 500 MW, equivalent to 10% of Western Cape consumption.

The restriction on criticism from within the PMT was in support of speaking with 'one voice'. At a mundane level, this was about avoiding contradictory information on the available supply and load-shedding schedules. This level of communication was then folded into public relations as Eskom centralised control of information more broadly to 'restore confidence'. Eskom's plan envisaged stakeholder support 'to elevate what seems to be an "electricity industry" [problem] to a "National Challenge"'.[14] In short, the sense of crisis would justify Eskom's national plans for new coal-fired and nuclear generation and be used to override criticism.

Load-shedding was to be governed by a set of 'common principles', including that the cuts should fall equally across all areas while priority should be given to supplying 'strategic' or 'sensitive' consumers such as the fuel refinery, hospitals and sewage works. Additionally, '[e]conomic hubs, such as the CBD, will not be shed if they meet their allotted savings targets, where practical'. Cape Town's central business district did not meet its targets but nor was it shed.

Demand-side management (DSM) savings on peak demand exceeded expectations. Ironically, the biggest saving was made by those who consume least while efficiency savings from commerce and industry were negligible.

The blackouts revealed the social nature of electricity consumption. People went from individualised consumers who simply assumed the availability of cheap and abundant electricity at the flick of a switch to understanding their consumption in the context of the city's functioning. It thus created a sense of common crisis and vulnerability. It even seemed that this crossed Cape Town's rigid class-divide as business and labour shared a number of platforms, but this larger solidarity was illusory. The middle classes could enjoy the camaraderie of crisis as they checked the load-shedding schedules and went to eat in restaurants where the supply was on. Lacking such easy mobility and free spending power, the working classes justifiably felt a strong sense of their class-specific vulnerability.

Prior to the blackouts, Eskom's DSM programme was remarkable only for its invisibility. The energy efficiency campaign was now made highly visible through TV and radio power alerts as well as extensive press coverage and appealed directly to the emergent sense of social solidarity with the slogan 'sharing the load'. Substantial savings of around 100 MW were made simply by turning things off. This was backed up by efficiency subsidies available for geyser blankets and low-flow shower heads. Households adopting electronic 'ripple' control – which enables electricity managers to switch off their geysers from a central control – could also claim subsidies.

In contrast, support for solar water heaters was excluded despite strong arguments from civil society that water heating in richer households is the largest component of residential consumption. Eskom claimed that, given the limited solar-industry capacity in Cape Town, they could not be installed fast enough to realise savings within the three-month crisis period. Environmental organisations responded with a call for immediate action to launch a huge drive for solar water heating and energy efficiency. They noted that research demonstrating the benefits had gathered dust at the DME for over ten years and suggested that efficiency should not be left to those with a vested interest in expanding electricity sales.[15]

An efficient lighting campaign made the biggest savings at 230 MW. It worked through free swaps of incandescent bulbs for compact fluorescent lights (CFLs). The bulk of these savings were made by poor households as over three million

Box 7.3   Mercury

Each CFL contains a minute quantity of mercury – as Sustainable Energy Africa pointed out in 2000.[16] Dumping millions of CFLs when they burn out thus creates a potential for groundwater pollution. Environmentalists again raised this issue as the five million CFL roll-out got into gear. The PMT responded by recommending a comprehensive disposal plan including a R5-million mercury-recycling plant. This plan relies on voluntary recycling from households in a context where municipal waste managers do not provide the infrastructure for recycling. Bringing these lamps to a mercury-capturing plant is thus highly unlikely but, if it does happen, it may just shift the problem. Mercury recycling plants, such as Thor Chemicals, have a wretched history of poisoning workers and local environments. Production appears equally problematic. In China, hundreds of workers in factories making CFLs for the European market have been poisoned.[17] Eskom has since recognised that mercury is an issue – which is an advance on 2006 – but its response appears limited to providing tips for householders. Some organisations are looking at more costly but even more efficient light-emitting diodes to avoid the mercury problem.

Coal-fired power generation also results in mercury pollution. Trace elements of mercury in coal are minute but, because massive quantities of coal are burnt, mercury emissions are significant. If CFLs have the effect of reducing the amount of coal burnt, it might be argued that there is an overall reduction in mercury pollution. Regrettably, this does not follow. Eskom is building new power plants as fast as it can and DSM is really aimed to prevent demand growing faster than Eskom can increase the supply.

CFLs were distributed door to door in the townships. In middle-class areas, the swaps were organised through retailers who distributed two million. Savings in poor areas have not been sustained, however, as blown CFLs are replaced with cheaper bulbs.

The poor were similarly targeted to swap electric for gas hotplates, including a gas bottle and two refills, while richer households could get a discount on gas appliances. Peak time savings were a modest 22 MW partly because poor households cook ahead of peak demand. Those of the poor who made the swap soon ran into trouble. First, rising crude-oil prices and the new demand for gas started driving up the price in mid-2006. The PMT proposed, without success, that the price be regulated or subsidised. Second, gas supplies ran short largely because it is produced as a refinery by-product. Refineries actively shape the market to fit demand to the ratio of their product streams so the market was already calibrated to the supply. With additional demand driven by the managers of the electricity crisis, the gas supply could not simply be increased – and was not. Despite supply problems, local research indicates that those who were introduced to gas now prefer it to both electricity and paraffin because they found it clean, safe and efficient. Most have retained their gas appliances even if they cannot afford to use them all the time (Mohlakoana and Annecke 2008).

Exempted from load-shedding, the business community in Cape Town's CBD exempted itself from 'sharing the load'. Eighty-eight businesses in the CBD were approached to undertake voluntary electricity conservation. Twenty-five were not interested, fifty-one expressed interest, only twelve pledged to efficiency measures and achieved just 4 MW saving. In the Western Cape beyond the CBD, twenty industrial and commercial customers took advantage of the 100% subsidy for energy-efficiency projects. Projects with a combined impact of 17 MW were approved but only 6 MW savings were achieved by the end July 2006. Implementation of the rest of the projects would extend over several months. Eskom's rule that only interventions yielding short-term gains should be considered was thus not applied to business.

More broadly, it is evident that voluntary measures did not work. It seems equally evident that learning from this will be avoided. The CBD did not come close to meeting the agreed targets. It was vigorously defended from the consequences of sharing the load-shedding by the Cape Town Partnership – a non-profit company set up by local government and business to promote the CBD.[18] The partnership argued that its energy-saving campaign was making good progress and would be undermined if load-shedding was implemented. The unspoken assumption here is that business would co-operate only if it was afforded privileged treatment.

Middle and working class Capetonians were incensed to see whole office blocks lit up at night. Many of these buildings are hard wired for profligate consumption because they operate on a single switch, meaning that the whole building is either on or off. Rental buildings are particularly problematic because tenants pick up the electricity bills but landlords decide on any retrofit. The effect is that the costs of high-energy buildings are transferred from landlords to tenants but, in the context of cheap electricity, business tenants did not notice. The Cape Town Partnership is

Box 7.4   The poor pay the price

Energy researcher Wendy Annecke argues:

> It wouldn't be an exaggeration to say that the Electricity Supply Industry
> has failed women in Africa . . . Women are often among the poorest of
> the poor, and we know that it is women who are largely responsible for
> acquiring energy on a daily basis to keep members of the household fed,
> clean and comfortable. To do this women juggle with multiple polluting
> and inefficient fuels: mostly wood, biomass and dung, but also kerosene
> (2006: 38).

Kerosene, or paraffin, is dangerous, smells bad and gives food a bad taste – yet
it remains the fuel of necessity for around two million South African households
who cannot afford anything more expensive. They pay the price of massive
externalities resulting from an energy policy and practice that does not address
the needs of the poor.

A Paraffin Users' Household Energy Summit, in June 2007, estimated 'costs
to households and to the economy due to paraffin related incidents . . . in the
region of R100 billion per year'. A 2001 report to the Paraffin Safety Association
of South Africa concluded that in 2000 at least 143 000 children drank paraffin,
at least 55 000 children contracted pneumonia after drinking paraffin, and at
least 4 000 children died from paraffin-induced chemical pneumonia. There
were at least 46 000 paraffin-related fires and 50 000 paraffin-related burns,
and 31 000 of these burns were the result of paraffin stoves exploding.

In densely packed shack settlements, fire spreads within minutes. Lives are
lost, people are injured and meagre possessions are destroyed. These include
documentation like identity books, leading to problems in accessing pensions,
health and other services. Recently, a safer paraffin stove was developed but it
sells for around R200 in Cape Town – four to five times the price of an unsafe
stove. The DME has failed to control the paraffin trade and the price more than
doubles from leaving the factory gate to being retailed to poor households in
unsuitable containers. Electrification does not solve the problem, as cooking
and heating requires more energy than the basic free energy allowance provides.

Sources: Paraffin Users' Household Energy Summit in June 2007, 'Final Declaration'; '2001
Report to Paraffin Safety Association of South Africa' in *Energy Management News*, Vol. 8 No. 2.
at www.eri.uct.ac.za.

now working to develop a model to 'incentivise' energy efficiency projects in multi-tenant buildings.[19] The real effect of this will be to protect the principle of voluntary measures, which will come at a considerable price in public subsidies.

Business did much better when they could see the money. Eskom paid premium rates – thought to be some ten times higher than normal industry electricity tariffs – to 33 customers who used their own backup generators during peak hours and so substituted about 58 MW. Many companies have since bought generators to protect against future outages. Most run on diesel and are indeed expensive to operate. They may substitute for grid electricity but at the cost of an overall increase in energy consumption and carbon emissions.

Keeping the CBD switched on may have defended Cape Town's world-class city ambitions, but the profligacy of the CBD contrasts starkly with conservation in the townships. It is part of the broader structure of discourse that defends cheap energy for industry and business in the name of national economic competitiveness and uses the need for conservation to justify prepaid meters and trickle-feed technologies for the poor. While these technologies restrict consumption, the need for energy is exacerbated by shoddy building and the neglect of environmental design in public housing projects.

Agriculture unplugged

The blackouts provided a detailed demonstration of the economy's dependence on external sources of energy and provided a test run, albeit in just the one dimension of electric power, of the implications of an energy future that, in Mandil's words, 'evolves from crisis to crisis'. The Western Cape Regional Chamber of Business claimed that, by the end of March 2006, the power failures may have cost the provincial economy as much as R8.9 billion, with businesses losing R5.6 billion and spending an additional R3.3 billion on generators and other equipment to help them manage the blackouts.[20] Large and small businesses reported lost production and damage to equipment. The Chevron oil refinery in Cape Town lost twelve days of production and a host of small businesses, from hairdressers to Internet cafes, closed during blackouts.[21] The modern rural economy is no less dependent on fossil energy and here we take a closer look at the impact of the blackouts on the farm.

Reliable electricity is a central input in the increasingly high-value export-orientated Western Cape agricultural sector. Cape farms were first electrified in the 1930s, when industrial agriculture integrated electricity into farming. In the 1980s and 1990s, Eskom renewed the drive to electrify agriculture as it tried to expand markets to soak up its excess generating capacity. Its agricultural arm, Agrelek, gave technical advice on how to electrify ever-more farming processes.

With the political transition, the electrification of agriculture put the leading Cape wine and fruit farms in a good position to respond, in exemplary fashion, to the policy of export-oriented production.[22] From 1998 to 2002 they experienced booming exports to Europe and rising profits as they linked into global production networks supervised by the Northern supermarket corporations. Thereafter, their profit margins were squeezed as the rand strengthened, global competition for access to the Northern markets intensified, competition from imports arrived on their home turf as the state stripped out protective tariffs, global over-production of wine created a glut on the markets, and input costs rose.[23] In short, their vulnerability at the subordinate end of the production network became evident and the squeeze on labour intensified.

Energy is critical to creating the 'cold chain' that carries fresh produce from across the world to the refrigerated display shelves of Northern supermarkets. The entire process is on a tight schedule defined by just-in-time delivery systems and, for fruit, the time from tree to ship is no more than 48 hours. Fruit is picked by casual workers in the heat of summer and taken to packing sheds where it is washed and rapidly cooled, packed and loaded on to refrigerated trucks. The trucks must meet refrigerated ships that are on tight turnaround times. Any delay incurs additional docking fees that are billed to the producer, not the buyer. The ships must also meet the trucks at the other end, with time penalties again imposed on producers. A thermometer inserted in each fruit box records its temperature throughout the journey and the box is rejected if it has exceeded the temperature limits at any stage. Finally, the supermarket may summarily cancel the whole contract if time or quality criteria are missed.

Timing is also critical to the wine farm harvest. White wines in particular are now drunk in the year they are produced, speed to market being crucial for profitability within the global markets structured by the Northern buyers. For quality, precision in production compensates for the time previously allowed for the wine to mature. Grapes must be picked when the acid balance is right, rapidly cooled in the cellar and the temperature precisely controlled by computerised systems throughout the production process.

Wine and fruit production are large-scale industrial operations of which the farms are but one component. Scale and the necessity for speed now dictate that a large pool of unemployed people should be available for seasonal work.[24] On a single Cape farm, something like 700 pickers may be employed while anything between 200 and 1 000 casual workers are employed in the packing sheds.

Blackouts brought the whole system to a halt. Without the cooling plant in cellars and packing sheds, harvesting had to be stopped. In the account of farmer organisation Agri-Weskaap, farmers had to pay workers for the day without getting the harvest. This may have been the case on some farms. However, 'most seasonal workers work on a piecework basis and are paid by the "basket" ', according to agricultural researchers Ewart and Du Toit (2005: 120). It seems unlikely that they would have been paid for more than they picked. Farmers also lost sensitive equipment. Winemakers, for example, reported that electricity interruptions fried their costly computer process-control systems. The larger threat, however, was that of losing market access.

In the event, the impact of the electricity crisis on farms was largely contained. Agri-Weskaap, according to CEO Carl Opperman,[25] anticipated further supply problems following the November 2005 blackouts and responded early and proactively. It administered a survey of the Western Cape's 6 000 farms to establish at what times – daily and seasonal – the power supply was most critical to the farming operation and when it was least critical. The responses were given to Eskom staff who analysed them and planned load-shedding on that basis. The unplanned blackouts, when the whole system crashed in February, were thus the most threatening interruptions.

In Opperman's view, farmers understand that they are tied to Eskom for the foreseeable future. Like other businesses in the Western Cape, they abandoned plans to sue Eskom but many were exploring ways to reduce their dependency. Some opted for conventional backup generators. In response to climate change as well as energy security, a small minority of the leading estates looked at their overall energy and carbon flows, seriously addressing energy efficiency and developing on-farm energy systems using low-carbon or renewable technologies.

These responses were motivated both by sensitivities in their export markets and by a real concern about climate change, not least because Western Cape agriculture will be severely affected by it (see Box 7.5). Thus, the Backsberg wine estate was declared 'carbon-neutral approved' according to 'The Carbon Standard', having reduced its energy consumption and offset its outstanding emissions by planting trees in a nearby township in partnership with the NGO Food and Trees for Africa (FTFA).

The Carbon Standard sounds both official and universal but was actually established by FTFA and implemented in a partnership between the NGO and transnational corporate auditors PricewaterhouseCooper. PricewaterhouseCooper's South African office was 'the first African company accredited to do carbon auditing' according to FTFA.[26] Its accreditation was based on the International Standard Organisation's newly developed standard for greenhouse gas reporting. This standard is one of several such initiatives, most notably the 'Voluntary Carbon Standard' established by the International Emissions Trading Association, The Climate Group, the World Business Council for Sustainable Development, and the World Economic Forum with the primary purpose of facilitating carbon trading. FTFA's Carbon Standard may be similarly used and PricewaterhouseCooper is actively engaged in facilitating the carbon market brought into being by the Kyoto Protocol of the United Nations Framework Convention on Climate Change (UNFCCC).

For Southern countries, trading is possible under Kyoto's Clean Development Mechanism (CDM). The rules require that a Northern organisation or business must invest in a Southern CDM project, which results in lower carbon emissions than a business-as-usual project. The Northern organisation is then credited with the carbon emissions that are held to have been saved. It can either sell the carbon credits or, if it is in danger of exceeding its own emission allowance, it can subtract the presumed saving in the South from its actual emissions in the North.

Backsberg rejected trading on these terms because it means that South African carbon rights are alienated to the North, but did expect to improve its brand position particularly in Northern markets. It thus pioneered the addition of climate change concerns to ethical trading, etcetera, at the producer end of global production networks. In the North, the supermarkets were already on to the management of these concerns. Not frying the planet is offered as one more consumer choice, mainly aimed upmarket, in a basket of ethical, quality and brand choices. Overall food-energy costs are rising, however. In Britain, the production, processing, distribution and preparation of food now consumes one fifth of total energy. Lucas, Shiva and Hines (2006) show that half the energy for transport is used within exporting countries and between exporters and Britain, and one third of the energy used for production, processing and packaging is expended in exporting countries.

Irrespective of the sincerity of individual initiatives, it is this larger dynamic driven by global capital that both creates poverty, as workers are outsourced, casualised or simply made redundant, and degrades local and global environments. And it is in the interests of global capital that the voluntary codes and standards are brought into being. Tree planting may bring its own benefits in particular cases, but offsetting carbon emissions just from agriculture would require more land than the planet has to offer. The question then arises as to whose land will be appropriated as the carbon market clamours for offsets.

> **Box 7.5   Climate change on the farm**
>
> Wine farmers in the Western Cape have had several meetings with climate scientists and are well aware of the threats. They expect that as the climate dries out they will grow varieties now suited to the dry margins of the winelands. Agri-Weskaap's Opperman thinks there should be more active agricultural research to help farmers deal with the coming changes but notes that national research capacity has been run down. His vision for responding to climate change is not limited to the Western Cape. He sees South African farmers expanding into the rest of Africa as the next option. It seems unlikely that African peasants will welcome this even if their governments do. Regulars at a Kalk Bay coffee shop are less sanguine about climate change as they contemplate the possibility that the railway line running along the coast will disappear under the water. Residents of the low-lying Cape Flats have more immediate reason for concern. Many settlements already experience regular flooding in storms and the intensity of storms is likely to increase.

THE LOGIC OF AN ELITE FUTURE

The Western Province has set a target for 14% renewable energy by 2015 – the most ambitious in the country. At the level of cities, Cape Town is leading the debate with plans to expand renewable energy and boost efficiency. Its sustainable energy policy aims to provide affordable energy for all while promoting the city's economic competitiveness.[27] It thus repeats the dual strategy noted above.

South Africa certainly has plenty of scope for expanding renewables but it should not be expected to sustain growth. Elmar Altvater comments that while 'life on earth remains dependent on the radiation of the sun . . . it is impossible to power the machine of capitalist accumulation and growth with thin solar radiation-energy' (2006: 9). This is because industrial capitalism works by debt financing driven by compound interest. The owners of capital, whether they are lenders or investors, require a profit that increases their capital. The enlarged sum must then be reinvested because the system cannot tolerate 'idle' capital: profit itself must make profit and the system as a whole must accelerate. This is achieved through two strategies: increased productivity within the production system or by 'making the other guy pay', that is through accumulation by dispossession.

The command of dense energy is essential to increasing productivity. At the global scale, the International Energy Agency's (IEA) projections of energy demand through to 2030 show what is required to maintain economic growth. Following the point of peak oil, there will still be copious quantities of oil to be had, but declining production will terminate growth in returns from rising productivity for the system as a whole. What is left is accumulation by dispossession. This strategy – enclosing people's resources and externalising costs on to people and the environment – pre-dates industrial capitalism and provided the initial capital for industrialisation, but did not then disappear. It has always subsidised the 'internal' return of profits from the production system, imposing untold misery on people in the colonies, the Third World and the global South. Since the 'signal crisis' of the US-led regime of accumulation, and with the triumphant turn to neo-liberalism in the 1980s, increasing returns from dispossession have compensated for diminishing rates of return from the production system.

The logic of the crisis of the regime of accumulation thus converges with the logic of the crisis of energy depletion. The dynamic of capitalism post-peak will intensify present processes. In contrast to Bush, the Obama administration has adopted a more inclusive rhetoric but has not abandoned accumulation by dispossession or the war on terror [28] because it cannot abandon growth. Nor should it be expected that any other nation state will spontaneously abandon growth for this is the foundation of legitimacy within the international state system that brought them into being.

Growth, however, is an impossible strategy in a shrinking energy system. The supposedly common good of abundant energy will therefore be 'transformed into a "positional", oligarchical or "club" good' (Altvater 2006: 13) with access regulated by price and violence. In spatial terms, this means it will be expropriated for the ever more exclusive benefit of the elite enclaves brought into being in the past two decades. Even as growth within these enclaves is sustained through the subsidy from dispossession, increasingly pursued in the mode of disaster capitalism, the enclaves of growth themselves will, in time, inevitably shrink. This, finally, is the lesson of the 2006 power outages in Cape Town. The incompatibility of economic growth with redistributive equity will grow ever more acute but the space of the rich city itself will shrink as energy becomes unaffordable even to the middle classes.

8

Future power

THE NATIONAL POWER CRASH in January 2008 appeared to come as a shock. The public might be forgiven for not anticipating it. The Western Cape blackouts were portrayed as a local matter while government and Eskom punted South Africa's cheap and abundant electricity to anyone with a few billion bucks and a plan for massive additional consumption. Major corporations were most forthcoming with such projects and utterly indifferent to energy efficiency. In September 2007, Eskom briefed government and business to expect load-shedding.[1] It nevertheless 'reacted as if it were caught unaware' when the lights went down, according to the National Energy Regulator of South Africa's (NERSA) report on the crisis (2008: 9). It approached the crisis with eyes wide shut, forgetting the Western Cape experience as one would a bad dream and making no active preparations for a major loss of power. Further, it allowed its coal stockpiles to decline throughout 2007, exaggerating a trend that began around 2001, even as it used more coal to run plant harder to keep pace with rising demand.

The bulk of Eskom's coal is supplied through long-term contracts from tied collieries operated by major mining houses while the remainder is supplied through short-term markets and trucked in. Eskom favoured emerging BEE companies for both coal and trucking and purposely ran down its stockpiles from 60 to 20 days supply to expand the short-term business.[2] At the same time, the tied mines supplied to the lower limit of their contracts, according to Eskom, as the big corporations focused on exports that were then returning rising profits.[3]

The weakness of Eskom's management had long been cushioned by a very large spinning margin. As the margin narrowed, plant was run harder, things broke down and what had hitherto been shrugged off as minor problems turned into major risks. Management could not see the difference. Nor, it seems, could any of the other major players in the energy system within South Africa. All were captivated by the faith of cheap and abundant power for industry. Bobby Godsell, a former Anglo executive brought out of retirement in 2008 to chair Eskom's board through the crisis, subsequently joked that 'South Africa has the cheapest energy in the world . . .we just don't happen to have any in stock'.[4] Opening Eskom's case for the 2009 price applications, he said the price must now cover the costs of expansion.

The national power crisis shifted the politics of energy. It exposed that politics to unprecedented scrutiny and provoked conflicting responses. It opened a space for public dissent but also confirmed the deep-rooted instincts of state and capital. The first part of this chapter looks at the immediate response to the crisis and the institutional fragility that it exposed and locates it in relation to the larger crises of the times – into which South Africa also walks with eyes wide shut. The second part looks at the contested future of power that is now under construction.

THINGS FALL APART

The response to the crisis was in marked contrast to that in the Western Cape. Cabinet declared a national emergency on 25 January promising 'vigorous and coordinated action' from what came to be called 'team South Africa'. The heart of the response was to be the Power Conservation Programme, intended to ration demand in the short term while Eskom recovered itself, with longer-term demand-side interventions to be fast-tracked. The pressure on Eskom's margins would only be relieved once the first of its big new coal-fired power plants came online in 2012. At the same time, Eskom's new build would be accelerated.

It was soon evident that 'team South Africa' was government, Eskom and corporate business, with the unions in the corridors and the rest of civil society not invited. Government established two structures to manage the crisis: the Forum of Energy Executives, composed of the state's energy mandarins and meant to co-ordinate government's response; and the broader National Electricity Response Team (NERT), which was chaired by the Department of Minerals and Energy (DME) and included business and labour along with government departments and state entities.

In the event, team South Africa barely held together. Despite the rhetoric, top-level leadership from government was not evident. Eskom muddled through the immediate crisis by imposing a 10% supply reduction on the big energy users, who co-operated more or less grudgingly, with load-shedding for the rest. As the threat of rolling blackouts receded after May, government lost interest and, following Mbeki's ousting, it abandoned NERT to the corporates. The DME chair did not appear at meetings and money for the management of the structure was unpaid.[5]

NERSA opened another channel for a rather more vigorous public contestation through its hearings on Eskom's applications for price increases. Debate on South Africa's energy future thus appeared to be confined to the issue of price. Participants, however, broached the questions of who was paying and for what. Eskom itself extended the theatre of dissent through its own ineptitude. It repeatedly failed to submit applications on time and then submitted interim applications with the apparent intention of provoking panicked decisions in its favour while holding out for doubled increases. Consequently, it repeatedly returned to NERSA with new applications, each of which was met by a storm of protest from all sectors.

Nevertheless, it did manage to wring out a series of increases over the five years to 2013, which totalled up to 137% above inflation. This is far less than it asked for and said was needed both to pay the rising costs of coal and to finance the borrowing for its ever-more expensive new build programme. The process pointed out its own vulnerability on funding and access to capital but also the vulnerability to which it exposes ordinary people and the country as a whole. Government meanwhile lounged on the sidelines, raising the odd cheer for Eskom but otherwise leaving a policy vacuum. In particular, it neglected to produce an Integrated Resource Plan (IRP) required by law to guide both Eskom's planning and NERSA's decision making. The Department of Energy (DOE, formerly DME) finally produced a paltry three-pager of dubious legal standing on the eve of NERSA's 2010 hearings into Eskom's multi-year price determination (MYPD) application.[6] It was immediately evident that, far from giving direction to Eskom, it had taken direction from Eskom's application. Policy on power, it seems, remains with Eskom and, as ever, at the service of energy-intensive corporate industry.

As if to confirm this, the DOE implicitly admitted that it was incapable of producing an IRP and, in February 2010, called in the minerals-energy complex A list – Eskom, Anglo American, BHP Billiton, Sasol, Xstrata and the Chamber of Mines – to do it for them. The existence of this 'technical committee' was revealed through leaks to the press, meetings were behind closed doors and civil society requests for minutes were refused.[7] Of course, committee members had to share 'proprietary' technical information and the IRP 2010 finally released for public comment was shorn of these details. It thus confirms that confidentiality is not an issue between these competitors but is an issue between corporate South Africa and the public.

Meanwhile, team Eskom was also falling apart. A boardroom tussle resulted in Godsell resigning from the board and CEO Jacob Maroga being sacked. Maroga sued for compensation of R95 million in lost earnings and benefits, exposing yet again the inflated remuneration of top executives. At issue was his failure to present a coherent funding model to the board and to renegotiate Special Pricing Agreements with BHP Billiton. Maroga claimed neither issue was within his powers: Eskom's funding crisis resulted from the extraordinary costs of the new build and the funding plan had to be negotiated with government. It was then subject to NERSA's decision on tariffs. 'Buying back' the power from Billiton would cost $800 to 900 billion and was unaffordable. The sum presumably indicates the long-term value of the contract to Billiton. The Special Pricing Agreements could not be renegotiated without political backing from government, which was not forthcoming.[8] If this was true, it would appear that the political backing then materialised. In April 2010, Eskom and Billiton announced that a new agreement was being negotiated and that it '*may* involve BHP Billiton assuming responsibility for the commodity pricing and currency exchange risks related to the contracts' [emphasis added].[9] The deal is claimed to be in the interests of both parties and of the public. The public, however, is unlikely to gain any insight into its putative benefits as the new deal will, like the existing deal, be 'confidential'.

Economic hit

The chaos of the Western Cape was repeated across the country as water treatment and sewage plants failed, small businesses floundered and large businesses installed backup generators. Energy Minister Buyelwa Sonjica gave the nation tips on energy saving and told people to stop whinging and go to bed early. The real focus was on big industry.

Industry claimed large losses from both the outages and rationing and threatened redundancies while labour rallied in defence of jobs. NERSA subsequently estimated that R50 billion had been lost to the economy.[10] Neva Makgetla, a former union economist in the Presidency, attributed a decline in GDP growth from 5 to 2.1% in the first quarter of 2008 directly to the power cuts. 'For most of its history,' she argued, 'South Africa has benefited from the tendency of precious metals prices to rise when the world economy faces a crisis.'[11] The power crisis undermined this benefit. She recommended that residential and commercial users conserve electricity in favour of the mines to restore 'hopes for renewed growth' – a proposal that suggested the continued subordination of other interests to mining.

That was July. In October, the commodity boom turned to bust. The gold price held up but, by the end of 2008, some 22 000 jobs were lost across the economy with mining (including gold) and metal industries leading the losses. Casualised workers were the first to go and it is doubtful that they were properly represented in these figures. Industry said it was working to protect 'permanent' jobs but massive job cuts were planned in all sectors including services. Mining giant Anglo American alone planned to cut 19 000 jobs. The flagship auto industry shed thousands of jobs and more were lost on the showroom floor as both export and domestic markets shrank. Government meanwhile represented its infrastructure programme as a 'countercyclical' stimulus to the economy. The construction industry is the immediate beneficiary of the programme, yet even here corporations were getting rid of workers. Murray and Roberts retrenched 3 385 workers despite rising profits. By the end of 2009, a million jobs had been lost across the economy and more followed as the 2010 World Cup stadiums were completed.

Recession saved Eskom. As smelters and mines closed or went on to short time, the spinning margin was restored from 5% in January 2008 to 14% in January 2009 – one point short of Eskom's target of 15%. By 2010, commodity prices were on the way back up and electricity demand followed. Eskom said it would reach 2007 levels during the year and, although it added another 1 000 MW of new build capacity in 2009, it anticipated a tightening spinning margin. Recession saved another margin too: that between global oil supply and demand.

The sweet spot

The economic crash exposed the hollow foundations of growth even as it demonstrated the vulnerability to its failure of those made dependent on it. Three years earlier, reporting growth of 4.4% for 2004, Trevor Manuel said the economy was 'hitting the sweet spot'.[12] It grew sweeter yet as the commodity boom pushed growth to around 5.5% in 2007, seemingly within striking distance of the 6% target. Yet the boom itself etched a corrosive insecurity into the fabric of the economy and particularly into the lives of poor people.

In terms of energy security, South Africa's vulnerability to peak oil and rising oil prices appeared to be offset by its considerable coal reserves, by the large proportion of fuel supplied by Sasol's coal-to-liquid (CTL) plant, and by its sizeable reserves of uranium. Nevertheless, imported crude oil still provides close to 70% of fuel for transport and most people and goods were and are moved by road. With very poor public transport and sprawling cities, the cost of working escalated dramatically. So too did the cost of shopping at car-dependent malls, which rely on road transport to bring in the goods. The growing and lucrative international tourist trade was also beginning to feel the effects of high and volatile fuel prices. South Africa's energy-intensive agriculture saw steeply rising input costs with little benefit from agricultural chemical production associated with the CTL process as Sasol fixed prices at parity with imports.

A modicum of energy security does not protect South Africa from the economic winds blowing through the global economy or that may be expected with peak oil. Oil is South Africa's largest import item and the rising cost was the key contributor to the growing gap between the value of imports and exports in 2007. Record trade deficits were covered by 'hot money' capital inflows as mining bosses started talking of a commodity 'super-cycle' and prices escalated. The strong rand mitigated the high price of oil imports but came at the cost of manufacturing, assumed to be the major creator of jobs, as South African products were priced out of export markets while cheap imports flooded into the local market.

Yet the commodity price boom was now being driven by the derailment of global capital. Various commentators blamed the oil-price spike to $145 a barrel on speculators but this was only part of the story. Money flooded into commodities because all other options looked increasingly dire. When commodities collapsed in mid-2008, the rand crashed to eleven to the dollar. In February 2009, *The Economist* marked South Africa's economy as one of the most vulnerable in the world as its exports dried up, the trade deficit ballooned and the prospects for investment seemed remote. It concluded: 'The rand, which has already fallen sharply, remains one of the most vulnerable emerging-market currencies'.[13]

The rand, however, defied expectations and rose sharply. First, the trade gap shrank as corporate South Africa cut imports of machinery and plant for expansion projects and indebted consumers stopped buying. Next, commodities recovered to around the 2007 levels for two reasons: China's massive stimulus programme partly substituted for declining Northern demand for commodities including iron ore, platinum and coal; and fund managers were once more buying into commodities as a least worst option to dollars and equities. Finally, the rand was supported by the 'carry trade' – money borrowed at zero interest in the US, Europe and Japan and invested in relatively high interest regimes in the South. This is a species of round tripping symptomatic of financialisation. It is initially paid for by Northern taxpayers through the bailouts of the banks but then siphons money out of Southern economies through dividends or profit taking on shares while also reducing the competitiveness of exports. It thus supplements global capital's pyramidal profits.[14] In sum, rand strength has been less a sign of the resilience of the local economy than of the weakness of the dollar and everything else. Meanwhile, the volatility of the rand in an increasingly volatile global economy remains a key point of vulnerability for South Africa's economy.

The oil price escalation through to 2008 also fuelled inflation. In response, the Reserve Bank lifted interest rates several times but, as Hendler et al. (2007) noted, this cure seemed as likely to provoke a recession and so prove worse than the illness. Indeed, it is a moot point whether the rising interest rate did not simply compound the inflation imported with crude oil. But while 'shocking inflation numbers' provided the justification for raising interest rates, it seems that the threat to the value of the rand from the record trade deficits was the real concern.[15] The imperative was to keep the hot money flowing in. Food price inflation was particularly severe and increases were steepest in poor urban and rural areas. Unregulated trading created a volatile market in maize, the staple of poor South Africans, and the price rose steeply from R500 a tonne in October 2005 to R1 300 in 2006 and over R1 900 in 2007.[16] On 29 September, Western Cape farmworkers on the minimum wage marched on parliament protesting that there had never been such hunger in the land. Their cry was echoed around the world as bread riots broke out in one country after another.

Booming the climate

The boom also produced an extraordinary intensification of greenhouse gas emissions beyond even the worst-case scenario projected by the Intergovernmental Panel on Climate Change (IPCC). Climate change, however, was not and is not a compellingly immediate issue in the minds of politicians. South Africa ratified the Kyoto Protocol covering the 'first commitment period' to 2012 as a 'Non-Annex I' country. As such it had no commitments to reduce emissions but is obliged to collect climate-relevant information, to report its plans for mitigation and adaptation, and to raise awareness. Ratifying Kyoto thus cost nothing but barely concealed government hostility to what it made out to be a Northern environmental agenda intended to constrain Southern development and keep the South in its subordinate place. Thus, in the lead-up to the 2002 World Summit on Sustainable Development in Johannesburg, South Africa's top environmental official told the parliamentary portfolio committee on environment that 'developing countries were "taken for a ride" in Rio with all the emphasis on environment and no focus on economic and social issues'.[17]

Government later professed to take climate change seriously. In 2005, it paraded six ministers at the first National Climate Change Conference. Speaking at the UN in September 2007, Mbeki berated the US for not taking climate change seriously and managed to sound like an environmental justice activist:

> To billions of the poor [the] linkages [between poverty, the environment and the use of natural resources] are real, the combination of their empty

bellies, their degraded environment and their exploited natural resources, for which they benefit nothing, defines a hopeless and heart-wrenching existence.[18]

However, his government's priority for economic growth remained absolute. It was widely expected that developing (Non-Annex I) countries would take commitments to reduce emissions during the 'second commitment period' beginning in 2012. As the international climate negotiations began to focus on the post-2012 deal, government participated more vigorously with the clear aim of avoiding such commitments. This was in continuity with South Africa's 2004 Response Strategy, which argued the priority for development over environment in terms of global equity. It suggested that 'the relocation of energy intensive industries from Annex I [developed] to Non-Annex I [developing] countries should be promoted' although it recognised that this 'may give rise to negative environmental impacts' and 'do little to alleviate the problem of unemployment'. Further, South Africa's export coal markets should be expanded and protected: 'Annex I parties . . . should initially concentrate on domestic actions that will not negatively impact on the market for fossil fuels from developing countries' (DEAT 2004: 7).

South Africa was also an avid supporter of carbon trading, treating the Clean Development Mechanism (CDM) as an alternative strategy to attract fixed direct investment and complaining only that Africa does not get its share. The first South African CDM project was an NGO initiative to install solar water heaters for people in Kuyasa, in Cape Town's Khayelitsha township. The project is repeatedly cited as proof of the benefits of CDM. But the big money is not in solar water heaters or in energy for poor people. In the polluted Vaal Triangle, the chemicals company Omnia calculated on creating 500 000 carbon credits per year by reducing its nitrous oxide emissions. Omnia got World Bank backing for trading the credits and expected to make around R60 million a year – not a bad return on a capital investment of R46 million.[19] Similarly, Sasol registered a nitrous oxide abatement project and 'expected to earn significant income' from carbon credits. The credits started to flow in August 2008 and the corporation is looking to develop several more CDM projects.[20] The logic is chilling: CDM is good business for polluters and the bigger the polluter, the greater the opportunity for carbon credits.

The perverse rewards created by Kyoto's trading regime played into the climate negotiations. Southern countries came under heavy Northern pressure to accept commitments on the Northern model of 'grandfathering' emissions rights. This means that future emission reductions are tied to the 'baseline' of historic emissions.

So those countries with the biggest baseline get the biggest share of rights to emit in the future. If South Africa had a conscious strategy for pushing up its greenhouse gas emissions in anticipation of future reduction commitments, that strategy would look exactly like what it was in fact doing before the power tripped out.

In the event, negotiations for the second commitment round collapsed in multilateral recrimination and distrust in Copenhagen in December 2009 and Southern countries will not be taking on binding commitments. South Africa, however, made a voluntary offer to reduce its emissions by 34% from baseline by 2020 and to start reducing emissions in absolute terms around 2035. Baseline represents the forecast for rising emissions if no mitigating action is taken. It was calculated in 2006 in the middle of the commodity boom and was not revised to take account of the bust. The offer has been held up as proof of South Africa's climate commitments and hence as justification for Eskom's new build. We will return to this in Chapter 10.

REMAKING THE COAL ECONOMY

If the blackouts lent a panicked urgency to the new build, the economic crash put the squeeze on Eskom's plans to fund it – particularly the capital plant imports – from international capital markets. Credit was drying up and the Wall Street credit ratings agencies – the watchdogs of global capital – put Eskom on 'negative watch'. In the 2008 budget, Finance Minister Trevor Manuel gave the utility a R60-billion 'subordinated loan' – effectively a capital injection – to shore up its balance sheet. But the ratings agencies were looking for a steep increase in the price of electricity to support repayment of loans for the expansion. Eskom applied to NERSA for a 60% hike but was granted 27%. Moody's then downgraded Eskom's credit rating by four notches so raising the cost of capital on international finance markets. The next day, news was fed to the media that Eskom was negotiating with the World Bank for a $5-billion loan.

The Bank was thus cast as saviour and Bank president Robert Zoellick drove the point home when he told the African Union that the loan was an example of scaled-up assistance to African countries affected by the financial crisis.[21] The image of the Bank as the friend-in-need to Southern countries contrasted starkly with its reputation for dictating structural adjustment programmes to supposedly sovereign countries in order to enforce debt repayment to global capital. South Africa itself had, in the Bank's own words, previously regarded it as an 'unwelcome suitor'.[22] Given South Africa's weight in Africa, Zoellick used the loan to signal that the Bank's political credibility was restored and/or that no country could afford to avoid it.

In the event, the loan was subsequently fixed at $3.75 billion. The Bank said it would bring financial stability to Eskom, support future economic growth, contribute to poverty alleviation, and help South Africa on to a 'low-carbon path'. This fits with the Bank's view of sustainable development and with the image it must cultivate to support its aggressive positioning as the world's leading broker of climate funding. The use to which the loan will be put also fits with the Bank's actual practice that is starkly at odds with this image.

The first news of the loan drew sharp criticism and opposition has since grown. It combines several strands in the justice movement: South Africans appalled by the social and environmental costs; Africans who argue that South Africa has already accumulated a 'climate debt' to the rest of the continent and see escalating carbon emissions as a threat to survival; and international and local groups opposed both to the World Bank's fossil agenda and to its use of debt to dictate policy in the South in the interests of global capital.

Pumping demand

Energy planning works on the assumption of growing demand and planning supply to meet it. The electricity system is primarily designed to supply power to large energy-intensive industries and mines that consume over 60% of power in most years. The 36 members of the Intensive Energy Users Group consume 40%. The very biggest users are the metal smelters supplied under long-term Special Pricing Agreements at well below the cost of production.

Speaking in 2006 of the supply constraints, Public Enterprise Minister Erwin said 'we were caught napping by our own economic success'.[23] Eskom was instructed to base its planning on ASGISA's 6% GDP growth target instead of actual growth projections of around 4%, a planning assumption that risked inflating the figures for future demand and over-building to meet them. Government policy of leveraging the 'competitive advantage' of cheap energy was calculated to ensure that actual demand did indeed inflate irrespective of the shrinking spinning margin.

A deal luring Rio Tinto Alcan to invest in an aluminium smelter at the Coega Industrial Development Zone was symptomatic. While smelters were shutting down in the North, largely because of the relatively high cost of electricity, South Africa baited the deal with a hefty energy and tax subsidy to win it from China and Brazil. As the deal was announced so too was the Developmental Electricity Pricing Programme (DEPP). Alcan was to be the first beneficiary and would use half the cut-rate power made available through the programme.

The rationality of this deal was questioned even by South Africa's growth-obsessed business press. A *Business Day* editorial asked: 'How far will the government go to attract foreign direct investment – and at what cost?' Government spending on Coega was heading towards R20 billion including R6.4 billion in high-voltage transmission infrastructure to supply the power for the smelter. 'As if that was not enough, the government sweetened the deal with a R1.93 billion tax incentive.' *Business Day* concluded:

> Essentially, South Africans will therefore be heavily subsidising the Coega smelter with cheap electricity at a time that they themselves will cough up considerably more for power – if they can get it. It's a lot to give away to a project that can in no way guarantee that Coega will become the industrial hub its creators dreamed of. [24]

Alcan's smelter was to produce 720 000 tonnes of aluminium a year from imported alumina and would require 1 355 MW. As with BHP Billiton's existing smelters, South Africa would effectively export energy provided through the DEPP at cut prices for at least 25 years. It was not clear if the DEPP replicated, or was additional to, the commodity-linked pricing deal already enjoyed by Billiton. Government, Eskom and Alcan used the usual alibi of 'commercial confidentiality' to conceal just how cut the price was. The environmental organisation Earthlife Africa went to court to force Eskom to reveal details of the deal in the public interest but was refused.

Local organisations also questioned the deal, arguing that the environmental and health costs would outweigh the benefits to the local economy. Greg Smith of the Nelson Mandela Bay Municipality Local Environmentalists said decision makers were 'stuck in the poverty versus environment scenario. It doesn't have to be like that. We don't have to destroy people's health to give them jobs'. Alcan would create only 1 000 jobs at a cost of R5 million each and 'at least 300 will only be available to highly skilled professionals, probably many from overseas'. [25] The pollution, however, would destroy other jobs and opportunities along with the resources of an environmentally sensitive area.

Government invested its prestige as well as big money in Coega. After a decade of trying to lure an anchor tenant, it evidently saw Alcan as a make-or-break deal and anticipated that other transnationals would follow its lead. On its own account, the Coega Development Corporation relied almost exclusively on 'cheap and reliable power' to sell itself to investors and it 'projected' that other investments, not counting Alcan, would result in the Industrial Development Zone consuming more power than the 810 MW consumption of the Nelson Mandela Bay metro.[26] Alcan pulled the plug on the deal in 2009 citing Eskom's inability to guarantee the power supply but without mentioning the crashing commodity markets. Since then, Coega Development Corporation has been scouting for an independent power producer (IPP) to build a plant in the Industrial Development Zone. More broadly, government is punting Thuyspunt, west of Port Elizabeth, as the site of a major new nuclear plant on the grounds of its proximity to Coega.

Coega was not alone. Cheap energy for the capital- and energy-intensive industries at the heart of South Africa's minerals-energy complex remained central to the state's strategy for growth in the 'first economy' and to Eskom's own growth strategy. Major expansions were either planned or in progress in the Mpumalanga platinum mines, at the Hillside and Mozal aluminium smelters, at Columbus Steel and ArcelorMittal, and at Sasol, while Indian conglomerate Tata had started construction on a high-carbon ferrochrome plant at Richards Bay. In each case, the corporations would be haggling over the electricity price and seeking to ensure that increases following from the costs of building new generating capacity were laid at someone else's door. The net result would be to lock in carbon-intensive economic growth for the foreseeable future.

The Cape Town blackouts in 2006 stirred Eskom to a renewed rhetoric on demand-side management (DSM). The DSM programme was introduced in 2003 to save 3 000 MW by 2013 and 8 000 MW, the equivalent of two new coal-fired power stations, by 2025. It consistently missed its targets for energy savings until the national blackouts compelled urgency. The 2013 target was then brought forward to 2011. For 2008/2009, Eskom claimed savings of 916 MW against a target of 645 MW and cumulative savings of 2 000 MW since 2003. The programme was focused on energy-efficient lighting, solar water heating and energy-efficient electric motors used in industry. Cancellation of the Rio Tinto Alcan smelter, by comparison, instantly knocked off 1 355 MW from forecast demand.

Cumulative savings on lighting seems more doubtful than Eskom claims. It includes the savings made during the Western Cape crisis but subsequent research has shown that households are replacing blown compact fluorescent lights (CFLs) with cheaper bulbs (Mohlakoana and Annecke 2008). The solar water heater programme appeared to end Eskom's and government's long-standing hostility to this most basic of technologies. But the 'roll-out' has done anything but roll. The programme provides subsidies in the form of a rebate following installation to offset high capital costs. These costs have more or less doubled since the introduction of the programme largely in response to Eskom's technical requirements. In its first year the programme supported the installation of just 1 400 units which hardly adds to existing sales of 35 000 units a year. The subsidy amount was increased substantially at the beginning of 2010 and now compensates fully for the increased price. In 2010, the Department of Trade and Industry (DTI) also got round to supporting 'green' industries. Its second Industrial Policy Action Plan (IPAP2) specifies support for the solar water heaters and aims to increase the market to 250 000 units a year with domestic production rising from 20 000 to 200 000 units over the next three years. IPAP2 notes that the industry is labour-intensive along 'the entire supply chain' (DTI 2010: 42) which rather begs the question of why it has taken so long to support it.

IPAP2 also aims to 'design and launch' a programme promoting industrial energy efficiency (45), a matter that has hitherto escaped its attention despite being nominally part of Eskom's DSM programme since 2003. Economic and industrial expansion is the DTI's *raison d'ére* and the basic assumption of Eskom's planning. The Jevons paradox therefore applies: energy efficiency expands overall energy demand in the medium to long term. This may contradict assumptions made in South Africa's energy planning that efficiency correlates to energy 'savings', but does not contradict either government or Eskom's priority for growth.

In their vision of the future, energy conservation – with the intention of reducing overall consumption – appears largely as a contingency. Conserving liquid fuels by transforming transport is scarcely on the agenda while Eskom's DSM is primarily driven by the power shortage. The utility expects to have restored its reserve margin of supply over peak demand by 2012. A second consideration is to save on the escalating cost of building new power plants. This, however, goes against the grain of Eskom's history of aggressive marketing to expand electricity sales. Even *Engineering News* observed the irony of Eskom 'having to champion efforts to curb consumption' and suggested that its new-found devotion to conservation might not survive once the new build programme had restored a comfortable spinning margin.[27] As with Eskom, government's record suggests that getting a return on its infrastructure investments will trump conservation as soon as an expanded power supply is secured and irrespective of any rhetorical devotion to climate mitigation. Its short attention span on NERT is symptomatic: conservation is an issue when the lights flicker out.

New build

Eskom CEO Jacob Maroga gave an update on the new build in January 2009 as shown in Table 8.1. These are the projects initiated since 2004 and most are either

Table 8.1   Eskom's new build.

	Technology	Name and location	MWh
Peaking Plant	OCGT	Ankerlig, Atlantis, Cape Town	2 080
		Gourikwa, Mossel Bay, Western Cape	
	Pumped storage	Ingula, Van Reenen, KZN/Free State	1 352
		Tubatse, Limpopo/Mpumalanga	1 500
	Wind	Sere	100
Total			5 032
Coal-fired base plant	Expansion	Arnot	300
	Return to service of mothballed plant	Camden, Ermelo, Mpumalanga	1 520
		Grootvlei, Balfour, Mpumalanga	1 170
		Komati, Middelburg/Bethal, Mpumalanga	955
	New coal	Medupi, Lephalale, Limpopo	4 764
		Kusile, Witbank, Mpumalanga	4 800
Total			13 509

Source: Eskom CEO Jacob Maroga: 'Presentation to the Media, 23 January 2009'.

completed or under construction and will add 17 000 MW of capacity by 2018. These generation projects are complemented by a major expansion of the grid transmission capacity particularly on the long lines to Cape Town and to the new demand centres such as the platinum basin and Coega.

Some 5 000 MW of this is peaking plant – either diesel-fired Open Cycle Gas Turbines (OCGTs) or pumped storage. The OGCT plants are all running and the Ingula pumped storage is under construction. Tubatse was postponed indefinitely as Eskom saw the economic recession reducing peak power demand. The Sere wind farm is not a peaking plant but was presumably too insignificant to be given a separate category. Eskom has since talked up its commitment to renewables, apparently under pressure from the World Bank, which needed a cloak, however thin, for its massive investment in coal. Sere is now due for completion in 2012.[28] Eskom has also committed to build a 100 MW concentrated solar plant as a pilot project.

The rest of the new build is base-load plant and all 13 500 MW is coal-fired. Of the mothballed plants now being returned to service, Camden is complete and several of the generating units in the other two plants have also been commissioned. The two new plants, Kusile and Medupi, are under construction and will be the third and fourth largest power plants in the world if completed. The first of Medupi's six units is to be commissioned in 2012 – although there are rumours that the deadline will not be met – with the others following at six-monthly intervals for completion in 2015. The first unit at Kusile was to come on line in 2013 with completion in 2016 but the project has been delayed for a year. Eskom planned to build a third such plant – known as Coal 3 – by 2018 but has cancelled it for want of funding. It nevertheless said that the plant is necessary and that IPPs must build the equivalent capacity.[29] Coal 4 was to follow later in the decade.

The draft IRP 2010 displays the minerals-energy complex perspective on future power. Integrated resource planning was introduced to shift planning from a one-dimensional focus on supply. However, the IRP is best understood as a traditional power expansion plan that justifies itself by projecting accelerated demand growth, largely driven by a major expansion of minerals processing, topped by a 30% spinning margin – double the international norm.[30] It has two components: the IRP itself covers the period to 2030 while a Medium-Term Risk Mitigation Plan (MTRM) focuses in on the immediate future.

Assuming high-demand growth, MTRM anticipates short supplies through to 2016 despite Eskom's new capacity and more active demand management. IPPs are seen as filling the gap. Several major corporations are negotiating subsidies to build or expand their own power production and 'well advanced . . . projects can produce between 1 000 MW and 1 500 MW by 2014'.[31] Sasol was first up, expanding its steam and power plant at Secunda from 320 to 600 MW to supply about half its power demand. No electricity will go to the grid but Eskom will pay Sasol above tariff rates and sell back at tariff rates. Anglo American and Xstrata are looking to build plants fired by the coal wastes heaped at their mines on similar terms.[32] Xstrata's project is specifically intended to power a new ferrochrome smelter.

Renewables are largely seen as the business of IPPs. In 2003, government set a very modest target of 4% electricity production from renewables by 2013 but did nothing to achieve it ahead of the power crash and the global depression. NERSA set renewable feed in tariffs – which pay a higher rate for each kWh produced – for several technologies in 2008. IPPs have lined up several projects but have been blocked in negotiations with Eskom over the cost of connecting to the grid amongst other things. MTRM sees IPPs installing 1 025 MW of renewables by 2014, less than 2% of capacity and about 0.6% of production.

Over the longer term, IRP 2010 shows total capacity expanding to 80 500 MW by 2030 and, from 2022, the new plant also replaces Eskom's older coal-fired stations which are due for decommissioning. Government now sees a niche role for renewables both to create jobs and to moderate the carbon intensity of the economy. The final IRP takes some account of declining costs and adds a further 17 800 MW between 2014 and 2030. It does not take adequate account of very large savings in water and associated infrastructure costs. Big base load remains the priority with the equivalent of Coal 3 and half of Coal 4 (6 250 MW) being built between 2014 and 2030. This implies that coal-fired power survives at least to 2090.

Nuclear power is held to provide the only viable base-load alternative. Nukes have long gripped the elite imagination. In 2007, government ministers talked up extravagant plans for 20 000 to 27 000 MW of new nuclear capacity by 2030. The bulk of it was to come from conventional pressurised water reactors (PWRs) while 25% was to come from Pebble Bed Modular Reactors. Eskom had already invited bids from Areva and Westinghouse to build the first PWR – a very large 3 500 MW plant dubbed 'Nuclear 1' – but baulked at the price when the bids came in and shelved the project. It has not revealed the price tag on the bids, but a good guess would be two or three times the R100 to R120 billion that Eskom had estimated and with plenty more room for escalation.

Government nevertheless insisted on pressing ahead with nuclear power and its ambition to develop the nuclear supply chain industry from uranium mining through to fuel fabrication. Instead of inviting bids for individual PWR stations, it invited the nuclear corporations to bid for the role of 'strategic partners' in its overall nuclear programme including a 'fleet' of PWRs. The IRP 2010 makes this a fleet of six new plants totalling 9 600 MW. Supporting these plans, DTI's IPAP2 observes: 'A future nuclear programme will cost in excess of R1 trillion. This will place enormous strain on the balance of payment and without an effective localisation programme will have severe consequences for the South African economy' (2010: 88).

If they pull it off, the conventional nuclear programme will displace the Pebble Bed Modular Reactor as the largest and most secretive industrial development

programme. Government sank several billions into developing this 'fourth generation' nuclear technology in which South Africa fancied itself a world leader. With
nothing to show for it, the programme was finally abandoned in 2010. Even the
skills necessary for localisation of the conventional programme have melted away.[33]

Government touts nuclear power as the means to reduce the extraordinary
carbon intensity of South Africa's economy. Given its ambition to establish a full
supply chain, the nuclear industry as a whole will scarcely mitigate emissions. Be
that as it may, it seems that government hopes to get financial transfers on the back
of climate change to pay for what it patently cannot afford. At the same time, it is
looking for cheap nuclear options from South Korea and Russia.

Overall, IRP 2010 results in the following energy mix in 2030: coal produces
65% of the supply, nuclear 20% and renewables 9%. The rest is supplied by peaking
plant, a little gas and imports. DSM displaces 3 420 MW capacity, less than half
Eskom's original target of 8 000 MW by 2025, equivalent to just 4% of the supply.

Coal and carbon

South Africa's carbon dioxide emissions for 2004 are estimated at 440 million tonnes
(mt) with Eskom accounting for around 45%. In the year to March 2010, Eskom
burnt over 122 mt of coal and emitted 224.7 mt of carbon dioxide. As shown in
Table 8.2, the coal and carbon figures have increased with rising production as
Eskom has run its plant harder to keep up with demand. Coal use and carbon
emissions per unit of production are markedly up even though higher 2010 emissions
also reflect historical under-reporting as earlier figures do not include emissions
from the diesel peaking power plants. Eskom still does not report methane emissions
but is reckoned to emit 2 267 tonnes (49 874 CO_2e) or close to 60% of national
methane emissions (Worthington 2009). IRP figures suggest that power system
emissions will peak at over 300 mt CO_2 in 2022 and then level off at 275 mt in the
later 2020s.

Table 8.2 Production, coal and carbon.

Year to March	2010	2008	2004	2000
Production (GWh sold)	218 591	224 366	206 799	178 193
Coal consumed (tonnes)	122 700 000	125 300 000	109 600 000	92 500 000
Carbon dioxide (tonnes)	224 700 000	223 600 000	197 700 000	161 200 000

Adapted from Eskom Annual Reports 2008; 2010.

Greenhouse gases aside, Eskom is a major league polluter of more local environments. Table 8.3 shows that its emissions of sulphur dioxide and nitrogen oxides have also increased in line with production. Only particulate emissions have been in any way mitigated and that only at some plants.

Table 8.3   Eskom's sulphur, nitrogen and particulate emissions.

	2008	2004	2000
Sulphur dioxide (tonnes)	1 950 000	1 779 000	1 505 000
Nitrogen oxides (tonnes)	984 000	797 000	674 000
Particulates (tonnes)	50 840	59 170	66 080

Adapted from Eskom Annual Report 2008.

Eskom has not installed sulphur scrubbers at any of its power stations. Medupi was planned without scrubbers on the rationale that there is a 'relative lack of pollution' in the Lephalale area as compared with Emalahleni (formerly Witbank) where Kusile is being built.[34] The Department of Environmental Affairs and Tourism (DEAT) in fact found that ambient sulphur dioxide standards were already being exceeded in the Lephalale area. Eskom's existing Matimba power station is the main source of emissions. The DEAT also found that people's health in nearby Marapong village – which houses miners and power workers – would be affected. Nevertheless, in 2007 it granted Eskom permission to build Medupi without scrubbers. Kusile was planned as the first South African power station with a full set of scrubbers. Pollution in Emalahleni, which the DEAT has declared an air quality 'priority area', is apparently adequate to justify the additional expense of R6 billion or more depending on exchange rates and cost escalations.

Eskom has since committed to retrofitting Medupi starting in 2018, apparently to comply with the World Bank's 'clean coal' agenda or because, as the Bank delicately puts it, the Waterberg 'airshed is in transition from an attainment to a non-attainment zone' and the DEAT may declare it an air quality priority area (World Bank 2009: xviii). That gives Eskom six years' unmitigated pollution. There's a catch, however. Scrubbers are water-intensive and Lephalale is dry. Although dry-cooled, Medupi will in any case require around six million cubic metres (mm^3) of water and the scrubbers will double that to 12 mm^3. This is about equal to the present consumption of both Matimba and the town of Lephalale that is already water-stressed. Water Affairs has promised to deliver the water through a series of 'augmentation' schemes but it remains to be seen whether Eskom's commitment will hold come 2018, or whether it then says that installing scrubbers is not feasible.

The coalfields of the Vaal and eastern highveld are now being depleted and the Waterberg, said to hold 50% of remaining reserves, is identified as the new frontier. Medupi is the first of a number of projects planned or mooted for the area. They include further power plants – whether built by Eskom or IPPs – and Sasol's Mafutha project as well as the associated mines and coal export ventures. The Department of Water Affairs (DWA)[35] projects water demand rising more than ten-fold to around 140 mm^3 a year. Re-plumbing the local rivers at an estimated cost of R10.5 billion barely covers the demand from Medupi without scrubbers. Further development requires 'augmentation' from the Vaal and implies a massive increase in water flow from the Vaal Dam and ultimately from the Lesotho Highlands Water Scheme, the Thukela and other catchments on the Drakensberg escarpment.[36]

The governments of South Africa and Lesotho have approved Phase 2 of the Lesotho Highlands Water Scheme – the construction of the Mashai Dam and water-transfer infrastructure. This will be the third major dam to be built and, like the others, will flood local people's best valley lands. The existing dams have already severely affected the downstream ecology of the rivers. The Mashai will add to the impact as Lesotho's rivers are drained dry.[37] On a wider scale, the Crocodile River augmentation would link the Limpopo into the national river plumbing that extends in the other direction to the Sundays River in the Eastern Cape.

The combination of projects lining up for investment in the Waterberg thus represents what David Harvey calls a 'spatial fix' on a grand scale. It involves not just the fixing of investment in the Waterberg itself but the massive infrastructure necessary to make them viable and to realise profits from them. There is a certain reciprocity here. If the area is not developed on a grand scale, then the investment in water infrastructure is not viable even on the narrow economic terms in which it is justified. Supplying the water to Medupi thus already assumes the rapid expansion of the area's coal economy and must be accompanied by the roads, rails, wires and pipelines to get the product out and, if Sasol's project goes ahead, by a whole new corporate-branded town.

Mining

Coal remains central to government's vision of South Africa's energy future. The industry's fortunes were boosted during the oil crisis of the 1970s, and it similarly benefited from rising oil prices in the 2000s. The price of coal rose sharply from $35 a tonne in 2003 to $65 in 2006, prompting speculation that coal 'may overtake oil as the best performing energy investment'.[38] In the first half of 2008, it spiked at $200 in European markets before crashing and then recovering along with the oil price.

Eskom is fed the cheap stuff. Most is produced under long-term supply contracts at 'cost plus'. This creates the economic base of the coal industry, which can then respond to market opportunities for higher-value exports and coking coal. Eskom's expansion requires a massive expansion in coalmining. In 2009, its planners said R100 billion must be invested in coalmining with some 40 new mines required by 2018, including 35 new mines devoted to supplying Eskom.[39]

Exxaro's Grootgeluk Mine on the Waterberg coalfield near Lephalale currently supplies 14.6 mt a year to the Matimba power station as well as coking coal for ArcelorMittal and coal for export. It is now being expanded to supply another 14.6 mt a year for Medupi under a long-term contract with Eskom. Sasol is currently conducting field tests on the Waterberg coal characteristics as part of its feasibility studies for Project Mafutha. The coalfield straddles the border with Botswana where Canadian corporation CIC is planning the Mmamabula power plant intended to export electricity to South Africa.

Emalahleni (formerly Witbank) on the Mpumalanga highveld has been at the centre of the coal industry since the late nineteenth century when it supplied fuel for the gold mines and subsequently for power stations. Eskom has contracted AngloCoal to supply the Kusile plant with 17 mt a year. Some coal will come from existing mines but the bulk will come from Anglo's New Largo project, described as a 'greenfield' development. Emalahleni's designation as an air quality priority area – or pollution hot spot – is well deserved. The mines add to the pollution from the cluster of power stations in the area. Apart from emissions from heavy equipment, opencast mines, mine tailings and old works are prone to spontaneous combustion. In some places, fires have smouldered underground for over half a century and the carbon emissions do not appear to be included in the national account.

The pollution of water is even more intense. The streams and rivers downstream of Emalahleni are ruined by acid mine drainage as described in Chapter 4. Nationally, over 100 mines (not only coal) are operating without water permits. Water Affairs Minister Buyelwa Sonjica told parliament that the department was 'negotiating' with them.[40] Eskom's mothballed plants are also located on the Mpumalanga highveld and bringing them back into operation has driven new mining development. The most convenient, and previously undeveloped, coal resource lies in the Mpumalanga Lake District at the source of three major river catchments – the

Vaal, the Usuthu and the Komati. A rash of mining applications has been waved through by the DME and some corporations have not waited even for its rubber stamp. Most of the coal deposits are small-scale and will be worked out in as little as five years. Acid mine drainage lags mine development by five to ten years. As the mines close, the rivers will be poisoned at the source.

With the rising price, the export channel was pushed to its limit. By 2007, South Africa was exporting about 72 mt of coal per year. Expanding the capacity of the Richards Bay terminal and the rail corridor leading to it to over 90 mt was the first priority in Transnet's infrastructure development programme. The cost was then put at R4.9 billion of Transnet's five-year capital spend budget of R47 billion. The Richards Bay terminal achieved its target of 91 mt capacity in 2010 but rail capacity lagged behind. Transnet is now targeting 81 mt on the coal line in the next few years. It is also expanding capacity on its iron-ore export line from the Sishen mines to Saldhana Bay and looking to develop a manganese export channel possibly through Coega. Transnet's 2010 five-year budget has increased to R95 billion, with R52 billion for rail expansions, and it plans to invite exporters to supplement that investment in 'private public partnership' deals.[41]

Europe remained the biggest export market during the boom but its demand contracted sharply in 2008. India and China took up much of the slack. To maintain economic growth of 8 to 10% a year, India plans a gargantuan expansion of its power system from 148 000 to 800 000 MW by 2030. In the short term, it intends increasing its coal-fired capacity from 77 000 to 127 000 MW.[42] India routinely runs short of supplies despite massive coal reserves and, assuming that it builds only a fraction of what it plans, its appetite for imported coal would seem insatiable. China is both the world's largest producer and consumer of coal and, in the last couple of years, has become a net importer on a grand scale.[43] In contrast to Europe, India and China are less demanding of quality. This suggests that expanding coal exports will increasingly be in competition with Eskom's low-quality demand. Whether to Europe or Asia, it seems doubtful that climate change diplomacy will penalise coal or energy-intensive exports any time soon.

The money

The World Bank's loan was complemented by several other loans more or less under World Bank management: two loans from the African Development Bank (AfDB) totalling $3.1 billion and a Clean Technology Fund loan of $250 million. Another $1.7 billion or so has been secured through the German and French export credit agencies (ECAs) from private European banks to fund the boilers and turbines for the coal plants.[44] This adds up to some R60 billion.[45] Medupi gets the bulk of the money – $3 billion of the World Bank loan and $2.6 billion from the AfDB. The World Bank loan allocates another $490 million to building a coal rail to the Majuba power plant that is currently supplied by road. The Bank calls this an energy-efficiency project. The remaining $260 million is allocated to the Sere wind farm and the concentrated solar power pilot plant. The Clean Technology Fund loan is also slated to support these two projects together with IPP wind projects, municipal and private sector solar water heating projects, and industrial energy-efficiency investments.

Treasury stands surety for the loans to Eskom. The World Bank required Treasury guarantees and Manuel's 2009 budget provided for R176 billion of loan guarantees, covering both development bank and private lending. The risk was thus shifted to the public purse.

Impressive as the figures were, a cavernous funding gap remained. From the first announcement of the new build in 2005, Eskom's five-year capital expenditure has risen in giant steps. Eighty-four billion rand, a staggering sum at the time, nearly doubled to R150 billion a year later. In February 2009, when Manuel recast the infrastructure programme as 'countercyclical spending' to stimulate growth, the spend to 2014 was put at R385 billion.[46] The big coal-fired plants were central to the escalation. First estimates were R30 billion for a big 'six-pack'.[47] By 2007, the price tags on Medupi and Kusile were put at R79 billion and R84 billion respectively. That has now escalated to R125 billion and R140 billion and it won't stop there.

In September 2009, Eskom submitted its delayed application to NERSA for the second MYPD. It asked for a 45% tariff increase in each of the three years from 2010 to 2012 and indicated that it would spend R638 billion in the five years to 2015. The application was greeted with the now familiar protest storm and, in November, Eskom submitted a revised application for a 35% annual increase. The five-year expenditure was then reduced to R500 billion while total spending in the ten years to 2015 would come in at R645 billion.[48]

On Eskom's account, these reductions followed from a review of its demand forecast. Reduced demand resulted from a more aggressive implementation of DSM and of the solar water heater programme in particular. But the most significant reduction was from the cancellation of the Rio Tinto Alcan smelter.[49] The lower demand forecast then allowed a one-year delay on Kusile and so shifted major spending to beyond 2012. This still left a R40 billion hole in Eskom's funding and the application indicated that Kusile was unaffordable without another round of major tariff hikes. Eskom also cancelled Coal 3 and put a two-year delay – from

2020 to 2022 – on the first nuclear plant coming on stream to shift massive costs out beyond 2015.

In part the price escalation reflected the general escalation of prices (steel, cement, etcetera) during the boom when Eskom was contracting. This was a 'seller's market' in which utilities globally were competing for construction projects. In 2009, Eskom hoped that some prices would come down in a buyer's market. There is little sign of it in Eskom's figures. Moreover, the big capital equipment is to be imported and the import bill will be around 45% of the cost. The value of the rand is thus likely to be more significant than any price reductions and Eskom's own demand for dollars will put considerable pressure on the currency. As Manuel put it:

> Lower consumer demand and the softer real exchange rate will dampen import demand in 2009, but infrastructure investment will continue to draw in capital goods. This will continue to generate a sizeable current account deficit, expected to average 6.7 per cent a year over the period ahead.[50]

If Eskom thought 45% tariff hikes would later make 35% seem reasonable, it miscalculated. Only government supported it, saying the lights would go out without the increase. Business and labour said the economy would go out with it. During the blackouts, they had both called for the new build programme to be expedited but they also agreed that Eskom's tariff demands would jeopardise a fragile recovery and threaten jobs. The mines and big industry, however, accepted that Eskom must recover costs. They thought 25% annual increases would be sufficient. On this cue, that is indeed what NERSA finally decided. Eskom's funding gap gaped wider. It said it was R190 billion short over the seven years to 2017.[51] Nearly R100 billion of this was for the new build already under construction and most of that was for Kusile. The remaining R90 billion was for projects still in planning – mostly the first phases of its nuclear ambition. It appointed JP Morgan and Credit Suisse to advise, respectively, on its funding plan and the sale of shares in Kusile.

COSATU rejected the increase and proposed the new build should be financed through a special tax on corporations. Community groups and environmental justice organisations were already on the streets at the NERSA hearings. They protested that increasing numbers of South Africans would be cut off, resulting in increased indoor air pollution from coal and paraffin with severe consequences for people's health. They also denounced the climate and other environmental impacts of the new build. They too rejected NERSA's award, noting that the 137% real increase in tariffs in the five years to 2013 was unaffordable to the majority of households. Following the NERSA hearings, and with the World Bank's executive board still to approve the loan to Eskom, they refocused their campaign to stop it and so pull the funding on the new build. Two critical questions were at the core of their concerns: cost recovery from whom and to pay for what?

Government was caught off guard by the intensity of the opposition to the loan and was clearly affronted at being challenged by civil society in an international forum. Energy Minister Dipuo Peters called opposition to the loan unpatriotic. Finance Minister Pravin Gordhan misrepresented the campaign as the initiative of Northern NGOs who were placing 'environmental concerns ... above the economic needs of South Africa'. South Africa, he said, had committed to reducing carbon emissions at the 2009 Copenhagen conference on climate change and it had 'a very clear plan' to do so.[52] The next chapter looks at the plan. The board finally approved the loan in April 2010 but with four countries abstaining.

Box 8.1   The World Bank's brighter future

Cost recovery is integral to the World Bank's view of sustainability. It claims that 'access to modern energy' is critical to its core mission of fighting poverty because it would liberate African people from subsistence chores and relieve women in particular of the burden of gathering wood or carrying water. Private investment is, in the Bank's view, the evident answer. Small investors could, for example, develop village-based energy systems using renewable technologies. Public-private partnerships, if not outright privatisation of utilities, would take care of larger investments. In either case, commercial terms are necessary to attract private investment, which in turn 'sharpens cost-consciousness and enforces payment discipline' according to a Bank paper put out for the World Summit on Sustainable Development and titled 'A brighter future? Energy in Africa's development'. It gets around the problem of how people without money will pay market rates by ignoring it. Not surprisingly, hardly any investment in village energy or renewables has taken place. As noted in Chapter 6, the Bank's actual projects have nothing to do with supplying local people or alleviating poverty but are overwhelmingly about getting the resources out to the global markets.

At the NERSA hearings, community groups testified that many households would be driven into penury by the increases demanded by Eskom. NERSA responded by introducing a rising block tariff, long called for by civil society, covering all residential consumers. This reduces the inequity within the residential sector as it works somewhat like tax bands – those who consume more pay higher rates. The distribution of the blocks is questionable however. Low-consumption households would still face a sharp and unaffordable increase in electricity costs and it is not clear how the scheme will be extended to those on prepaid meters. With only four blocks, the top block captures most suburban and many township households and adds no extra penalty for truly heroic consumers.

On the other side, as NERSA finally confirmed after years of evading the issue, industries with long-term Special Pricing Agreements were exempt from the tariff increases. Their very substantial share of the costs of the new build is thus transferred to tariff customers. This is rather rich coming on top of the news that Eskom made a R9.5 billion loss on 'embedded derivatives' – code for the link between the price of electricity to BHP Billiton and the international aluminium price – in 2008/2009. How much Billiton profited from this remains secret.[53]

Moreover, new questions were raised about actual prices charged to Eskom's energy-intensive tariff customers. In March 2010, Eskom told parliament that 138 large industrial customers would pay increased tariffs but that both the actual tariff paid by each of them and the tariff increases were confidential.[54] Earthlife Africa shows that Eskom sells to them at very little over cost or at below cost. In 2008/ 2009 it made a R3.2 billion 'operating loss' on top of the embedded derivatives loss. The average selling price across all customers was 24.97 against average costs of 27.63 cents/kWh. For 2007/2008, Earthlife estimated average tariffs at 11 to 14 cents/kWh for Billiton and 15 to 19 cents/kWh for other big users, as compared with 38 cents/kWh for suburban consumers and 45 cents/kWh for people on prepaid meters.[55]

Eskom argues that it is cheaper to supply bulk electricity to big users. This may be so but largely because the entire electricity infrastructure is designed to deliver to their needs. It hardwires social power relations into the technology of electrical power. The new build reproduces that bias: it is required by industry, not households. Government now says that the era of cheap power is over. Yet the wall of secrecy protecting energy-intensive corporations from public scrutiny strongly suggests that cheap and abundant power for industry remains at the core of South Africa's strategy for international competitiveness while households are made to buffer the cost. To be sure, industrial tariff customers will face a sharp hike. This is offset by rising power prices elsewhere in the world and it may be anticipated that the increase is calibrated to keep the costs to power-hungry corporations comparatively low. Since the largest consumers are listed in the major financial centres, the profits reaped from cheap energy are returned to global investors to invest wherever in the world will yield the highest return. When full power is restored, South Africa will again offer cheap power to attract foreign direct investment by transnational corporations. The capital for such investments will be from a global pool of profits, which includes the profits originally produced by operations in South Africa.

Box 8.2   Coal stain

The World Bank's loan specifically excludes the boilers and turbines for Medupi. The African Development Bank (AfDB) loan is specifically for these components only. The AfDB is to all intents and purposes the World Bank's less scrutinised branch in Africa and the two loans were clearly co-ordinated. The reason for this split in funding is that Eskom awarded a R40 billion contract to Hitachi Power Africa to supply the boilers for both Medupi and Kusile. They will be made by Hitachi Europe which is located in Germany – hence the German export credit agency loan.

Chancellor House, an investment company set up to fund the ANC, is Hitachi Africa's accredited BEE partner with a 25% shareholding. The ANC consequently gets a very large rent off the deal. At the time that the boiler contract was awarded, Valli Moosa was both chair of the Eskom board and on the ANC's National Executive Committee. The public protector, not hitherto known for making findings that discomfort the ruling party, found that Moosa's conduct was improper in that he did not manage the conflict of interests appropriately. Prior to this finding, ANC Treasurer Matthews Phosa admitted the conflict of interest and said that Chancellor House would withdraw its stake in Hitachi. It did not do so.

The World Bank's procurement rules prohibit lending to projects that benefit a political party. The comfortable arrangement with the AfDB, which operates under less stringent criteria, was patently a subterfuge to circumvent the rule. The major European countries and the US are members of the AfDB as they are of the World Bank. It must be assumed that they knew very well what the game was. Once the matter was splashed across the international media, it seems that some heavy diplomacy followed. Within days of the vote, Phosa again promised that Chancellor House would sell the shares but was immediately contradicted by ANC General Secretary Gwede Mantashe.[56] Chancellor House has subsequently said that it has no intention of selling its shares.

TERMINAL LOGIC

Despite the government makeover following the ANC's Polokwane conference, the executive has consistently reiterated its support for Eskom's tariff applications. Public Enterprise Minister Barbara Hogan sounded much like her predecessor Alec Erwin as she argued that the new build would stall and the country black out if the utility's demands were not met. This view was shared by the World Bank and other investors. According to the Bank, 'effective pricing and cost recovery are key for achieving financial sustainability for [South Africa's] electricity sector'.[57]

Pricing is also a condition for private capital investment and opening up the sector to transnational power corporations. Production costs from Eskom's new coal plants will be far higher than from existing plants in part because of the costs of paying off the debt. New private plants will similarly need to pay off the capital and, in addition, return a profit to the investors. Negotiations between IPPs and Eskom, as the 'single buyer' of their electricity, have mostly foundered on the question of price.

Price is, of course, capitalism's basic approach to DSM although recession proved rather more effective. Eskom attributes its 2009 operating loss to reduced sales and increased coal costs. Its 2009 interim price application noted the potential for higher prices to cut into demand as a critical risk. The failure of economic recovery would result in an additional drop in sales and shrinking revenues would further destabilise operations. This clearly points to the corporation's dependence on expanding volumes and to the limits of its conception of DSM. It also indicates the other side of 'effective pricing'. If the price increase retards economic recovery, then Eskom is cutting at its own revenue base. It thus raises the possibility that cost recovery and expanding sales have become incompatible.

This incompatibility will be exacerbated into the future. The major economic actors focused on price but not on what they are paying for. They did not question either the need for the new build nor its base in coal. Eskom made the rising coal prices central to its argument for a higher tariff and argued that 'the true economic cost' includes 'the cost of increasingly scarce primary energy and the cost of shifting to cleaner and renewable electricity generation technologies'.[58] It did not mention the external costs of pollution.

Government and World Bank statements notwithstanding, Eskom's traditional aversion to renewables remains evident in the limited ambition of the IRP 2010. Instead, the new build ties power production to coal for the next 50 to 60 years on the logic that South Africa's energy-intensive economy relies on base load that cannot be met by renewables. The limits of the global capacity to expand oil production have been obscured by the recessionary collapse in demand. Even in the absence of economic recovery, it is doubtful that supply will meet demand much beyond 2012. Coal will then once more follow oil prices up even if the coal supplies can be expanded. The moment of peak oil marks a terminal point in the logic of the regime of capitalist accumulation.

A second terminal point is visible in the economic crisis. Eskom calls it a 'downturn' while government acknowledged in 2009 that South Africa was in 'recession'. Neither word was adequate to the moment. The world was entering a major depression. In contrast to the recession of the 1980s, which was managed to restore the political power of the US, the managers of global capital have lost control. Investors run from pillar to post to find a safe haven – now into US bonds, now into emerging markets, now into commodities. The result is increased economic volatility. Peak oil plays into the crisis. At the first sign of 'green shoots', the oil price spikes as investors rush in only to strangle the shoots. There may be more booms and even bigger busts to come but the global political and economic order will not survive the next few decades.

Increasing the spinning margin is no doubt essential. Building R140-billion power stations in anticipation of supplying yet more energy-intensive corporates with cut-price power is hardly a sensible way of doing it. In taking on the debt, the Treasury is making a double bet: that future economic growth, and the continuous expansion of the energy system, will more than cover repayments; and that the rand will hold its value. Otherwise the debt becomes a trap as it did for many Southern economies in the 1980s. Neither bet looks good. Moreover, in conventional economic terms, they pull in opposite directions. Because the debt must be repaid in dollars, growth must be led by exports to earn the dollars. A high rand value, however, will suppress exports of everything but commodities while also reducing the rand value of commodities. Either way, the corporations listed in the global centres will take the benefit of higher dollar prices while the squeeze is put on the local operation and specifically on wages. In October 2010, the Treasury made its choice in a game of double or quits. It doubled the guarantee on Eskom's debt to R350 billion rather than call it quits on Kusile, which, it appears, would not otherwise be funded. On top of this, it indicated that it would inject a further R20 billion of equity into Eskom.

The third terminal point is the ecological crisis. The costs are now escalating at all scales, from the local consequences of pollution and the destruction of 'ecological services' to the global consequences of climate change. The regime of accumulation founded on growth is not compatible with addressing climate change. While the global managers have thrown stupendous sums of money at saving the economy, losing it now presents the best prospect of inadvertently saving the climate. Anderson and Bows come to the reluctant conclusion that a 'planned economic recession' would be necessary to avoid warming, not of 2 °C, but of more than 4 °C (2008: 18).

The World Bank, deeply involved in climate negotiations and financing as it is, is not the institution to support the drastic change in direction that is required. The South African government's own assumptions are not very different from the Bank's. The new build is, after all, a home-grown idea. It was nurtured in an economy that is based on cheap labour and cheap energy. For big industrial users, but not for people, it provides the cheapest power in the world. This is the competitive advantage that has made the country one of the world's most carbon-intensive economies. The managers of SA Inc are determined to retain the advantage. In doing so, they are recreating the logic of an economy that is internally subordinated to the interests of the minerals-energy complex and externally subordinated to the imperial market. This is the economic model that the Bank set out to save with its loan.

9

Driving climate change

THE PEOPLE ON THE STREETS of Copenhagen chanted, 'Change the system, not the climate'. Down the road, behind the police cordons and a world away in the Bella Centre, the official delegates to the 2009 climate conference did not hear them. Negotiating a serious response to climate change was not the real agenda. It is precisely for that reason that heads of state felt compelled to 'emphasise our strong political will to urgently combat climate change . . .'.[1] Strong political will was certainly evident in Copenhagen. The purpose of all major parties was to defend their respective interests in the global accumulation of capital. In their vision, this is what is meant by 'development'. Their disagreements reflect the conflicts inherent within that agenda. In particular, they reflect the dominance of the North and the subordination of the South within the orders of capitalist development.

The official meeting finally abandoned efforts at an agreement on the second commitment period within the United Nations Framework Convention on Climate Change (UNFCCC) process. It produced instead the 'Copenhagen Accord', a political agreement made in a back room by the US with the 'Basic' countries – Brazil, South Africa, India and China – and then endorsed by the leaders of 26 countries at a 'non-meeting' that lesser countries were told wasn't happening. The Danish chair of the final plenary of the official process tried to gavel agreement on the Accord without discussion and before some delegations had even seen it. When that failed, arm-twisting and threats followed. Finally, the Accord was not adopted but merely 'noted' and its substance – or lack of it – was greeted with dismay. Copenhagen broke up in disarray but, one year of heavy diplomacy later, national delegates meeting in Cancun, Mexico, applauded themselves as they effectively adopted the Accord. They had, they said, saved the UN process but, some added, sacrificed the climate and people (Khor 2010).

Copenhagen marked the end of a pretence. The international process, or some international process, will of course continue. It will be a ghostly charade, no longer a pretence but now the pretence of a pretence, made necessary because the world's leaders cannot announce their failure nor admit the futility of a process that refuses to address the central issue: the capitalist economy over which they preside cannot be reconciled with a credible response to climate change. As La Via Campesina leader Josie Riffaud put it: 'Money and market solutions will not resolve the current crisis. We need instead a radical change in the way we produce and we consume, and this is what was not discussed in Copenhagen'.[2] The organisation concluded that the only way forward was with people's movements.

This chapter gives a brief critical review of the history of the negotiations in order to explore the common interest that lies beneath the conflict between the parties. It then looks at South Africa's Copenhagen offer to see how it stacks up against the research that is said to underpin it. A critique of that research, the Long-Term Mitigation Scenarios (LTMS), wraps up the chapter.[3]

FALSE DEALINGS

The UNFCCC and the Kyoto Protocol were negotiated under the sign of the Washington Consensus. They make governments responsible for implementation while private sector corporations are made the agents of implementation. This agency, however, is voluntary and supposedly driven by the carbon market brought into being by states.

The Convention recognises that developed and developing countries have 'common but differentiated responsibilities'. This principle is meant to secure developmental equity between North and South and its inclusion was seen by Southern countries as a major victory. The Convention thus recognises first that Northern countries are responsible for the bulk of emissions to date and are better resourced to implement the agreement, and second that Southern countries have a priority for development. It then emphasises 'sustainable economic development' within an 'open international economic system' – meaning a capitalist system – and allows that all countries will define sustainable development in line with their own development priorities. It distinguishes between Annex I (developed) countries and 'non-Annex I' (developing) countries, with the former taking tougher commitments and supporting the latter with financial and technology transfers.

The UNFCCC initially relied on voluntary reduction targets for Annex I countries. No one volunteered. A binding agreement was therefore called for. Kyoto is based on a proposal put forward in 1997 by the US under Bill Clinton and sets up emissions trading. This followed an earlier Brazilian proposal, rejected out of hand by the US, that Northern countries exceeding their reduction targets should pay a fine into a common pot that could then be used to finance projects in Southern countries. The US proposal also displaced European proposals for a carbon tax.

The US essentially proposed a cap-and-trade scheme similar to one that it claimed had successfully reduced sulphur dioxide emissions in the US. Larry Lohmann (2006) has since shown the claim was in fact dubious although no one at Kyoto had the information to dispute it.[4] Nevertheless, no one really believed cap-and-trade would work to reduce global carbon. The proposal was finally adopted for two distinct reasons of political expediency. First, the preferred 'market mechanism' of European Union (EU) negotiators was to tax emissions but, because any EU tax requires a consensus of all member states, trading offered a politically easier route (MacKenzie 2007). Second, it appeared that trading was a precondition for US agreement. Having imposed its preferred system, however, the US exempted itself from abiding by it. The Clinton administration avoided putting it to Congress for ratification and, under George Bush, the US actively rejected Kyoto.

The Bush administration rejected Kyoto on the grounds that it was unfair for Northern countries to take commitments if Southern countries did not. The US, of course, knew Southern countries would not accept this. From the start, they have refused commitments until the North demonstrates real reductions. They argue that Northern countries developed on the back of high emissions and still produce the majority of emissions. They also suspected, with some justification, that the North was using the climate negotiations to lock in economic dominance by blocking development in the South.

Following the US withdrawal, the EU took over as the champion of Kyoto and of 'multilateralism'. It led a series of negotiations culminating in the adoption of the Kyoto Protocol at Bonn in 2001. The US nevertheless maintained a strong presence in these negotiations where it used its outsider position, and the bait of its possible ratification, to weaken the agreement. It thus substantially shaped the outcomes of negotiations while exempting itself from the rules being negotiated.

Kyoto set mandatory emission reduction targets – the would-be cap[5] – to be achieved in the 'first commitment period' (2008 to 2012). It specified targets for each Annex I country of between 8% below and 10% above emission levels in the 'baseline' year of 1990. This worked out as an overall reduction for all Annex I countries of about 5%. NGOs under the umbrella of the Climate Action Network International welcomed mandatory targets. Although the targets were woefully inadequate, it was argued that they would be ratcheted up in successive five-year commitment periods. Thus, in the 'second commitment period' beginning in 2012, it was expected that Annex I countries would take on tougher targets while Non-Annex I countries would also take mandatory reduction targets. The targets themselves, however, were founded on the deeply inequitable principle of 'grandfathering' by which historic inequalities are enshrined and projected into the future.

Kyoto set up carbon trading through three 'flexible mechanisms': Emissions trading allows Annex I countries and corporations that exceed their reduction targets to trade their surplus allocation with other Annex I countries that do not meet the targets; Joint Implementation projects enable investors in one Annex I country to invest in projects that produce less emissions than a business-as-usual project in another Annex I country and to claim 'carbon credits' for the difference; the Clean Development Mechanism (CDM) works in the same way except that the investors must be from Annex I countries and CDM projects must be located in Non-Annex I countries.

The stated objective of CDM was to support sustainable development in Southern countries while reducing the costs to Annex I countries of meeting their reduction targets. Thus, Northern polluters could invest in 'clean development' projects in the South and claim carbon credits known as 'Certified Emissions Reductions' (CERs). Alternatively, they could buy CERs produced from CDM projects on the market. The explicit reasoning behind this was first that the costs of meeting targets would be unaffordable to Northern economies and second that reductions would be cheaper in the South. It is thus founded on unequal development – that is, on economic, social and environmental injustice.

Wolfgang Sachs (2005) concluded that negotiators 'were charged with protecting economic growth and not the climate' to which end Kyoto embodies three strategies: Northern obligations are transferred to the South and East – through CDMs and Joint Implementations; obligations are discharged through sinks – that is, through 'carbon-offset' projects mainly located in the South and again funded through CDM;[6] and negotiations are framed to focus on the economic tailpipe and exclude discussion of driving interests in the engine room.

The effects of trading on carbon emissions are predictably dismal. The EU set up its own internal emissions trading scheme, which has delivered profits to polluters and traders without reducing emissions. This followed the over-allocation of give-away emission rights to big corporations, notably the power utilities, effectively lifting the cap right off the corporate heads and leading to a collapse in the carbon price. It was then proposed to ratchet down the cap and also to auction emission rights to corporations rather than give them away. Both proposals were the subject of intense political wrangling. A minor portion of credits (7%) have been auctioned in Britain and Germany since 2008 and these two countries have since resisted the formation of a single European auction system.

The crash in commodity prices similarly crashed the carbon price. European industry slumped, energy consumption shrivelled, corporate revenues dwindled and the creditors came knocking at their doors. What they had in surplus was carbon credits that were sold off to plug the holes in their balance sheets. Got free, they produced pure profit at whatever price. Credits dropped from around 33 euros to under 10 euros and have since traded at 12 or 13 euros. The only reason the price didn't disappear off the bottom of the charts is that corporations were allowed to store up pollution rights by rolling their credits over from one year to another. The cap evaporated.

CDM has an equally inglorious record. It invites players to 'game the system' and they have embraced the invitation. But even if the rules are followed, the carbon accounting is based on a series of fictions and false assumptions, particularly in respect of sinks. For Southern countries, CDM has simply created a new arena of competition for foreign direct investment. Real or not, the carbon credits are subtracted from the Northern country's total carbon count and must logically be added to the Southern country's count. This is fudged. Thus, Sasol includes its CDM projects in its strategy for reducing its greenhouse gas emissions. So it takes the money from selling CERs but still reports the carbon reductions that are simultaneously claimed by the buyers.[7]

Box 9.1   Trading targets away

Britain's climate change bill was put to parliament in 2008. It requires that the country's CO_2 emissions are cut by 60% by 2050 with legally binding interim targets every five years. This, however, is at odds with its energy policy. Environmental journalist George Monbiot observes that an obscure government briefing note shows how the contradiction will be resolved:

> It explains that, during the latest stage of the bill, the government 'remov[ed] the quantified limit on the use of internationally traded credits in meeting the UK's targets'. In other words we could buy the entire cut from other countries . . . But there are three problems. The first is that we are exporting

emissions that are difficult to address and importing, through carbon trading, the easiest and cheapest cuts.

The second is that while the emissions we export are certain and verifiable, the cuts we buy through carbon credits are often fraudulent. For example, as the writer Oliver Tickell documents, 96% of the carbon credits from hydroelectric dam construction were issued after construction had begun: the dams would have been built without the carbon market, so no additional cuts have been achieved. Around 30% of all carbon credits comes from the sale of trifluoromethane cuts by Chinese and Indian companies making refrigeration gases. Many of them are still producing this pollutant only because they make so much money from cleaning it up: the carbon market pays them 47 times more for these cuts than the gas costs to remove.

Behind these problems lurks a much greater one, which is mathematically impossible to resolve. You can trade your way out of trouble when the cut you are trying to achieve is a small one. But when the global cut required to prevent two degrees of warming is 60 or 80 or 90%, then every rich nation must reduce its emissions by roughly the same amount. Otherwise half the world would have to buy credits equivalent to 180% of the emissions produced by the other half.

The government will have to impose some kind of cap on carbon trading. But I bet it will be set high enough to cover any failures in domestic policy, as measured by the rigged accounting methods civil servants use. This means that successive governments will have no legal incentive to change their energy policies. The carbon trading provision torpedoes the useful content of the entire [climate change] bill.

Source: George Monbiot, 'Traded away', *The Guardian* (London), 24 July 2008.

The second non-commitment

The Fourth Assessment Report of the Intergovernmental Panel on Climate Change (IPCC AR4) made it clear that the world is running out of time. Following its release in 2007, the political classes appeared genuinely alarmed at the prospects of climate change. The EU committed to a 20% unilateral reduction by 2020 and said it would up this to 30% if equivalent reductions were forthcoming from other Annex I countries – meaning the US. It gave itself generous space to meet these targets through trading CDM and Joint Implementation credits. In the US, Bush's denial of climate change became politically untenable and even ExxonMobil gave up on this line.[8] German proposals at the 2007 G8 for more stringent Annex I targets were, however, rejected by the US, which proposed an alternative agreement between major emitters, implying that the US won't move without China and India taking equal commitments.

China specifically ruled this out while announcing its own climate action plan, claiming that 'China will not tread the traditional path of industrialisation, featuring high consumption and high emissions. In fact, we want to blaze a new path to industrialisation'.[9] In the world's most polluted country, this seemed a somewhat belated ambition. China has since become the world leader in building renewables but remains the world leader in new coal projects and has overtaken the US as the world's largest emitter of CO_2. The US still emits four times more per person but China's per capita emissions are now above the world average and well above any credible per capita carbon allowance. On the basis of historical emissions per person, Chinese academics argue that the country has used only 28% of its 1900 to 2050 carbon budget whereas the US has used 320% and most other Annex I countries are also in deficit.[10]

The Bali conference of the parties, in December 2007, was to initiate the series of negotiations to agree on the carbon regime for Kyoto's second commitment period. Here, the US appeared increasingly isolated as nation after nation castigated it for blocking progress. In fact, the US was merely repeating the negotiating tactic that had proved so successful at Kyoto. Once more it made its own participation in 'the Bali roadmap' the defining issue at the conference, holding out till the last moment in order to create an appropriate level of desperation and exhaustion, before appearing to concede. It then joined the 'global consensus' on the Bali roadmap, but the roadmap now appeared without even the targets that were held to be Kyoto's saving grace. At the same time, Bali entrenched carbon trading. National delegates, North and South, brooked no questioning of it. South Africa's environment minister spoke for the consensus view when he said there was no going back on carbon trading. The result: trade but no cap.

Nevertheless, the world's assembled politicians applauded the US 'return to multilateralism'. At the same time, the US attempted to keep open alternative negotiations between big polluters and so to circumvent the multilateral UNFCCC process. There were no takers as Bush limped off stage but the strategy was driven through under President Barack Obama. The Copenhagen Accord was the result. As George Monbiot remarked, negotiators would have done better to sign a blank piece of paper.[11] While small island states and Africa had argued that global warming should be restricted to 1.5 °C, the Accord aims for 2 °C and avoids any mention of the scale of reductions required to meet it. Instead, it invites each country to set its own target, effectively restoring the voluntary pre-Kyoto approach, endorses carbon trading and links it with sink offsets through 'Reducing Emissions from Deforestation and Forest Degradation' (REDDs).[12] In response to Southern demands for financial transfers, US Foreign Secretary Hillary Clinton triumphantly held out a bag marked $100 billion. There was nothing in it. It was a photo opportunity promise and the excuses for breaking it are written into the Accord: 'This funding will come from a wide variety of sources, public and private, bilateral and multilateral, including alternative sources of finance'. No obligations for the US or for Europe there. And it came with the threat that money would be made available through the Accord – those who did not sign up would not be eligible. Promises on technology transfer, another key Southern demand, are similarly empty – as they have been since the signing of the UNFCCC.

This is what the delegates adopted in Cancun. By then, the sum of country 'pledges' made under the Copenhagen Accord, assuming they stick to them, implied global heating of 4°C. The agreed 2°C target is thus meaningless. Carbon trading saturates the text but is no longer supported even by the fraudulent logic of Kyoto because the agreement dispenses with the very notion of a cap. Cancun did establish a 'Green Climate Fund' under UN control, providing a glimmer of light that was immediately snuffed out by putting the still empty fund under World Bank management. It also established a 'technology mechanism' but, at the insistence of the US and over Bolivian protests, avoided discussion of intellectual property rights, the single most important block to technology access. Further, Cancun recognised carbon capture and storage (CCS) as a valid mitigation technology eligible, amongst other things, for carbon trading credits.

The international consensus was already in tatters on the way to Copenhagen. The Europeans gave up on Kyoto as the US refused to buy into it. Southern nations then rallied to 'save Kyoto' as they saw the North wriggling out of binding commitments while shifting the burden of emission reduction onto the South. Northern nations responded that without commitments from 'major economies' there was no hope of mitigating climate change. Southern commitments were indeed anticipated in the second commitment period but this assumed that the North would have already proved its credentials through serious reductions in the first period. Most have effectively missed their modest targets – but for the recession, all would miss it – and Canada simply tossed them in the bin because its targets were incompatible with developing the tar-sands. 'Binding' has proved anything but. Besides, the entire logic of Kyoto was to shift the burden south.

Southern countries are clearly justified in treating Northern negotiators with suspicion. Bad faith at the World Trade Organisation (WTO) negotiations is echoed in the bad faith of Kyoto and Copenhagen. At the start in 1992, major Southern countries essentially treated climate change as a Northern scam aimed to prevent them climbing the ladder of development. They have since used the principle of 'common but differentiated' responsibilities merely to refuse significant commitments. In doing so, they signed over to the North the power to define the global response to climate change in its own interests. Kyoto's one achievement was to institute the carbon-trading regime. As one trader emphasised to negotiators, 'the market' cares only about the price of carbon, not about carbon reductions or sustainable development (Lohmann 2006: 296). As a response to climate change, Kyoto was bankrupt from the beginning. As an expression of the financialisation of global capital, it remains exemplary.

Carbon bubbles

To recall, financialisation was one of two strategies to compensate for the squeeze on profits consequent on over-accumulation. Spinning financial assets based on debt through ever-more complex derivatives required the suspension of state regulation in favour of market regulation to facilitate systemic fraud. Carbon trading similarly turns the state's regulatory function over to the markets in order to create new profit streams out of thin air. This does not, however, imply the retreat of the state. To the contrary, the profits of the market depend on the state's creation of property rights, which are not in carbon as such but in 'avoided' carbon emissions. They are, curiously, rights in what is not. Within the European trading system, you can sell the carbon you did not emit against what the state gave you the right to emit. In the CDM, what you did not emit is calculated against what you would have emitted in the business-as-usual world without the CDM project – that is, it is calculated against a story of what will not be. As Lohmann (2006) argues, the CDM 'saves' carbon against that fictional alternative future but has no relationship to total carbon emissions. For example, CDM projects producing energy from landfill gas in South Africa are held to displace energy from coal-fired plants but do not relate to shutting down the equivalent amount of existing coal-fired energy.[13]

The scope for gaming the system in this market is, as Monbiot put it, wide enough to drive a Hummer through.[14] More broadly, the hallmark of global financialisation is stamped on the carbon market. It creates new instruments of profit from trading, innovative accounting or outright fraud to compensate for declining profits from production. Those profits then join the global pool of capital and must be reinvested wherever the profiteer can find the best return. The profits of trading must either go into expanding production – irrespective of whether that is in a coalmine, a wind farm, a Hummer plant or a perfumery – or into the bubble economy.[15] Carbon funds grew fast. By 2008, $12.5 billion was invested and being spun out through new derivatives according to Nicholas Hildyard (2008). The good from Copenhagen-Cancun is that it deflated the market. Durban may well be accounted a success if it provides air for that new bubble demanded by investors. It is a market that awaits its Enron.

The profits of financialisation are not cost-free. Although counted as 'value-added', Ben Fine argues that they are in fact 'value-subtracted' from the economy (2008a). They represent a transfer of wealth from poor to rich in all countries and from South to North globally. This appropriation relates financialisation to the second, and interlinked, strategy used to compensate for declining profits of production: accumulation by dispossession. Both externalisation and enclosures have been documented as direct results of carbon trading. Externalisation is visited on fenceline communities living next to polluting plants that can buy carbon credits to stay in business. It is also visited on the neighbours of dodgy CDM projects. Durban's Bisasar Road landfill gas-to-energy project is a case in point. According to the original proposal, it burns 'dirty, low calorific value gas' because this gas is too dirty for use elsewhere and the costs of cleaning it were reckoned to be too high for profitability.[16] Enclosures are most often associated with sink offset projects. Most involve planting trees and need land that is cheapest and where people's rights are not recognised. In Uganda, land in the Mount Elgon National Park has long been disputed. In 2003, local people ripped up 400 hectares of exotic eucalyptus trees planted to offset emissions from Dutch power stations. Subsequently, they won a court ruling recognising their rights to the land. Ironically, these farmers then started planting their own tropical fruit trees but these trees will get no carbon credits because no Northern investor has rights in them.[17]

Theatre of conflict

The climate conflict was also played out in the politics of fossil-fuel funding prior to the World Bank decision on the Eskom loan. Just as Copenhagen was reaching its anti-climax, the US Treasury issued a guidance note saying that, to assist developing countries mitigate carbon emissions, the Bank should not fund coal projects except as a last resort. This looked very much like carbon imperialism.

The US itself produces about a quarter of the world's coal-fired power and it provides large subsidies to domestic coal production.[18] Indeed, in 2007 the US industry looked to be booming with 150 new coal plants planned. The boom was headed off by three factors: an unexpected reduction of gas prices consequent on the shale gas boom; the recessionary contraction of demand; and a vigorous campaign by environmentalists who blocked planning permission for over 130 plants. Further, the guidance was sent to World Bank president Robert Zoellick on the day that the US announced to the Copenhagen conference that it would commit to reducing carbon emissions by a mere 4% by 2020 – as compared with the 40% that is the minimum requirement from Northern countries if even the inadequate 2 °C target is to be met.[19] The note thus embodied the US intention to impose responsibility for carbon reductions on the South.

Southern country representatives on the Bank board objected strongly to the use of the World Bank as an instrument of US power. At the same time, they defended the use of Bank funding for fossil energy and specifically coal-fired power stations, which they justified as necessary for 'achieving poverty alleviation and economic growth'.[20] This is what India's climate justice movement, in a memo criticising their government's approach to climate change negotiations, describes as 'hiding behind the poor'.[21] The easy association of growth and poverty alleviation ignores the rank dispossession and growing inequality that accompanies economic growth in all countries. As the memo puts it, this development 'has witnessed the exploitation of natural resources, the greater displacement of adivasis and other forest dwellers, intensified exploitation and continued pauperization of the urban poor, casualisation and contractualisation of labour, and the promotion of consumption by and production for elites'. The poor are mostly left worse off than before. Even where their income improves on the conventional measures, the gains are lost to cost recovery, to health costs imposed by pollution, to the loss of resources including land and water, and to the increased cost of access to amenities previously provided as public goods. And increasing numbers of poor people are already feeling the harsh impacts of climate change.

As in South Africa, World Bank coal-power projects are primarily designed to supply industry, not people. The industries in turn are mostly geared for export and controlled by transnational corporations. The goods are then consumed primarily in developed countries. The Southern countries themselves compete vigorously for Northern corporate investments as industries move South looking for the cheapest energy, labour and environmental regimes while the North rigs the rules to keep profits, cheap goods and strategic resources flowing North.

The US drove the policies that resulted in this global restructuring of industry and sent the Bank to impose them. It now calls for carbon savings in developing countries while depending on them to produce carbon-intensive goods on the cheap for the home market. Similarly, developing countries defend carbon-intensive production in order to produce those goods while calling on developed countries to reduce consumption. But neither side in fact wants what it is asking for – and it is certainly not what they ask for in the WTO negotiations. It is difficult to avoid the conclusion that the guidance note was intended to provoke the Southern reaction.[22] It keeps the conflict over climate change on the boil and conceals the deeper common interest.

At the climate negotiations, the US is justifiably seen as the spoiler in chief but, commenting on the 2008 G8 meeting, Walden Bello remarks that all Northern governments 'hang on to the position that economic growth can be "decoupled" from energy use' and energy decarbonised with a magical technofix or two. This is a symptom of 'growthmania', a concept he takes from environmental economist Herman Daly:

> Growthmania . . . is a cultivated ideological predisposition that serves as a protective shield for global capitalism. Capitalism is an expansive mode of production, and it can only reproduce itself by continually transforming living nature into dead commodities. This is essentially what growth is all about. This is why ever-increasing consumption is so central to the engine of profitability that drives capitalism.
>
> The G8 – the directorate of global capitalism – is trying hard to avoid just such radical controls on growth, consumption, profits, and the market that a viable strategy to stave off the looming climate catastrophe will necessitate. Voluntary cuts, technofixes, and carbon trading are desperate efforts to prevent the inevitable.[23]

But growthmania is not the preserve of the North. In the name of 'development', Southern governments are equally determined to defend economic growth and, as Bello observes in another article, have shown a determination to catch up with the North at whatever cost to the environment and to people. He disputes that this elite view represents the South's perspective on the environment and documents the emergence of growing environmental and other movements resisting 'a model of growth that has failed both the environment and society'.[24]

The disputes between Northern and Southern governments are the product of a long history of unequal power relations but they conceal a common interest in a dysfunctional climate regime. As Bello puts it:

> When the Bush administration says it will not respect the Kyoto Protocol because it does not bind China and India, and the Chinese and Indian governments say they will not tolerate curbs on their greenhouse gas emissions because the US has not ratified Kyoto, they are in fact playing out an unholy alliance to allow their economic elites to continue to evade their environmental responsibilities and free-ride on the rest of the world.

A dysfunctional climate regime allows each to use the other as an alibi for inaction or failure while rallying the home crowd in support. It also provides an alibi for the world's middle classes. In the European context, Monbiot observes that politicians 'know that inside their electors there is a small but insistent voice asking them to try and to fail. They know that if they have the misfortune to succeed, our lives would have to change. They know that we can contemplate a transformation of anyone's existence but our own' (2006: 213).

Several countries refused to let the Copenhagen Accord pass in the final plenary at Copenhagen. They did so both in defence of the UN process and on substantive grounds. US envoy Jonathan Pershing subsequently confirmed that the US would sideline the UN process in favour of negotiations between the largest polluters. He accused four Latin American opponents of US imperialism – Bolivia, Venezuela, Nicaragua and Cuba – of blocking agreement on the Accord because they saw the process 'not so much as a solution to climate change, but in fact as a mechanism to redistribute global wealth'.[25] This misrepresents the scale of opposition to the Accord. Nevertheless, it touches on the heart of the issue. In response to Copenhagen, Bolivian president Evo Morales called a People's Conference on Climate Change and Mother Earth Rights in the city of Cochabamba, renowned for the social movement that defeated the attempt to privatise Bolivia's water. Opening the conference, he put the issue bluntly: 'Either capitalism lives or Mother Earth lives'.[26] Yet even here, the state initially tried to close down the dissident 'Table 18' where local movements fighting mine developments pointed to the conflict between Bolivia's promotion of privately owned extractive industries and the conference theme of people living well with the earth. It may be hoped that the movements of which Bello speaks will take charge of the processes that emerge from the conference, that it is the sign of a growing power to shape the world's response to climate change.

For its part, Bolivia has kept faith with Cochabamba, attempting to insert key demands into the negotiating text and, in a manner that can only be described as heroic, standing alone against the abject agreement at Cancun. Word from insiders to the negotiations is that several delegations wanted to support Bolivia but were silenced by their political bosses. Morales was the only head of state who could not be bullied or bribed into ordering his delegation into acquiescence. In breach of UN procedure, which requires consensus, the agreement was gavelled over Bolivia's expressed objections.

SOUTH AFRICA'S OFFER

The South African offer at Copenhagen was hailed as a bold initiative. It is said to be founded on the LTMS, a study commissioned by the Department of Environmental Affairs and Tourism (DEAT) in 2006. The World Bank and government itself have repeatedly cited both the offer and the LTMS as proof of South Africa's climate commitments. Medupi and Kusile, they say, are already factored into the promised reduction so there can be no objection to these plants on climate grounds.

The Copenhagen offer is for a 34% 'deviation' below baseline by 2020 and 42% below baseline by 2025. The baseline represents the projected increase in emissions assuming 'business as usual' so these cuts are intended to trim the rate of growth of emissions. Emissions are then to level off: 'With financial, technology and capacity building support from the international community, this will enable South Africa's greenhouse gas emissions to peak between 2020 and 2025, plateau for approximately a decade and decline in absolute terms thereafter'. The offer is conditional on 'an ambitious, fair, effective and binding multilateral agreement under the UNFCCC and its Kyoto Protocol' being finalised at Mexico in December 2010.[27] As the South African negotiators have repeated several times: no money, no deal. And without a deal, no action.

The LTMS put forward two scenarios: 'Growth without Constraints' (GWC) and 'Required by Science' (RBS). These two scenarios produce top and bottom lines for emissions through to 2050 with 2003 as the starting year. Four 'strategic options' look at ways of bending the top to the bottom line. Each is more ambitious than the last. The first three options are called 'Start Now', 'Scale Up' and 'Use the Market'. Together they 'only get South Africa two thirds of the way' to RBS (DEAT 2007: 20).[28] The fourth strategy, 'Reach for the Goal' attempts to close the remaining gap.

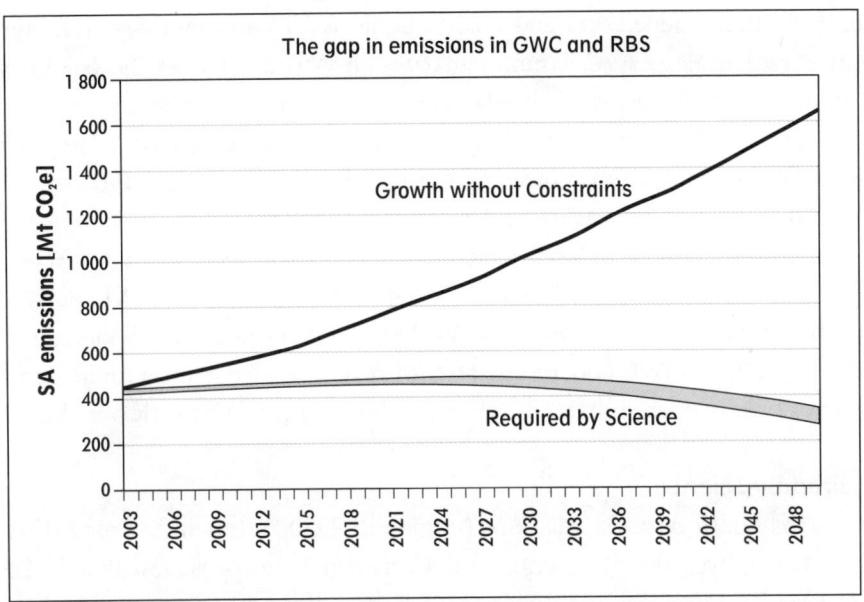

Figure 9.1   GWC and RBS carbon emissions to 2050.

Growth without Constraints

GWC extrapolates from present trends. It assumes no action to mitigate climate change and sees greenhouse gas emissions rising four-fold from 440 mt in 2003 to 1 600 mt by 2050. It shows emissions of about 750 mt of CO_2e per year in 2020 and 870 mt in 2025.[29] This is the baseline for South Africa's offer and seems to imply emissions targets of 495 mt in 2020 and 504 in 2025. This depends, however, on how government rigs the numbers. Civil society demands for clarity were initially met with silence but, eighteen months later, DEAT officials finally presented an unusually candid take on the numbers and suggested 'a new expression of our objectives' to take account of the 'error range' in the GWC projection.[30] The 2020 target could then be put at between 418 and 571 mt and the 2025 target at between 412 and 599. Since they also confirmed that current emissions are around 540 mt per year – and Medupi and Kusile alone will add another 60 to 70 mt – it seems clear that it is the higher numbers that count.

The scenario reflects the assumptions that have shaped actual policy – before those assumptions tripped on the power outage and the global depression. In GWC, industrial policy remains focused on energy-intensive industries while coal and nuclear electricity generation and coal-to-liquid (CTL) and crude-oil refining all expand dramatically. GWC assumes that South Africa achieves the 6% growth target, that climate change does no damage, and that oil, water and other resources are available to meet ever-expanding demand. This, as the LTMS notes, is not realistic. So the reductions on offer are measured against a theoretical construct of a future that will not materialise under any circumstances.

The power sector expansion, if it is fully realised, will keep South Africa on the high emissions path described by GWC well past 2020. Indeed, GWC puts 2025 generating capacity at 60 000 MW (DEAT 2007 TR: 51) while Eskom puts it around 80 000 MW. Either GWC underestimates the capacity needed to meet future demand growth or Eskom overestimates it, as it did in the 1980s, and is already over-building.

Required by Science

RBS shows South Africa's emissions peaking in 2020 and then declining (DEAT 2007 SD: 10). By 2050, the country emits between 30 and 40% less than in 2003. The LTMS Technical Report shows that whether 30 or 40% is achieved depends on the date and level of peak emissions as shown in Table 9.1.

Table 9.1   RBS parameters for peak emissions.

Peak year	Peak level Mt CO_2e	2050 / 2003 Reduction %
2016	463	40
2020	473	35
2026	483	30

Adapted from DEAT (2007 TR: 117).

Emissions rise before the peak but the rate of increase, starting from 2003, is considerably slower than in the GWC scenario. Since 2003, actual emissions have increased more or less in line with GWC, only slightly moderated by recession. With 2008 emissions well above even the 2026 peaking figure, RBS is already blown. Getting back to it would require an early peak followed by a much steeper decline in emissions than the scenario envisaged. The Copenhagen offer does not come close. Citing the LTMS as evidence of South Africa's climate commitment is thus disingenuous.

The next question is whether RBS itself is adequate. This scenario is defined by the goal or target. It is not derived from the economy but from the logic of climate change as set out by the IPCC AR4 produced in 2007. Having set the target of a 30 to 40% reduction, it then works backwards to identify what needs to be done to reach it. The target band is calculated on the assumption that the world must reduce emissions by 50% by 2050 and that Northern countries make reductions of 80%, so allowing more modest reductions in Southern countries. However, the 50% global reduction is at the bottom end of the range of 85 to 50% reductions, which the IPCC says is necessary to keep temperature rise within the 2 to 2.4 °C range. Similarly, the LTMS says that global emissions must peak in 2015 whereas the IPCC says emissions must peak between 2000 and 2015.[31] Finally, by bundling South Africa with the South in general, the LTMS gives it a free ride on the really low emissions from least developed countries.

The IPCC report was itself a conservative assessment of the pace of climate change. The reports are produced under the scrutiny of governments who cavil over findings that might imply some qualification to their economic interests, and they cannot take account of the latest studies. By the time of its publication such studies showed, amongst other things, that actual emissions are running higher than the most pessimistic of earlier IPCC projections, that the expected impacts of climate change are happening earlier than expected, and that a 1 °C rise is already dangerous. The implications are: first, a 2 °C rise may soon be inevitable but will not prove 'tolerable'; second, ecological feedbacks are already kicking in and picking up pace; and third, reductions must be considerably more ambitious than the IPCC says (see Box 5.1 in Chapter 5).

It must be concluded that the LTMS's early target for South Africa's emissions to peak in 2016 is cutting it fine and, even if 2 °C is accepted, the RBS bottom line of a 40% reduction by 2050 is still too high. Limiting the rise in temperature to 1.5 °C, as demanded by the small-island states that will otherwise drown, would require that carbon emissions are pretty much shut down within the next decade.

That said, RBS 'imagines a post-carbon world very different from ours, one that is therefore difficult to describe in detail' (DEAT 2007 SD: 11). Part of the reason why it is difficult to imagine is that the energy planning model could not find a way to meet projected future energy demands while at the same time meeting the RBS target. The Technical Report concludes:

The RBS climate target cannot be met within this framework. This suggest[s] that either one need[s] to redefine what is realistic (eg, re-considering the extent to which mitigation options can be achieved 'realistically'); or the analysis needs to be conducted outside of the confines of a constrained modelling approach (DEAT 2007P TR: 16).

The modelling tool in question is Markal (Market Allocation). It was developed by the International Energy Agency (IEA) and is widely used by planners to model possible energy futures. 'The model is demand-driven, in that it starts from projections of useful energy demand' and is designed to match supply to growing demand at the least cost (12). It assumes first, that supply *can* match projected demand – Markal does not admit shortage – and second, that investors and consumers make 'rational choice' decisions based on costs. For the reference case (GWC), it assumes that development is a continuation of present trends: 'For instance, energy efficiency is only increased in line with historical trends' (13). To define mitigation actions within the first three strategy options, the model is 'constrained' by external criteria, such as a limit on carbon emissions or a target for energy efficiency or renewable energy, imposed by the modellers. It then finds the least-cost options within these parameters. These mitigation actions are nicknamed 'wedges' because, represented on a graph, each action shows a rising wedge of carbon savings over time. As ever, the 'saving' is measured against business as usual and represents a reduced rate of increase rather than a reduction of emissions.

The LTMS story of development

Although the economic costs of RBS cannot be modelled, LTMS finds this scenario 'more robust' than GWC. Table 9.2 shows the conditions in which each scenario can survive, as presented by LTMS.

That GWC is entirely unrealistic has been noted above. In addition, LTMS observes that both scenarios will fail if climate impacts become unmanageable but adds that GWC actively contributes to that outcome. This point is well made but the other assumptions shown in the table are less convincing. With the exception of peak oil, they refer primarily to two international negotiation processes: the UNFCCC and (implicitly) WTO processes.

Taking the assumptions relating to RBS:

- International climate consensus reached and effective: The LTMS does not consider the possibility that a global consensus is reached but is not effective.

Table 9.2 Conditions under which the scenarios are plausible.

GWC is only robust if:	RBS is only robust if:
• International climate consensus collapse/fragment • Technologies not developed or don't flow freely • Oil cheap and abundant, no carbon premium on coal • Fragmented trade systems, bilaterals and free for all	• International climate consensus reached and effective • International flows of appropriate technology/finance • Peak oil arrives, oil scarce and expensive, coal premiums • High degree of trade integration and globalisation

Source: DEAT (2007 SD: 12).

The agreement in Cancun, however, confirms that ineffectiveness is the condition of consensus.

- International flows of appropriate technology/finance: The UNFCCC already promises technology transfers from North to South. The promise is contradicted in the WTO process and has been dishonoured. Intellectual property rights, leveraged on and designed to sustain unequal power relations, remain at the core of the global technology regime.
- Peak oil arrives, oil scarce and expensive, coal premiums: Peak oil may drive greater energy efficiency and technology innovation. Already, however, it is driving greater energy and carbon intensity and dirtier production. There is no reason to think that governments or corporations will focus innovation driven by peak oil only on low carbon options rather than wringing the last drops of liquid from fossil fuel by whatever means possible.
- High degree of trade integration and globalisation: Globalisation and expanded trade have hardly contributed to reduced carbon emissions. The WTO agenda is itself largely determined by those with power and in their own interests. Indeed, the challenge to those interests by the more powerful Southern states is the primary reason for the failure of the Doha Round. The Southern challenge itself, however, is trapped within the same calculus of power that dictates the position of Northern countries: it is about competition for the rewards, in political and economic clout, of growthmania.

The LTMS does not in fact make an argument as to why it finds these four conditions significant. It appears rather to take its cue from the IPCC Special Report on Emissions Scenarios (SRES). IPCC has proved critical in terms of understanding the biophysical processes driving climate change but is a good deal less convincing when it comes to the social and economic systems because it cannot address power relations or talk about capitalism. It sees patterns of inequality but cannot take account of the interests that create, and are sustained by, inequality.

SRES develops four 'scenario families' through to 2100. These are not predictive but designed to 'explore' the relationship between the main drivers of emissions, which SRES identifies as economic growth, population and technology development.[32] All scenarios portray a wealthier world with reduced inequality, though some less than others. Those scenarios with high levels of 'global convergence' produce the best results for equity and, depending on technology choices, the best long-term results for emissions reductions. Scenarios describing a heterogeneous world in which societies are regionally or locally oriented end up, in 2100, with larger populations, greater inequality and higher emissions than the equivalent globalised scenarios. The growing inequality actually produced by globalisation thus far does not appear to inform the underlying assumptions.

IPCC naturalises 'five stages of economic development', which LTMS assumes will shape future GDP growth in South Africa (TR: 25, 26):

- First, the pre-industrial economy, in which most resources must be devoted to agriculture because of the low level of productivity;
- Second, the phase of capacity-building that leads to an economic acceleration;
- Third, the acceleration itself (about two decades);
- Fourth, industrialisation and catch-up to the 'productivity frontiers' prevailing in the industrialised countries (about six decades);
- Fifth, the period of mass consumerism and the welfare state.

This harks back to the development theory constructed in the US in the 1950s and 1960s to provide economic mechanisms for extracting resources from the Third World as a substitute for the direct colonial control of European imperialism. In the context of the Cold War, that theory provided a justification for US global hegemony in conflict with the Soviet Union and held up capitalism as the model for all to copy. This is what defines the politics of 'catch-up'.

As the LTMS applies the five stages to South Africa, the whole of the twentieth century seems to disappear from the history of economic development. In one brief reference to the apartheid legacy of inequality, it echoes Mbeki's metaphor of the dual economy. It then concludes: 'South Africa could be described as being an accelerating economy (stage 3)' (TR: 26). This fits with ASGISA's growth objectives that LTMS explicitly takes into account. However, whereas 'governments would like to project a continuously high GDP growth', an examination of 'other developed regions of the world' shows that 'GDP growth increases, reaches a peak and then declines' (TR: 25). GDP growth is therefore projected to increase over the next ten years or so to peak in about 2020 and then decline over time, 'flattening out around 3%' by 2050 (27). Following the peak, South Africa presumably enters the 'industrialisation and catch-up' stage and is on the way to the more equal society implied by the fifth stage of 'mass consumerism and the welfare state'.

There are several problems with this story of GDP growth. It is constructed with reference to developed and industrialising states: post-war Europe and Japan, South Korea between 1965 and 1990 and China since 1980. Except for China, all these countries were on the frontiers of the Cold War and received massive US support in the age of global Keynesianism. The neo-liberal Washington Consensus since 1980 has been considerably less expansive. Countries where growth and industrialisation failed, or which were de-industrialised, are not mentioned. Indeed, economic growth in South Africa itself was running at 6% in the 1960s – just a little below Korea's 7% and certainly good enough to qualify as an accelerating economy. Black workers and their families saw little benefit then and the working classes will see little benefit now.

China's industrialisation, on the back of massive peasant dispossession and pitifully low wages for workers, signals a major shift in global power relations. It does not, however, leave much space for industrialisation (stage 4) elsewhere. In the triangular ordering of the global economy, South Africa's dependence on resource extraction is confirmed. Far from promising stages 4 and 5, this reinforces the centrality of South Africa's increasingly capital-intensive minerals-energy complex.

LTMS is undoubtedly right to challenge government's assumption of constantly high growth. But the smooth graph of its own projection is embedded in a narrative of development that equates development to GDP growth. This narrative is now exhausted and will be shredded in the coming crisis. If anything, this adds urgency to finding a path to, and beyond, RBS for reasons other than climate change. It indicates the need for a wholesale transformation of power relations within which a different logic of development can be articulated. This cannot be led by the interests of corporate capital centred on the minerals-energy complex that have shaped development to date. If such a transformation is to be achieved, then people must organise for it and create the social movements that can bring it into being.

Strategic options

The LTMS process was somewhat closed and, if not exactly secret, certainly very discrete. This was justified on the grounds that Cabinet needed to be able to make an 'objective' assessment of it free from political pressure. Participation in a 'Scenario-Building Team' was by invitation. Participants were selected 'to cover' different stakeholder groups and economic sectors and for their 'known technical expertise' on climate change. They were supposedly there as individuals but are mostly referred to as 'stakeholders'. The Scenario Building Team was heavily weighted in favour of government and industry.

The first three strategic options, and the wedges that compose them, were modelled 'bottom-up' according to the Technical Report: 'Stakeholders defined mitigation actions [wedges], which were then modelled by the research teams' (10). The wedges are thus defined by economic sectors as they exist now and by comparison with the way they are expected to grow in GWC. In general, they also rely on data provided mostly by participants from government and industry and on their assumptions about what is or is not feasible. The effect is that each wedge is bound to the dominant interests in the relevant sector and, for the most part, the emission savings are made without jeopardising those interests.

These first three options are composed of wedges that are modelled for costs, emission reductions and economy-wide impacts. They fall into three categories: energy supply, energy use and non-energy sectors (industrial process emissions, waste and agriculture). Because it stalled the planning model, the fourth strategic option has a different logic from the first three.

'Start Now' is rather modest. It includes a set of wedges that save money – or at least cost very little – over time. The major wedges relating to energy use are industrial energy efficiency, more people using public transport in preference to private cars, and energy efficiency in cars. By 2050, the energy supply in this option is 27% renewables, 27% nuclear and 27% 'cleaner coal'. The remaining 19% is traditional dirty coal. Renewables and nuclear, which is assumed to emit no carbon, provide the big wedges. 'Scale Up' starts costing money. It adopts all the wedges from 'Start Now', extends some and adds some. The extended wedges are for renewables and nuclear energy, which each provide 50% of electricity in 2050. This requires much larger generation capacity to compensate for the variability of some renewables. Capacity is 180 000 MW in 2050 compared with 120 000 MW in GWC.

'Use the Market' might more accurately have been titled 'Economic Measures'. Taxes on carbon emissions and subsidies for renewables influence prices to effect decisions across the economy. A rising carbon tax – from R100/t CO_2e in 2008, to

R250 in 2020, and then to R750 after 2040 – produces the largest single wedge modelled by LTMS. Coal-fired power starts closing down after 2025 as existing stations reach the end of their life. New build is biased to renewables because of the subsidy, with 118 000 MW installed by 2050 while nuclear adds 25 000 MW. Total installed capacity is 150 000 MW. In liquid fuels, coal-to-liquid (CTL) is phased out but five more crude refineries are built. The carbon tax has a smaller effect on demand 'than one would expect in reality' and industry and transport emissions continue to rise. In transport, this is because 'other options are limited' (DEAT 2007 SD: 19).

'Start Now' produces a saving to the economy but this is pocketed by the rich. The poor are worse off. The cost to the economy comes in at about 1% of GDP in 'Scale Up' and 2% in 'Use the Market'. Here, however, the poor do slightly better as the rich carry most of the cost. The LTMS does not show it but these results reflect the bias inherent in GDP and other macro-economic indicators.[33] A percentage increase in the wealth of the rich contributes more to GDP growth than a percentage increase in the wealth of the poor and a 'saving to the economy' is effectively a saving to the rich and to the corporations in which they invest. This is what happens in 'Start Now' where savings are largely produced through energy efficiency.

Industry dominates energy demand, has neglected efficiency and is responsible for a high proportion of emissions. Industrial energy efficiency therefore makes for the biggest wedge on the demand side as boilers, fans and pumps, etcetera are made more efficient. The largest transport savings come from more efficient vehicles, including limiting the market for SUVs and so forcing a shift to smaller cars. The LTMS does not draw the conclusion that the conspicuous consumption of the rich – which adds to GDP – is at the expense of everyone else as well as the environment.

Nor does it take account of the 'Jevons paradox', explained in Chapter 5, that energy efficiency leads to an overall increase in energy use within a capitalist economy. The present Eskom crisis demonstrates both that a limit on energy supply may push efficiency and that efficiency is meaningless if, as Bobby Godsell quipped, there is no energy in stock. On the other hand, a radical reduction in demand makes meeting it relatively easy. As a first step, the campaign against the World Bank loan to Eskom called for phasing out the supply to the aluminium smelters, so cutting demand by 10%.

The greater cost to the economy in the next two strategic options hits the class of investors, corporate and individual, and must therefore curtail economic growth.

This is not a result that is wanted and the LTMS gets around it in two ways. In 'Scale Up' it assumes financial and technology transfers under the UNFCCC. In 'Use the Market' it proposes that the costs of the carbon tax should be offset by government recycling the revenues through the economy. How that revenue is distributed or invested would then become a critical political issue.

While cabinet has 'adopted' the LTMS, it is not clear what exactly that means. Moreover, there is some dissonance between the LTMS and policy as it is revealed by the actions of the state. Thus, the modest 'Start Now' shows renewables producing 27% of electricity by 2030 while the latest revision of Integrated Resource Plan (IRP) 2010 indicates 9%.[34]

Eskom itself is rather more excited by 'clean coal' technologies than by renewables. In fact, Eskom has been researching most of these technologies for well over a decade and long before it felt constrained to recognise climate change, let alone to develop a 'climate strategy'.[35] For the most part they are simply the latest coal-burn technologies given a green spin. Some are mature technologies being applied in South Africa for the first time. Medupi and Kusile, for example, will be supercritical steam generators and this is expected to improve the energy conversion efficiency from 35% to around 40%.[36] Others have yet to be proved internationally. Thus, Eskom has a long-running research and development programme on Underground Coal Gasification. The original motivation was to access energy from coal in situations where, for geological reasons, mining is not possible. It is thus primarily intended to expand the usable coal resource.[37] Environmental benefits are nevertheless claimed relative to the impacts of mining.

Carbon capture and sequestration is the one technology that responds specifically to climate change. Government and the big corporations have grasped this technical fix to get coal off the climate hook. Kusile is to be 'CCS ready' although Eskom's technical supremo, Steve Lennon, admits that 'no one really knows what that is at the moment'.[38] LTMS is distinctly cool towards CCS on power plants and allows just 2 mt CO_2 per year saving from it – which scarcely seems worth the cost. It allows 23 mt per year for CCS from Sasol's Secunda plant but remains sceptical for three reasons: 'South African geological conditions are not favourable for CCS'; there is a severe penalty on conversion efficiency; and the technology is not proven (DEAT 2007 TR: 81). Nevertheless, at the behest of government and with sponsorship from Eskom, Sasol and Anglo American, a 'CO_2 Storage Atlas' has been prepared by the South African National Energy Research Institute and the potential talked up.

The favoured option for 'low-carbon' generation was always nuclear power.

For the generator, nuclear is indeed a low-carbon technology since carbon emissions associated with nuclear are in construction, mining uranium, fabricating fuel, disposing nuclear waste and finally, in decommissioning plant. Most of these emissions will be attributed to someone else's carbon account. Eskom's 2008 Annual Report shows nuclear saving about 70 mt of CO_2 in 2025,[39] which is more or less in line with the LTMS 'Scale Up' projections. With government now talking of 9 600 instead of 20 000 MW of nuclear capacity by 2030 there is evidently a large hole in Eskom's climate strategy. While promoting nuclear power, neither Eskom nor government seem particularly bothered by the contamination on the West Rand discussed in Chapter 4.

In more optimistic times, Eskom liked to boast about the scale of capital investment in the new coal plants, comparing the cost of Medupi and Kusile with that of China's Three Gorges Dam. These investments now carry a real risk of bankrupting the country. The nuclear fantasy will escalate the risk. If a white rabbit is plucked from the hat of climate funding, it will be just that – a conjuring trick diverting attention from who gets stuck with the bill. And the initial investment will be compounded by rising fuel prices.

The proponents of renewables see these massive investments as wasted resources. South Africa has very good renewable resources, particularly for solar energy, which have hitherto been ignored. Moreover, the development of a renewables industry is within the scope of South Africa's capacities. There would now be a functioning industry if, in the last decade, renewables had been supported on the scale of the Pebble Bed Modular Reactor. LTMS notes that renewables create more jobs, particularly for lower-skilled workers, than conventional fossil energy. In contrast, nuclear leads to an overall loss of jobs and most are high-skilled (DEAT 2007 TR: 141). Holm et al. (2008) identify several other advantages for renewables, including that several technologies are composed of many small-scale units that can be built relatively quickly in response to actual demand rather than uncertain projections of future demand; and that, being widely distributed, they reduce transmission costs and minimise the risks of grid and plant failures.

'Reach for the Goal'

The fourth strategic option does not follow the logic of wedges and, like RBS itself, the economic costs of 'Reach for the Goal' cannot be calculated because it relies on 'unknown technologies and behavioural change' (DEAT 2007 SD: 21). LTMS proposes four sets of actions for this option:

1. New technology: Investing in technologies for the future
LTMS calls for 'aggressive' research and development. Technologies should be identified for their potential to reduce emissions, particularly in the carbon-intensive energy system, and on the level of risk and the potential for international technology transfers. These last criteria assume that successful new technologies are those that are developed and adopted globally rather than in isolation. Technologies are also seen as integrated into systems and interacting with 'human behaviour': 'An example would be a decentralised grid, in which citizens can generate their own electricity and pass surpluses back to the grid' (SD: 21).

2. Resource identification: Searching for lower-carbon resources
LTMS proposes further investigation on importing hydro power from the Congo and importing gas from the southern African region to substitute for coal.

3. People-oriented measures: Incentivised behaviour change
'Changes in social behaviour, whether driven by policy, education, or awareness, may yet prove to have large scale and low cost mitigation effects' (SD: 22). In fact, most of the changes proposed are systemic: changing the pattern of urban development, including reducing distances between home, work and amenities; shifting to public transport; localising food production and consumption, implying a major effort around urban agriculture as well as enabling rural markets; responding to urban population (including urbanisation) growth 'and high commodity expectations'; and 'greening' towns. People's behaviour is located within systems: however 'aware' they are, they can't get on a bus that isn't there.

4. Transition to a low-carbon economy: Redefining our competitive advantage
'Perhaps the most difficult but most fundamental approach to mitigation would be to shift South Africa's economy away from its energy-intensive path' (SD: 23). LTMS argues that policy still defines South Africa's competitive advantage around energy-intensive sectors. This must change and moving to a low-carbon economy must be integrated into industrial policy. This implies redirecting investment, removing incentives designed to attract energy-intensive investments and using the money to promote low-carbon sectors. Strategies to support energy-intensive industries and workers to make the transition would also be needed. LTMS suggests that such measures would support what is a natural process: 'Over time, most economies shift from primary and secondary sectors to tertiary sectors. South Africa's GDP has already shifted significantly from mining through manufacturing to services'.

With the exception of point 2, this seems to represent a radical departure from present development policies. Whether it is or not, however, depends on how it is read. The assumptions behind the LTMS's larger story of development – the five stages leading to a service-based economy – seem present in Cabinet's adoption of the strategy that commits to redefining competitive advantage and shifting to 'a climate-friendly path as part of a pro-growth, pro-development and pro-jobs strategy' and an industrial policy promoting 'sectors using less energy per unit of economic output'.

Yet the shift to services is not necessarily what it seems. First, many services are dependent on energy-intensive industries as with Sasol's chemical engineering design services. Second, the development assumed by the LTMS is not universal but reflects geo-political power and ranking. The major Northern economies have outsourced production, along with the carbon emissions, and it is through services that they retain control of production. Finally, Fine (2008a) notes that financial services are now held to account for 20% of South African GDP. He argues that this is a symptom of financialisation and does not represent the banks' contribution to the economy but the economy's contribution to the banks. Further, the banks themselves are closely tied to the major corporations at the heart of the minerals-energy complex. One might add that carbon trading falls under financial services and is similarly symptomatic of the parasitic financialisation of the global economy.

The LTMS doggedly abstracts its analysis from social power relations, but the first real test of government commitment to RBS must lie in its readiness to confront the power of the minerals-energy complex to shape development and appropriate the proceeds. To date, it has acted to entrench the state's own interest in the minerals-energy complex. The 'people-oriented measures' imply major social and economic investment. This will not be made within the terms allowed by the minerals-energy complex or, indeed, by capital more broadly. As the evidence from Cape Town shows, economic growth driven by global competitiveness and pro-poor development are not compatible.

If the people-oriented measures are to mean anything, they must define the shift to a low-carbon economy. These two sets of actions cannot be treated independently of each other. Resources currently devoted to capital- and energy-intensive growth in the service of capital must be redirected to create an economy in the service of people. This will not be accomplished unless the process starts with people and is about people taking control of their economies through the process. Similarly, it should be recognised that technologies are not neutral but that they embody social power relations. Investments in new technologies must therefore be conceived as integral to this process. A decentralised grid based on many small generators, for example, embodies a very different set of social relations to a centralised grid based on a few giant plants while nuclear power entrenches the political power of the minerals-energy complex and the security establishment and, beyond that, of the imperial powers.

Out of bounds

RBS is uncosted because it cannot be achieved within the confines of current planning models. The assumption that informs these models is that economic growth constitutes the central organising principle of development. This is not because growth is needed to alleviate poverty but because it is needed to reproduce capital. This is what determines the bounds of realism in planning and it is this realism that has produced the crisis of climate change, the crisis of peak oil and the political and economic crisis gripping global capital.

The wedges were developed within the bounds of planning realism. Some wedges indicate pathways leading out of these bounds but others are based on interests vested in the present realism. In particular, LTMS energy modelling assumes ever-increasing demand. It is this that could not be reconciled with the carbon reductions required by science and it is this that will not be reconciled with declining global oil production following peak oil. While the LTMS sees that RBS takes the country into uncharted realms, it retains growth as the central organising principle of development. It then takes the definition of national competitiveness as the fundamental issue. National competitiveness, however, becomes necessary because it is a function of growth. Rather than redefining national competitiveness, what is needed is a radical redefinition of what is meant by development and who defines it.

First, the central organising principle should be sustainable development founded on economic, social and environmental justice. This means a commitment to growing human solidarity and equality as well as a relationship to the environment that enhances rather than degrades the functioning of ecosystems both for their intrinsic value and for the 'eco-services' they provide. Put differently, it implies that people recognise themselves as a living part of earth's ecology. This does not imply that economy and production are unimportant, but that the economy must serve people, as may be inferred from the 'people-oriented measures', rather than people serving the economy.

Second, peak oil implies either a compelled shift to economic localisation or the exclusion of ever-more people from the shrinking enclaves of elite development.

The choice for localisation follows from the choice for justice and is essential to any serious programme to avoid catastrophic climate change. This implies that national resources should be focused on supporting people's capacities to direct local development.

Third, if we are to address climate change, the energy system must be transformed as a matter of urgency. The resistance to the new build programme and the focus on energy-intensive industries and development creates the beginnings of a movement with that purpose. Overall, energy systems including power generation should be localised and placed under people's common control. Maintaining a level of national and regional grid capacity will remain important and this capacity should be provided by renewables. An aggressive programme of renewable energy, including solar water heaters, should therefore be prioritised. Supporting the capacity for local production of renewable energy components should be made central to industrial development policy.

Fourth, the transition to a different energy and development order will require energy inputs from the declining fossil-fuel system. If these investments go into the declining system, they will represent a permanent loss. In the period between now and the latest credible peak emission target date of 2015, fossil-fuel resources should therefore be used to build the new system.

Fifth, food is the most basic form of energy for people and the food system must be thoroughly transformed to enable people to define and take control of production and consumption and hence of their own futures.

10

Change coming

AGRICULTURE IS OFTEN represented as a backward sector of the economy. In the post-war period it was progressively industrialised, with state support, throughout the world and this process has been accompanied by a depopulation of the rural landscape as machinery and chemicals increasingly replaced labour. Investments and growth in local economies were associated both with job-shedding and with a concentration of land ownership as smaller commercial farmers fell behind in the technology race. In most countries, the better part of the value of state support was captured by large corporations. Now, with the withdrawal of state support in Southern countries, the market is capitalising on those earlier investments and the process of depopulation is intensified.

What has happened on the farms is now happening in the factories. The industrial landscape is being depopulated. Workers are being swept out of both farms and factories and they and their families are forced to find shelter where they can in increasingly impoverished urban settlements. The service sector – particularly tourism and call centres – is now looked to for jobs. Services are, however, very diverse and people are already being replaced by computerisation in many areas such as financial services. Indeed, call centres themselves exemplify this. As they centralise functions such as bookings and information in one place, they destroy local jobs in other places both within South Africa and across the world. Globally, they destroy more jobs than they create and they pay less per job. In the name of productivity, that is precisely what they are meant to do.

The post-war promise of full employment within the Fordist regime of production, backed by the safety net of the welfare state, held good only for the First World. It was the outcome of struggles between Northern social movements and the managers of capital but it was underpinned by the transfer of resources from the Third World. Since the 1980s, under the neo-liberal regime, employment and welfare in the industrialised countries have been leached away. As financier Warren Buffet is reputed to have said: 'If there is such a thing as class war, my side is winning'. In the Third World, full employment was the political privilege afforded only to certain categories of people – whites in South Africa. For most of the rest, labour was coerced into the factories at pitiful wages and only as it was needed.

Across the Third World now, the workers abandoned to poverty are being joined by people whose subsistence economies subsidised industrialisation to much the same effect as the 'reserves' and 'homelands' did in South Africa. These economies staggered under the burden and the environmental resources of land, water and biodiversity that maintained them, and were previously maintained by them, eroded away. This erosion has undermined the resilience of local environments, making them ever-more vulnerable to the stress of drought and flood. On the other side of the fence, industrial agriculture has also undermined environmental resilience, destroying forests and wetlands, compacting soils under the weight of machinery and substituting chemicals for organic fertility. Thus it was, in the drought of the early 1990s, that thousands of people abandoned what remained of production in the homelands and thousands of farmworkers lost their jobs, creating a 'pulse' of migration to towns and cities.

These disasters are still called 'natural'. They get less natural all the time. It is no longer just that environmental resilience is destroyed on the ground, but the climate itself is made more erratic and extreme under the impact of industrial emissions. The people who flee these unnatural disasters mostly join those already crowded in urban shack settlements located on land that is not valued by the urban real estate market. This land is often in flood zones, on steep slopes or in polluted areas, and the shack settlements are inadequately served, if at all, with water, energy, sanitation and waste removal. Here, people face a new round of environmental disaster from contaminated floodwater and mudslides, from periodic outbreaks of disease, and from the fires that repeatedly burn through them.

A third of the world's urban people now live in slums, mostly in the Third World, in old working-class areas drained of income as well as in shack settlements. Almost all the growth in the world's population is being absorbed into the ranks of the urban poor while the population in rural and richer urban areas is, or soon will be, declining. Both the absolute number of people and the proportion of the world's population living in slums are therefore rising fast. At the same time, inequality between rich and poor countries and between rich and poor people everywhere has been taken to an extremity as the proceeds of economic growth are appropriated almost exclusively by the rich. Thus has the plight of the wretched of the earth come full circle. Those dispossessed in the countryside and forced into, or to subsidise, the work of urban industrialism are now also dispossessed in the cities to which they were and are being driven.

Contrary to the spurious arguments of the World Bank and major powers, the environmental impacts associated with poverty are not caused by poverty. Poverty is as much a sign of unsustainable development as environmental degradation and for the same reason: both are produced by the working of the economy that concentrates wealth in the hands of the few – and particularly in the hands of fewer and fewer corporations who are then able to decide where and how to reinvest it and so determine the future of development. Under the rule of neo-liberalism, the practices of capital at the periphery have thus returned to its centre and the system as a whole now feeds on its own entrails.

FOR PEOPLE'S POWER

It may be tempting to trust that peak oil and economic depression will achieve the necessary reductions in carbon dioxide emissions. That depression is the best hope for credible emission reductions in the elite future is the ultimate expression of the meaning of unsustainable development. The politics of elite power remains inseparable from capitalist growth. As that power begins to fail, it is ever-more brutally imposed every day on people and their environments and it will finally cost the earth. Those who are represented as the leaders of the world cannot conceive another way and cannot confront the challenges of the time. Time is short: the world is on the cusp of runaway climate change – the point at which natural feedback loops such as the melting of ice become more significant than industrial emissions. Time is short: the growth machine is running low on fuel while the elites remain determined to suck the last drop of oil to preserve an impossible political order. It is necessary that a different order and logic of politics come into being.

The elite energy agenda creates resistance everywhere and everywhere people create the possibilities of new life. This chapter tries to convey something of the creativity of people's responses to crisis. It recounts people's call for food and energy 'sovereignty' – the demand that people must control the resources necessary for life, that they take the power of decision making and, in so doing, that they make a future fundamentally different to that planned by the elites. How they respond is different in different places. In Cuba the state followed and supported the people's lead. Elsewhere, people are consciously anticipating peak oil. In very many places, crisis is visited on people by those who speak in their name but put themselves at the service of elite power. Resistance is then made an immediate necessity of life but it is always accompanied by a vision of another world. It is in these actions that hope resides.

New life

In the way industrial economies are described, agriculture appears almost as a residual sector contributing a minor proportion of GDP and invariably bracketed from the rest of the economy. Thus, major economic indicators such as employment are often qualified as 'non-farm'. Services and industry are what matter. But this conceals how much of services and industry finally rely on what is produced on farms and, indeed, how much of what is now defined as industry – processing food, fibres and timber – was once part of the farm or household economy.

It now takes about ten calories of fossil energy to produce one calorie of food energy. This includes the chemicals and machines, the fuel and electricity necessary for industrial farming, food processing and packaging, transport over ever-greater distances and refrigeration all the way along the fresh produce supply chains to the supermarkets. In Britain, getting food to the plate consumes 20% of total national energy. At the wealthy end of global production networks, a brief blockade of fuel depots in 2000 revealed how vulnerable Britain's food supply is to fuel shortages. Within days, the supermarkets shelves were emptied of even such basic commodities as bread. At the poor end of the networks, many permanent as well as seasonal farmworkers can scarcely afford the price of a loaf.

The global elite has long claimed that industrial farming is necessary to feed the growing population of the world. Proponents of organic agriculture have long since recognised that this argument is really a cover for promoting the interests of corporate agriculture – in much the same way that the bias against renewables reflects the interests of big oil. University of Michigan researchers Ivette Perfecto and Catherine Badgley have recently refuted the elite claim. They calculate that organic farming in developed countries would produce 92% of what industrial farming produces but, in developing countries, it would produce 80% more than industrial farming. They specifically refute the much-repeated claim that organic farming cannot overcome the loss of nitrogen fertilizers.[1] A 2007 report from the UN Food and Agriculture Organisation (FAO), long an advocate of industrial agriculture, found organic farming superior in terms of food security, productivity and local economic returns while it reverses the ill-effects on the health of workers and consumers and on the environment.[2]

These findings chime well with the Nyeleni Declaration on food sovereignty adopted in February 2007 by organisations of peasants, pastoralists and artisanal fisher-folk, amongst others, from 80 countries meeting in Mali. They declared themselves 'ready, able and willing to feed all the world's peoples'. They defined food sovereignty as:

> ... the right of peoples to healthy and culturally appropriate food produced through ecologically sound and sustainable methods, and their right to define their own food and agriculture systems. It puts those who produce, distribute and consume food at the heart of food systems and policies rather than the demands of markets and corporations.
>
> It defends the interests and inclusion of the next generation. It offers a strategy to resist and dismantle the current corporate trade and food regime ... prioritises local and national economies and markets ... promotes transparent trade that guarantees just income to all peoples and the rights of consumers to control their food and nutrition ... ensures that the rights to use and manage our lands, territories, waters, seeds, livestock and biodiversity are in the hands of those of us who produce food.
>
> Food sovereignty implies new social relations free of oppression and inequality between men and women, peoples, racial groups, social classes and generations.

This agenda resonates with the call for local control of resources and energy sovereignty made by communities affected by the fossil-fuel industries. In September 2005, civil society activists met in opposition to the agenda of the World Petroleum Congress in Sandton, Johannesburg. This agenda proclaimed the oil elite's intention to shape the world's energy future. Mindful of the scale of human and environmental atrocity associated with big oil's activities all along the production chain, as well as the consequences of climate change, the activists responded that 'another energy future is necessary'. They endorsed the conclusions of *The groundWork Report 2005*, that the oil elite's power 'is neither stable nor inevitable and that it is always and everywhere contested and renegotiated' and that the potential for people's energy 'lies in connecting the promise of renewable energy sources and technologies with social movements struggling for deep transformation of the way the world works' (2005: 121).

In September 2006, member organisations of Friends of the Earth from 51 countries adopted the Abuja Declaration. It took up the theme that 'another energy future is necessary' and linked it with the idea of 'energy sovereignty'. It observes that struggles for economic, social and environmental justice are linked through their common resistance to the elite economic and political order and calls for the co-ordination of 'energy struggles around the world by adopting a global strategy for resisting environmental degradation, destruction of local livelihoods, and rights abuses associated with corporate controlled energy sourcing and consumption globally'. It resolves that another energy future is necessary based upon:

- Abandoning the belief in export-led growth in favour of servicing local . . . needs;
- Restructuring the price and production of energy;
- A new approach to restructuring ownership of the energy regimes; and
- Abandoning the mistaken dichotomy between development and environment.

It declared support for 'community struggles towards energy sovereignty and democratic control of natural resources that will be the basis for alternative fair and just trade regimes that link producers with consumers, eliminating corporate led control of our energy systems'. It particularly noted the role of women in those struggles and said that they should be 'fully involved in all negotiations over energy production and allocation of natural resources'.

In summary, it called on governments to:

- declare a global moratorium on new oil and gas exploration and development;
- terminate neo-liberal trade agreements and economic policies that strip people of their entitlements to basic resources and lead to their impoverishment;
- enforce strict environmental standards and redirect the super-profits of the oil multinationals to clean up the mess;
- repudiate Joint Venture Agreements between governments and corporations and replace them with agreements between governments and local communities;
- resolve the Niger Delta crisis through democratic dialogue; and
- support decentralised, democratically controlled and sustainable energy systems using clean energy like wind and solar energy.

The struggle to make the vision of food, energy and resource sovereignty real faces formidable opposition. It is a struggle that is carried on in different ways and different circumstances by millions of people across the globe and for many it is a matter of life and death. There is no guarantee of success but, as *The groundWork Report 2005* put it,

... even if these social and environmental justice movements do not succeed against the enormous power of the current regimes, and the descent into a post-fossil-fuel (and post-US empire) era of uncertainty and collapse continues, then the spaces of self-reliance and local democracy created through such struggles will emerge as the only viable basis for rebuilding a new world (groundWork 2005: 121).

Yet the moment of crisis is also a moment of opportunity and hope, as the Cubans showed in their extraordinarily creative response to an energy crisis that provided a preview of peak oil.

Cuba

Just as Taiwan and South Korea were subsidised by the US on one side of the Cold War frontier, Cuba and North Korea were subsidised by the Soviet Union on the other. When the Soviet Union collapsed, these two countries found themselves isolated and under virtual siege by the US. Both were dependent on subsidised oil from the Soviet Union and both had adopted the centralised high-input agricultural systems that mirrored the technologies of the green revolution. North Korea's oil imports were slashed by 60%. Although well endowed with coal, the power and transport systems failed because the country was cut off from technical support to maintain its Soviet-built infrastructure. Agriculture collapsed. Industrial agriculture had produced enough grain for the country's needs and the rigid North Korean regime attempted the impossible project of sustaining it. Famine wracked the country and over three million people are thought to have died.

In the three years between 1989 and 1992, Cuba's petroleum imports were more than halved and fertilizer and pesticide imports dropped by 77% and 63% respectively and, without oil, there was no feedstock for its own agricultural chemicals industry. In addition, Cuba was heavily dependent on food imports, which were also cut by half as the country's foreign exchange dried up. Thus began the 'special period' in Cuba. People's food consumption was dramatically reduced and people did go hungry, with the number of under-nourished people rising from 5 to 30%, but there was no famine. By 1995, Cuba had transformed agricultural production and restored adequate levels of nutrition.

Prior to 1990, Cuban agriculture was dominated by state-owned estates expropriated from capitalist plantation producers during the 1959 revolution. Ownership aside, agricultural ecologist Peter Rosset (2000) observes that the management and technology regimes were similar to corporate agriculture in

California. The estates were water- and energy-intensive, producing sugar and other cash crops for export to the Soviet Union at five times world market prices and exchanged for energy imports. A small peasant sector – co-operatives and individual farmers – produced 40% of food on 20% of the land.

Unlike North Korea, Cuba turned to organic agriculture supported by massive household, animal and human waste recycling. The first initiative was taken by urban people who cleared waste land for gardens and learnt how to grow food organically using the methods of permaculture, and by peasants who responded to rising food prices and used animal draft for ploughing and transport. The state followed their lead. Amongst other things, it supported the establishment of farmers' markets to ensure that the value was not captured by intermediaries but returned to producers, it focused its formidable scientific expertise on issues such as organic pest control, and it supported a draft animal breeding programme. With this, the urban agriculture movement 'exploded to near epic proportions' (Rosset 2000) while peasants acquired a new status in society and thousands of urban people migrated to the country to take up farming. In contrast, the large estates floundered as the management regime precluded the relationship to the ecology of the land required by organic production. Most of the estates were consequently broken up and later turned over to the workers.

The transformation of energy appears less decisive. Cuba has small reserves of very low-quality crude oil and has expanded both production and exploration through product-sharing agreements with independent oil corporations and, more recently, with the Petróleos de Venezuela and China's Sinopec. In 2002, Venezuela agreed to supply crude and fuel to Cuba on very favourable terms and so relieved the shortage. More recently, it entered a partnership with Cuba to upgrade the mothballed Cienfuegos Refinery with production of 65 000 b/d starting up in December 2007. Cuba's two other refineries are old, dirty and dangerous.

For power, most Cubans continued to rely on the grid. Most power plants were designed to use heavy fuel oil. Lacking this fuel, the state instead used the crude oil from its production sharing agreements. As well as polluting the neighbourhood, this wrecked equipment and resulted in plant failure and extended blackouts in 2004 and 2005. In 2006 government responded with the 'energy revolution'. Diesel generators were installed in 116 of Cuba's 169 municipalities and linked through the grid to restore the power supply at a cost of $1 billion.[3] The state developed some renewable generating capacity from wind and from bagasse.[4] It also made a concerted effort to light up off-grid areas, particularly schools and clinics, with photovoltaic (PV) electricity. Energy efficiency and conservation was critical to the energy revolution and Cuba dramatically reduced petroleum consumption in the two years to 2009.

Not quite paradise then but, in 2006, the conservative World Wildlife Fund rated Cuba the world's most sustainable country as measured by ecological impact and human well-being (WWF 2006: 19). The experience illustrates some key points. First, technologies are not neutral but embody relationships of power and, as for sustainable agriculture, so for sustainable energy and sustainable production in general, the institutional relations of production are critical. High-energy large-scale production in all sectors generally requires centralised management irrespective of ownership. Energy efficiency can certainly be improved within this regime and niche-market renewables can be established – but within limits. Ultimately, these systems are not compatible with sustainable development or social justice.

Second, the cheap-energy, cheap-food, cheap-goods regime of production claims to represent the interest of consumers. This ignores that consumers are also workers. Cheap goods are the product of cheap labour and polluting production and part of the arsenal of managing labour demands. Energy-intensive agriculture is justified on the basis of productivity but small-scale production is generally more productive per hectare than industrial agriculture unless the machinery and chemical inputs are heavily subsidised. Peasants and small farmers are not being wiped out for lack of productivity but for want of power in a market constructed by and for corporate production and the extreme concentration of economic power secured by these means. Similarly, fossil-fuel energy is made to appear more economic than sustainable-energy technologies through massive subsidies, including the externalisation of environmental and social costs.

Third, Cuba was and is an authoritarian and patriarchal state. The path of sustainable agriculture was forced on it as a matter of national security. The alternatives were famine or allegiance to Washington. Refusing dependence on the global regime of accumulation, Cuba was also excluded from development aid and famine relief, which elsewhere acts as a palliative for 'market failure'. Yet, however imperfect, and unlike North Korea, Cuba's political elite appears to retain the egalitarian spirit of the revolution. During the special period, the state imposed equal food rations on all citizens irrespective of status. According to Pat Murphy of the Community Solution (in Morgan 2006), this prevented a situation in which competition for declining resources would lead to social disintegration and instead created the essential basis for social solidarity. Cubans felt that they were facing a common crisis together. The role of the state has been decisive – both in making the conditions of dependence on the Soviet bloc and in supporting and building on the creativity of people's response to the crisis. Since the alliance with Venezuela has partially restored cheap-energy supplies, it may be that the Cuban state will choose to revert to high-energy agriculture and so repeat its earlier dependence on the Soviet Union.

Yet it appears that people will defend what they have created. The 'green revolution' involved everybody in a way that resonates with the aspirations articulated in the Nyeleni Declaration on food sovereignty. It ended the regime of the 'passive consumer' for food as for energy. It changed people's imagination of the world, creating a new sense of social identity in which people see themselves making their own future and remaking in their daily lives the social solidarities previously made over to the paternal state. Reflecting on the experience, Cuban lawyer Rita Pereira comments that 'we can be happy with less' and she sees the potential of peak oil as 'a time for sharing, for cooperation, for solidarity. Maybe we'll have a better world' (in Morgan 2006).

From the point of view of the state, it seems that sustainability is the last resort rather than the first choice. The South African state has not faced a crisis of last resort, although many of its people face a crisis of survival on a daily basis. The state's policy choices are for integration into the global circuits of capital accumulation and climbing the ladder to what that system defines as higher value production.

Anticipating peak oil

The assumption that energy expansion is never ending informs energy planning in most countries. Cuban town planner Miguel Coyula remarks that countries dependent on imported oil are not seriously thinking about alternative energy and are, in effect, 'just planning for the next week' (in Morgan 2006). People's initiatives at local level are challenging this. In the North, a growing movement is drawing together the permaculture and localisation movements to confront both peak oil and climate change.

This movement makes several basic assumptions that distinguish it from those discussed in the renewables section in Chapter 5. First, renewables cannot compensate for declining fossil energy enough to power never-ending accumulation and economic growth. Peak oil thus implies a radical restructuring of economies including enforced localisation. People and governments therefore need to plan for an 'energy descent' or, in Heinberg's phrase, to 'power down' (2004). This would certainly include an expansion of renewable energy systems and the construction of local mini-grids as well as a heroic drive for energy conservation. Local production of food and other goods would produce local livelihoods and radically cut the energy now used to transport goods around the world. People would still trade but only after satisfying local needs. In this vision, the ideal future is a democratically directed, locally centred, steady state economy compatible with natural energy flows that cannot be expanded.

Second, powering down is also necessary to limit global warming. Even if oil production declines faster than anticipated, it is unlikely to match the very steep decline in carbon emissions required now. Using more than a fraction of the remaining oil, gas and coal is planetary suicide. The elite's proposal for the technical fix of carbon capture and sequestration is at best a massive gamble because it cannot be shown that it will work on the scale required – that is, that 80% of global emissions can be safely sequestered. More likely, it is a cynical pretence at action designed to save fossil corporations rather than address climate change. Nuclear energy similarly fails to address climate change and leaves an additional toxic heritage. It also adds to the threat of nuclear weapons proliferation and multiplies the excuses for the US to embark on imperial wars.

The key challenge for this Northern movement is seen to be persuading people that the energy descent is both doable and desirable. The alternative is the impossible politics founded on economic growth to which the elites are wedded. Given the inevitable failure of growth, and the collapse of the market in jobs, this implies a descent into chaos and fascism as people seek scapegoats for their pain.

The movement is therefore supporting people's planning processes, which consciously anticipate peak oil by developing local 'energy-descent plans'. The process builds the local movement and creates the popular support and pressure for local councils to adopt the plans. In Britain, this process has been captured in the idea of 'transition towns' advanced by permaculture designer Rob Hopkins (2006). Just as Pereira considers peak oil an opportunity for a better world, so too does Hopkins argue that 'life with less oil could, if properly planned for and designed, be far preferable to the present' (undated: 5). Anticipation of peak oil and the enforced localisation of economies is thus embraced as an opportunity for reviving local democracy and relations of mutual solidarity and creating local livelihoods founded on environmentally sustainable practices.

Green activists Caroline Lucas, Vandana Shiva and Colin Hines give a wider dimension to this vision, arguing that the rules of global trade are designed to prevent both local and democratic control of economies. Further, the organisation of the global economy supported by these rules has the effect of 'making poverty inevitable' (2005). Localisation in the North is thus the other side of the coin to ending the plunder of the South.

The concept of an energy transition poses the question of how to use the resources now available to society as a whole to create the basis of a future energy regime that will provide for people but not for profit. Environmentalists have commonly argued that a sustainable-energy regime will create more jobs than the centralised fossil-energy regime.

Urbanist Mark Swilling adds that a policy priority for ecological sustainability ahead of economic growth will create greater social equity even if this choice is accompanied by economic recession. The 'consumption city' can be, and should be, replaced by a 'sustainable city' in which improving living standards are decoupled from rising (and unsustainable) resource consumption. Such a city, composed of sustainable neighbourhoods

> . . . generates more energy than it consumes, generates zero waste (both liquid and solid), meets most of its basic food requirements from local sources, requires little or no fossil fuels to transport people, and releases minimal amounts of CO_2 into the atmosphere. The 'sustainable neighbourhood' helps to rebuild eco-systems and mitigates the risks associated with the rising costs of fossil fuels as these non-renewable resources run out (Swilling 2007: 5).

In south Durban, people believe that the new round of industrial modernisation being pushed through by the City authorities is not only destructive but that it assumes a future which will not materialise. Being founded on planning realism, it does not in fact prepare for the real future. The South Durban Community Environmental Alliance (SDCEA) notes that the City's spatial development plan stakes the future on 'accelerated growth based on large-scale investments, where attracting capital is a first priority'. Thus far, this strategy has resulted in job-shedding and dispossession. This 'should indicate a need to rethink the social and ecological relations of production'.[5]

The City is planning for the continued expansion of container traffic through the harbour and on the roads, of petrochemical industries, of international tourism and of exports of energy-intensive products through the new international airport. This will lead to over-investment leaving stranded assets and a legacy of intensified pollution as the global economic recovery fails and oil supplies decline. Indeed, the assumption that investments in 2010 football World Cup infrastructure would be justified by a major boost to tourism has already foundered.

Durban is already feeling the impact of climate change. Sea levels are beginning to rise, storms are more intense but also less frequent and the temperature is rising.

SDCEA observes that the City is beginning to take the issue seriously and welcomes its focus on adaptation and its recognition that 'eco-services' are essential 'to the future functioning of the city'. Mitigation, however, is entirely absent from its planning as the City continues to 'invest in development that is reliant on fossil fuels'. SDCEA calls on the City to recognise 'that unsustainable oil dependent industries must be phased out within the next 20 year planning period'.

In August 2009, SDCEA organised a hearing on climate change and poverty. The recurring theme of people's testimonies was that the City treated them with contempt and ignored their needs. Where they were consulted, plans were already predetermined and in some cases construction was already underway. Thus, the route for Transnet's multi-fuel pipeline avoided rich areas and was taken through poor and densely settled rural communities on the urban edge. The people were alerted by SDCEA and not by the authorities. In the words of a participant from the area: 'They are taking the pipeline through our gardens. What will happen when there are leaks and explosions? Why are they taking it through our area? They don't talk to us, they don't care about us, because we are poor'.

The demand for people-centred development, starting with democratic participation in the decisions that shape people's lives and futures, is at the core of SDCEA's project. It suggests an inversion of the planning process: that it should start with people at the local scale and work outwards to meet the world rather than starting with the assumption that corporate capital shapes development and the local must be fixed for its demands. The crisis of capital sharpens the question of politics in these and a thousand other struggles in South Africa and around the world. They are a part of what Harvey calls the struggle for 'the right to the city':

> To claim the right to the city in the sense I mean it here is to claim some kind of shaping power over the processes of urbanization, over the ways in which our cities are made and re-made and to do so in a fundamental and radical way (2008: 2).

In the neo-liberal period since the late 1970s, the making, unmaking and remaking of the city, and of the hinterland it makes of the country, has been driven by global finance capital. It has impoverished people in the country as much as in the city while creating globally connected enclaves of 'world-class' affluence. The battle for the city cannot therefore be a parochial affair but it is also always the struggle in each country district, town and city. It challenges trade unions and social movements to join forces in struggle and to respond to the question of the future. If capital is terminated in the struggles that intensify over the next decades, what will be the base, to succeed the corporation, for organising production and doing so democratically and without laying waste to the planet?

Everyone's village

Aamar Gram, Tomar Gram, Shobar Gram: Nandigram, Nandigram.
My village, your village, everyone's village: Nandigram, Nandigram.

The chant was taken up across India after police killed fourteen people at Nandigram, West Bengal, in March 2007. The people were part of a massive protest against the enclosure of their land which the state wanted for 'development' to sustain economic growth at over 8%. Nandigram was to be one of the many Special Economic Zones being set up around India in areas close to mineral resources, to energy sources and to ports from which export goods can be shipped. In most, people are stripped of their rights to land and fisheries as the transnational corporations move in and are granted exemptions from tax and from labour and environmental regulations. For writer Amit Sengupta, the chant links the 'thousands of farmers, dalits, tribals, being forcibly displaced to benefit big business and big projects, in Kashipur, Kalinganagar, Bastar, Punjab, Dadri, the Narmada valley, Tehri Garhwal'.[6] And it links these protests to the long history of resistance to dispossession stretching back to the great 1857 rebellion against the British Raj.

It is a chant that might echo around the world as states and corporations collude in appropriating people's rights in order to maintain profits and growth and, when they resist, assaulting the people themselves. In the Niger Delta, people's struggles against big oil's despoliation of their land and water have locked in a large proportion of oil production; in Ireland, the people of Rossport are blocking the state's appropriation of their land in favour of Shell which wants it for a gas pipeline; in Colombia, peasants and small-scale miners are resisting removals designed to benefit AngloGold Ashanti subsidiary Kedahda; in Britain, activists gathered at a 'climate camp' next to Heathrow Airport protesting government's expansion of airports; in China, local peasant and worker rebellions are now a daily occurrence.

In South Africa's northern province of Limpopo, platinum mining boomed to meet world demand for catalytic converters among other things. Reducing car emissions is no doubt an environmental good so long as the world is dependent on cars. As with much 'green consumerism', however, the environmental and social costs are relocated up the production chain. AngloPlatinum, part of the giant Anglo American, topped groundWork's 2006 Corpse Award for worst corporate practice.

It was nominated by the Mapela people for 'removing communities from their ancestral land, stealing peoples' resources and gagging voices of resistance'. The corporation claims its rights on the basis of agreements obtained by bullying and buying off leaders who, according to the community, had no mandate from them.

By mid-2009, 'approximately 15 000 residents [had] been forced to relocate and at least 10 000 more [had] lost most or all of their farming land'.[7] People were removed to a dusty relocation village where they say there is no adequate infrastructure for energy and water and no livelihood prospects. More are threatened with removal and are resisting through legal and direct actions. Their fields have been literally enclosed with security fencing by the mines, their water supply destroyed and their houses rocked by mine blasting. People have repeatedly asserted their rights to their land, taking down the fences, ploughing the land and forming human chains in front of the bulldozers. The police have consistently enforced the rights claimed by the corporation, arresting people and using rubber and live ammunition to break up protests. In September 2007, the people of Maandagshoek detained mine officials who 'illegally entered Maandagshoek community land' despite warnings to keep out. Community leaders then called the police to arrest the officials:

> When police arrived (in large numbers) they instead refused to open a case and indicated they would arrest all the community . . . members present. Not surprisingly, people legitimately resisted and clashed with the police. Chief Isaac Kgwete and [Maandagshoek Development Committee] Chair, Michael Kgwete, were beaten and arrested and then charged with robbery, public violence and kidnapping . . . The situation in Maandagshoek today is reminiscent of the old apartheid days when mining corporations did exactly as they pleased to any community and were protected by the police and the government.[8]

Land is not the only resource in question. The development of these huge opencast platinum mines contributes significantly to the expansion of national electricity demand. It also drives the construction of dams on Limpopo's rivers. This is a generally arid area susceptible to periodic droughts and the water table has been sinking for some decades due to heavy extraction by commercial agriculture. The new dams are justified in the name of 'delivering' water to people. The thirsty mines, however, take precedence. Already, the indications are that in dry years there will be little left for the people or for the legally required 'ecological reserve' needed to sustain the rivers.

The enclosure of people's resources precedes the second form of dispossession: the externalisation of environmental costs. In many cases, there is a direct continuity between these forms of dispossession as pollution poisons people's resources – their land, crops and water – and diminishes their livelihoods. Finally, pollution is an assault on people themselves. After long years of campaigning, the people of south Durban have forced official corroboration of the health impacts of living in the neighbourhood of two of South Africa's largest refineries and of several hundred smaller smokestack industries.

The *South Durban Health Study* confirmed that the transgression of people's constitutional right 'to an environment that is not harmful to their health or well-being' is systemic: it is built into the economic fabric. And whereas the state is obliged by the Constitution to enforce and promote the realisation of this right, it has in fact protected and promoted corporate polluters in its efforts to 'grow the economy'. For people living in South Africa's pollution hot spots, demonstrating the health impacts has been integral to a larger campaign to force government to withdraw the extraordinary rights it has granted to corporations and to take responsibility for the devastation that it has promoted in the name of development. This campaign has seen some success with the enactment of the new law on air quality and, after years of neglect, more determined regulation of polluters in some areas. Thus, the unrestrained freedom to pollute in south Durban has been curtailed and routine emissions from the refineries reduced. Incidents, however, are still part of the everyday reality of life in the shadow of the chemicals industry throughout South Africa.

Even as the regulatory system is being tightened in some respects, it is being loosened in others, particularly in respect of planning permissions. Thus, the Environmental Impact Assessment (EIA) process was 'streamlined' in 2005 on the rationale that it was delaying 'development' and so inhibiting South Africa's drive to 6% growth. Development is always about the future and specific projects are details of a picture of the future to be built. Along with other planning laws, EIAs are a way of securing the appearance of consent to the elite future but they carry the risk of opening that future to contestation. The focus on the detail of single projects to the exclusion of the broader implications and cumulative impacts narrows the scope of contestation and thus manages the risk. Thus, the vigorously contested EIA for South Africa's Pebble Bed Modular Reactor was restricted to the demonstration plant alone and excluded the implications of the broader plan for twenty to thirty plants – with twenty times the nuclear fuel, transport and waste while the risk of accidents would multiply faster than the number of plants.

In the Philippines, by contrast, resistance to plans to build new coal-fired power stations has resulted in government declaring the island of Negros a fossil-free zone. Local activist Romana de los Reyes comments on the necessity of vigilance in defence of this declaration because the state 'conveniently forgets the fossil-free policy' when lobbied by investors.[9] In a context where much of the generating capacity is supplied by private-sector independent power producers (IPPs), democratic governance is the first casualty of corporate proposals for new coal plants. Proponents have typically campaigned for the projects by securing the sponsorship of powerful politicians at national and local levels ahead of community consultations required by Philippine law. These elite figures have then not hesitated to use the coercive power of the state to suppress resistance, including posting armed police outside public scoping meetings to exclude opponents. The opposition, by contrast, campaigned from the bottom up to build a broad movement founded on local churches and people's organisations and including elements of the local elite, particularly professionals. Using the procedural requirements of planning laws to focus action and build the movement, they reclaimed democratic decision making from corporate subversion.

This confrontation of campaigning styles was also accompanied by the confrontation of information and images of the future. In the Philippines, as in South Africa and everywhere else, 'host communities' were promised new jobs and 'multiplier effects' that would create rising local prosperity. Indeed, it is generally the case that the larger the project, the more exaggerated the benefits. 'Only time exposes the empty words of the coal plant proponents,' observes De los Reyes (2006: 1). The proposed future, however, is mostly already the experience of people elsewhere. People living with coal power stations told those where new plants were planned what the reality is like. They told of lost health, lost land and polluted water. The told of plants wilting in the fields and orchards, animals dying and fisheries in decline. They told of the daily battle with coal-dust and ash in their houses, on their dishes and on their washing. They told of the smell and the noise. They told of the lost beauty of their land. They said: 'Do not allow the coal plant to be built in your area. You will end up suffering like us' (quoted by De los Reyes 2006: 15).

Industrial Development Zones are the South African equivalent of India's Special Economic Zones. Coega is the most ambitious Industrial Development Zone to date and the state has sunk billions into it, claiming that it will create thousands of jobs, boost economic growth and contribute substantially to eradicating poverty in the impoverished Eastern Cape province. The benefits to Alcan were obvious and confirmed by the very secrecy that obscured their precise value. As in the Philippines, the local campaign of resistance disputed the promised benefits but foresaw heavy costs to local communities, including the litany of environmental impacts from just the aluminium smelter. With that project cancelled, the Coega Development Corporation's hopes now rest with PetroSA's proposed refinery. Their vision entails the construction of a new pollution hot spot bought at considerable expense to the public purse. The local activist group Nelson Mandela Bay Municipality Local Environmentalists has proposed environmentally friendly alternatives for the area but have been disregarded by government.

Organised public opposition has persuaded the conservative German Chancellor, Angela Merkel, against trying to revive the nuclear industry. The annual ritual of this opposition is the turnout of hundreds of thousands of demonstrators to obstruct the delivery of radioactive waste for disposal in Gorleben by literally putting their bodies on roads and railway lines. This opposition has radicalised the residents of what was a parochial region, changing their attitude both to nuclear energy and to public participation in national decision making.

In 2007, Eskom identified several potential sites for the Nuclear 1 pressurised water reactor (PWR) plant. The communities expected to 'host' the plant went on to high alert. Veteran anti-nuclear campaigner Mike Kantey attended public meetings in these areas and reported that 'each and every site has an active anti-nuclear lobby' linked to each other 'through an informal anti-nuclear network'.[10] Public debates in Thuyspunt, close to Port Elizabeth and the nearby Coega, questioned the fact that alternative sources to nuclear power were not being discussed. In the Northern Cape, there were well-founded fears that the Vaalputs low-level nuclear dump will eventually be made to take high- and low-level waste from the expanded nuclear industry. The community of Komaggas in the Northern Cape said they would not allow Eskom to go ahead with its plans to build a nuclear power station on their land. Andy Pienaar, a community representative, threatened: 'I think from here on we are going [to] shut these people out of the community and we are going [to] make every effort to make sure that they do not erect a power station at Brazil or Schulpfontein for that matter'.[11] Residents resisting the new uranium enrichment plant planned for Pelindaba organised themselves into the Pelindaba Working Group, also part of the new nationwide anti-nuclear alliance. Government's nuclear ambitions have not diminished with the cancellation of Nuclear 1. Although the outcome of Eskom's EIAs seems predetermined, they have provided the focus for continuing opposition.

The blatant ambition of agricultural corporations such as Monsanto to take control of food systems through genetic engineering and patenting of seeds has similarly run up against serious resistance in Africa, Europe and Asia. European consumers reject it because of concerns about food safety while African and Asian farmers understand that it is a threat to their very ability to produce food.

Struggles in tension

South Africa remains a sharply divided society. The historical divide, where race and class were virtually synonymous, is now cut across with the emergence of a black middle class and increasing stratification within the working class. For many white people, environment has been primarily about nature conservation. For many black people, environment was an elite white concern. However, most of those who live on the fenceline of polluting industry are black and working class and are increasingly resistant to carrying the costs of pollution.

These divisions are always at issue in any attempt at building broad campaigning coalitions and, on this ground, the elite representation that the environment must be played off against development – 'balanced' is the usual word – has traction. David Harvey (2005) argues that, during the 'golden age' of post-war capitalism, the exploitation of labour was the primary means of accumulation and this created a working-class politics for 'expanded reproduction'. The working-class gains made then have been severely eroded. Moreover, these gains were largely confined to the First World and were not shared by Third World workers. Now inequality is growing in all countries and the promises of development ring hollow. Yet they retain great power because there is no evident escape from dependence on capital: if there are no jobs on offer, then there is nothing but scavenging scraps from the world's overflowing rubbish dumps. Yet the numbers of those made destitute through the enclosures and externalities of accumulation by dispossession grows every day, while the potential for expanded reproduction within industrial capitalism shrinks and will collapse with peak oil.

Nevertheless, the tension that Harvey sees in contemporary struggles for justice between demands for expanded reproduction and demands for an end to dispossession grows sharper but also more ambiguous. Thus, for example, there is an evident tension in Nigerian demands for cheap petrol while the delta and its people are trashed. As argued in *The groundWork Report 2007*, the demand is not only about the price of transport but also resistance to oil corporations running off with windfall profits while people pay for it in higher costs. Similarly, in South Africa, a series of strikes have won wage increases above government's inflation target but below the actual rate of inflation, particularly for food, experienced by most workers.[12] For economists, this raised the threat of 'second round' inflation.

For workers, it was about barely keeping up with the cost of living. They had, moreover, seen corporate managers walking off with record multimillion-rand bonuses in 2006. Few economists saw inflationary dangers in these inflated rewards.

This dynamic will be terminally destructive in environmental, social and economic terms. It is rooted not in the demands of workers and consumers but in the massive appropriation by the elite classes that has created gross inequality globally and in most countries. At present, scarcity for some and plenty for others is engineered through the markets. As scarcity of energy, food and other goods becomes absolute with peak oil, and assuming that capitalist accumulation survives for some years past the date of peak oil, this dynamic will intensify. The contrast with the food rations that imposed equality, at least in the bare means of life, in Cuba could not be sharper.

Equality is no longer only about a utopian hope or the survival of the poor. It is now also a matter of planetary survival. The fundamental problem with Kyoto is that it evades this necessity and, to the contrary, promotes growing inequality and environmental injustice through its carbon-trading regime and the grandfathering of rights.

Movement

Given the urgency of climate change, many civil society organisations have concluded that those who now hold decision-making power – in the state system and capital – must be persuaded of the necessity for change. As corporations proclaimed themselves 'part of the solution', this strategy met with the dubious success of Kyoto, winning targets for carbon reductions, which, it was hoped, would be made more rigorous over time, but at the cost of acquiescing to the carbon-trading regime.

That the governments who claim to speak in the name of citizens, and the corporations who are given such powers by the nation states, must be confronted with the devastation of the future that they are bringing into being is clear. But the tactic of persuasion misses the logic of the power that it would persuade. As the Retort group argues:

> . . . right at the heart of capitalist modernity . . . has been a process of endless *enclosure*. The great work of the past half-millennium was the cutting off of the world's natural and human resources from common use. Land, water, the fruits of the forest, the spaces of custom and communal negotiation, the mineral substrate, the life of rivers and oceans, the very airwaves –

capitalism has depended, and still depends, on more and more of these shared properties being shared no longer, whatever the violence or absurdity in converting the stuff of humanity into this or that item for sale (2005: 193–4, original emphasis).

Retort goes on to quote war apologist Thomas Friedman: '. . . the hidden fist that keeps the world safe for Silicon Valley's technologies to flourish is called the US Army, Navy, Air Force and Marine Corps' (195).

In his compelling critique of carbon trading, Larry Lohmann argues:

Defining the climate crisis . . . as a problem to be solved through indefinite capital accumulation, state subsidies for large corporations and consultants, transnational capital flows, international trade and national 'development', makes it almost impossible to connect top-down emissions targets with support for effective actions at the local level (2006: 349).

For activists in the environmental justice tradition, the issue is not only about what decisions must be made, but about where decisions are made and who makes them. It is, in short, about the order of power in society.

Environmental justice requires a radical redistribution of rights from private capital and corporations to people. Across the world, the daily struggle over rights takes place every day at innumerable locations and on many fronts. The tactics used vary according to circumstances. In many cases, particularly in the South, it involves a direct clash as people's lands and resources are invaded and enclosed. In other cases, as with the transition towns, it involves softer tactics but still works for a transformation in power. This too will finally lead to a confrontation of power for it should not be expected that capital will relinquish its rights in people's lives without a fight – even if the fight is on the edge of the precipice of its self-destruction.

Central to this approach is solidarity with community and popular resistance to enclosure of people's commons and to the global institutions supporting enclosures, including the International Monetary Fund, World Bank, WTO and, indeed, Kyoto. Lohmann warns against 'silver bullet' solutions such as carbon trading that play into the hands of such powerful institutions and try to avoid the messy stuff of 'democratic political organising and an uphill political struggle' (337). Thus, on the question of the subsidies enjoyed by fossil-fuel corporations, he argues:

Powerful enough political movements could shift [subsidies] towards a coherent programme of, for example: renewable energy development; community-based planning for lower-carbon lifestyles; support for local movements protecting land, forests and smallholder agriculture; better insulation and heating; promotion of public debate and exchange on climate change; and just treatment for those who would otherwise suffer from the transition to less carbon-intensive industry, including fossil fuel workers and the poor (331).

This raises the fundamental question of whether the broad justice movement can indeed create 'powerful enough' movements. This may hopefully prove to be the real significance of Copenhagen. In the spirit, although not always the practice, of the World Social Forum, such movements would have to be created without replicating the vanguard activism that has hitherto provided the means of unified mass mobilisation but has also repeated, and often exaggerated, the authoritarianism that it seeks to undo. Such movements must therefore honour the specificity of local struggles and respect the leadership of local people in those struggles. Solidarity that turns to the colonisation of local struggles to the benefit of movement leadership is no solidarity.

In 2005, Oilwatch issued an open-ended invitation for dialogue with others to create 'powerful enough' movements. It is a powerful statement of the need and it does not duck the difficulties. And it may be observed that capital itself is linking together the crises that it engineers in people's lives. Thus, the turn to biofuels has the effect of repeating the invitation to dialogue between the movements for energy and food sovereignty.

Never before have the limits of the current development model based on hydrocarbons been so clear or close. Never before has the relationship between oil and the networks of power that control the world been so clearly understood, nor have the relationships between oil and the main causes of misery that affect humanity been so evident . . .

For the Southern part of the world, the oil model has meant the perpetuation of inequitable exchange, technological dependence, indebtedness, and impoverishment. The ecological debt between North and South, which began during the colonial years, rose with unequal economic and ecological exchange.

We have accepted separately each one of these aggressions. Or worse still, fought among ourselves: inhabitants of one country fighting against another, oil workers against indigenous communities, people from the North against those from the South, the poor of the cities against indigenous and peasant peoples . . . those that propose against those that criticize . . . And the list goes on and on.

What are the organizations and networks with whom we can start a positive collaboration in the fight against the oil civilization? What are the social, local and global movements that cannot be ignored in our efforts? What are the international agreements and programs that can best help us in this process? What are the new initiatives that we could and should devise?

To answer these and other needs, Oilwatch is inviting sympathetic networks to initiate a joint dialogue on our struggles and launch a global campaign against a civilization based on oil.

We invite you to share your opinions, comments, suggestions and ideas, to build a new path together . . . where we can reflect each and every one of our struggles. This way, each and every one of our battles will gain a new dimension.

Oilwatch, 16 September 2005

Box 10.1 Enough

Climate change has revealed our home on the planet to be fragile. We have reached the limits of what the planet can absorb in waste and pollution, and still remain liveable. 'If all the countries of the globe followed the industrial model, five planets would be needed to provide the carbon sinks required by economic development' (Sachs et al. 2002: 19). We have only one. The situation invites us to think about what is enough.

Continuing greenhouse gas emissions above the safe limit, with enough knowledge of the consequences and the means to change, is no less than climate crime. There is a difference between what carbon emissions, from cows and rice paddies releasing methane, are needed for subsistence and for necessary industrial production, and that which is superfluous, including elite travel and transport in search of the cheapest sweatshop labour. While it may not always be easy to draw the line between 'necessary' and 'criminal' activities, there is such a difference, embodied in the time-honoured and multifaceted concept of enough.

Enough means poor people must be able to enjoy more of the earth's riches, and rich people must not endanger everyone by consuming more than their share.

Enough – besides being a spiritual goal of not only material self-sufficiency but also happiness in oneself – also promises a series of non-monetary rewards: community, time, health and a clean conscience! In the Jo'burg Memo, prepared for the 2002 World Summit on Sustainable Development in Johannesburg, Wolfgang Sachs and his co-authors propose reducing consumption through 'wealth alleviation' instead of the inept meddling of 'poverty alleviation'. Better still would be the eradication of wealth that sustains unequal power relationships and starvation in a world of potential abundance.

'Enough' is not only a restraint. It is also autonomy in that, when enough is achieved, the overwhelming majority of people will not need anything from the capitalists because they will have reversed the enclosure of resources. 'Enough' is the hope that people can throw off dependency as the power of the current rulers erodes so that a fair sharing of the resources of the planet and those created by people becomes possible.

In the words of Peter Kropotkin:

That we are Utopians is well known. So Utopian are we that we go to the length of believing that the Revolution can and ought to assure shelter, food and clothes to all – an idea extremely displeasing to middle-class citizens, whatever their party colour, for they are quite alive to the fact that it is not easy to keep the upper hand of a people whose hunger is satisfied (1913: 69).

Notes

INTRODUCTION

1. 'People's action for corporate accountability', *groundWork Newsletter*, September 2002.
2. The Gini coefficient is a somewhat rough measure and the use of different datasets results in different numbers. Thus, Leibbrandt et al. (2010) show it rising from 0.66 to 0.70 between 1993 and 2008. There is general agreement, however, that inequality was rising over this period (Seekings 2007).
3. This figure was produced by the Presidency using a method developed by Servaas van der Berg, an independent analyst who has generally come to optimistic readings on poverty and inequality using 'controversial methods' (Seekings 2007: 5).
4. This is a classic case of 'technological enclosure': water is available to those with the most powerful pumps. The concept of enclosure is discussed under 'The global scale of ecological debt' in this chapter.
5. SAMWU Media Release, 20 February 2008.
6. 'Water contamination cited in E. Cape child deaths', SAPA, 30 April 2008.
7. The *Digest of South African Energy Statistics*, produced by the Department of Minerals and Energy (DME), is intended to give timely and accurate information on energy. Timely it is not. The first edition, which came out in 2005, and the 2006 edition give stats up to 2004. These are the latest available on the DME website. There are major discrepancies between its figures and those of the *Energy Outlook* produced in 2002. The DME said this is the result of improved information, minor changes to methods of collecting data and what appear to be substantial changes in assumptions used in allocating energy use to different sectors. In general, the *Digest* makes South Africa's energy performance, and the energy sector's environmental performance, look better than the *Energy Outlook* did.
8. A joule is a basic measure of energy. A petajoule is 1 000 000 000 000 000 joules and 3.6 PJ is equivalent to one terawatt hour (TWh), or 1 000 000 000 kilowatt hours (kWh), of electric energy.
9. Apart from Vanderbijlpark, ArcelorMittal has plants at Vereeniging, Newcastle and Saldanha.
10. Two plants are at Richards Bay. The Mozal plant outside Maputo is not included in the South African statistics but is supplied by Eskom. It consumes more power than the rest of Mozambique.

11. Treasury, Discussion paper for public comment: 'Reducing green house gas emissions: The carbon tax option', December 2010, p. 16.

12. In March 2011, the Department of Environmental Affairs estimated total greenhouse gas emissions of 540 million tonnes per year. This is up from 440 mt/y in 2003 and suggests per capita emissions of around the OECD average of 11 tonnes.

13. The rubbish figures are discussed in detail in *The groundWork Report 2008*. The CSIR report issued several cautions on the data: it relied on information from corporations, most of whom knew little about waste, had no waste information, and would only speak on condition of confidentiality because they thought the information might involve 'commercial risks' (1992: 5).

CHAPTER 1

1. William D. Hartung and Frida Berrigan, 'Militarization of US Africa policy, 2000 to 2005', factsheet prepared March 2005, World Policy Institute at www.worldpolicy.org.

2. Ba Karang, 'Africom and the USA's hidden battle for Africa', posted at *The Hobgoblin Online Journal*, 19 March 2010.

3. See *Green Left Weekly*, 11 May 2005, at www.greenleft.org.au.

4. See www.newamericancentury.org.

5. Naomi Klein, 'The rise of disaster capitalism', *The Nation*, 2 May 2005.

6. Shalmali Guttal, 'Reconstruction: An emerging paradigm', *Focus on Trade*, No. 107, February 2005, Focus on the Global South.

7. Quoted by Naomi Klein, 'The rise of disaster capitalism', *The Nation*, 2 May 2005.

8. Seminar given at Rhodes University: 'From Washington Consensus to Beijing Consensus', 2 May 2007. This figure represents state borrowing. It does not include equity investments in US stocks, profits and royalties returned to US corporations from foreign investments, never-ending Third World debt repayment or the corrupt collusion of Northern banks and Southern elites by which capital flight from the South is organised to the benefit of global finance capital.

9. Walden Bello, 'Chain-gang economics: China, the US, and the global economy', posted at Counterpunch, 1 November 2006.

10. In 2006, they were spending as much as $1.22 for every $1.00 that they earned.

11. Yale economist Jacob Hacker quoted by Ambrose Evans-Pritchard in 'Spending spree over as Americans walk tightrope without safety net', *The Telegraph*, London, 6 February 2007.

12. Video: *Enron: The Smartest Guys in the Room*, directed by Alex Gibney, HDNet Films, 2005.

13. Presentation: 'Finance and investment factors' to International Forum on Globalisation: Global Economic Transitions Project Seminar, 23–25 February 2007, London.

14. See Manuel Castells (2000a: 141) for evidence that these institutions were taking orders directly from the US Treasury.

15. Martin Wolf, 'Do not learn wrong lessons from Lehman's fall', *Financial Times* (www.FT.com), 15 September 2009.

16. Jeremy Thomas, 'Cats get fat on a diet of junk stocks', *Sunday Times* (Johannesburg), 9 August 2009.

17. Jack Grubman, Citigroup executive, quoted in Brenner 2003.

18. In October 2008, the derivative exposure of US banks alone was 'estimated at a breathtaking $180 trillion . . . three times the gross domestic product of all the countries in the world combined,' according to Ellen Brown, 'Financial meltdown: The greatest transfer of wealth in history', posted at www.opednews.com, 18 October 2008.

19. There is a special irony here. Johnson was part of the IMF hit squad sent out to discipline Indonesia during the Asian crisis and so support the transfer of Asian assets to Wall Street. He was not wrong about crony capitalism in that country but has evidently mistaken his part in the theatre of global capital.

20. Cited by Paul Krugman in his op-ed column in *New York Times*, 18 July 2008.

21. Janet Bush, 'Monetary jihad?' *The New Statesman*, 4 October 2004. See also Paul Harris, 'Petro-dollar or petro-euro?' *ASPO Newsletter*, No. 29, May 2003.

22. Robert Fisk, 'The demise of the dollar', *The Independent* (London), 6 October 2009. All parties predictably denied the story but Fisk is too good a journalist not to take seriously.

23. This is not, of course, a hard and fast divide. Some economists follow the conclusions of the geologists while some geologists take the industry position.

24. The reserve is the total amount of oil that it is possible to pump from the ground. The resource is the total amount of oil calculated to be in the ground. Typically, less than 40% of the resource can be extracted.

25. The list includes geologists, academics, investors, retired Middle Eastern officials, and even Volvo Trucks. Most have spent a lifetime in the field.

26. In general, oil analysts tend to privilege oil as the key driver of the broader economy. Rising or falling oil prices are thus seen as dictating economic health rather than as symptoms of broader political and economic relations. This view will become more convincing after peak.

27. Steven Mufson, 'OPEC says it will cut 1.2 million barrels a day', *Washington Post*, 20 October, 2006. Oil tycoon T. Boone Pickens describes Saudi surplus capacity as 'good for asphalt'.

28. Interview with *Le Monde* reported in *The Oil Drum: Europe* (www.europe.theoildrum.com), 28 June 2007.

29. Steve Connor, 'Warning: Oil supplies are running out fast', *The Independent* (London), 3 August 2009.

30. David Strahan, Guest Commentary, *ODAC Newsletter*, 28 August, 2009. In 2010, the IEA acknowledged that conventional oil production had already peaked around 2006.

31. *Financial Times*, 24 January 2006.

32. See Hirsch 2007.

33. *The Times* (London), 'World "cannot meet oil demand"', 8 April 2006.

34. Cited by Chris Martenson, 'It's official: The economy is set to starve', posted at www.energybulletin.net, 24 November 2010.

35. This suggests a much more rapid rise in sea levels than indicated by the Intergovernmental Panel on Climate Change Fourth Assessment Report (IPCC AR4 2007). McKibben (2007)

points out that the latest findings are not included in the IPCC report because the politics of its production keep it some years behind current research.

36. Sulphur dioxide aerosols reflect heat outwards and so reduce warming. Particulates visible as 'brown haze', on the other hand, absorb heat and so increase warming.

37. Roger Harrabin, 'How climate change hits India's poor', *BBC News Mobile*, 1 February 2007 at http://news.bbc.co.uk/2/hi/south_asia/6319921.stm.

38. The link was made by Louis Pasteur, Robert Koch and Ignaz Semmelweis in the second half of the nineteenth century.

39. Cremation was indeed the word used for incineration in the nineteenth century.

40. This represents a much smaller proportion of the population as fewer people live in wealthy households than in poor households.

41. Abahlali baseMjondolo, 'Statement on the xenophobic attacks in Johannesburg', 21 May 2008.

42. Abahlali baseMjondolo, 'Statement on the xenophobic attacks in Johannesburg', 21 May 2008.

43. Kennedy Road Development Committee (KRDC) Emergency Press Release, 27 September 2009.

44. Statement by Sibusiso Zikode, 'The ANC has invaded Kennedy Road', 29 September 2009. See also, Kerry Chance, 'The work of violence: A timeline of armed attacks at Kennedy Road', School of Development Studies Research Report No. 83, University of KwaZulu-Natal.

CHAPTER 2

1. Eskom was Escom – the Electricity Supply Commission – until 1987 when the Afrikaans acronym was preferred. Iscor was taken over by Mittal and renamed Ispat Iscor in 2004, then Mittal South Africa in 2005 and finally ArcelorMittal South Africa – following Mittal's takeover of European steelmaker Arcelor.

2. The estate was named after Lewis and Marks's Zuid Afrikaansche en Oranje Vrijstaatsche Kolen en Mineralen Vereeniging – the South African and Orange Free State Coal and Mineral Mining Association.

3. VOCs are also reported as 'non-methane hydrocarbons' in corporate reports.

4. Sasol Sustainable Development Report 2005: 58.

5. ArcelorMittal South Africa Sustainability Report 2008: 19. See also Mittal Steel South Africa, Annual Reports 2004 and 2005.

6. Eskom Annual Report 2005: 120.

7. The researchers can verify these symptoms from personal experience, even from relatively brief exposure during visits to the area.

8. Sasolburg Community Working Group Minutes, 31 August 2005. Ppb is parts per billion. See Sasolburg Community Working Group Minutes 28 June 2006 for 2005/2006 incidents.

9. groundWork and SAQMC letter to DEAT, 2 December 2006. DEAT made some recommendations to Sasol but there was no indication that their implementation would be monitored or, indeed, that the corporation was obliged to implement them.

10. Calculated from the daily average for February and March 2006, including raw and potable water, given in Sasolburg Community Working Group minutes for 26 April 2006.

11. Calculated on ArcelorMittal's reported consumption of 2.82 kilolitres per tonne of steel produced.

12. Speech by Mrs L.B. Hendricks, Minister of Water Affairs and Forestry, on Vaal River Eastern Sub-System Augmentation Project (VRESAP) function for the commissioning of part of the project, 1 April 2009. The minister noted with apparent satisfaction that Sasol uses '11.5 litres of water to produce one litre of petrol and Eskom requires 1.5 litres of water for every kilowatt-hour'.

13. Mittal Steel Vanderbijlpark and the Environment, brochure, 2006.

14. Annual Report 2004: 64. Pickle liquor is usually sulphuric or hydrochloric acid used to clean scale and oxides from steel.

15. *Beeld*, 17 January 2006.

16. Tony Carnie, 'Workers march with coffin', *Mercury*, 26 February 2008.

17. Sasol press release, 6 June 2006.

18. The apartheid work regime is described in *The groundWork Report 2006*. Karl von Holdt (2003: 33) gives a detailed account of *baasskap* in the iron and steel industry.

19. See Mittal South Africa Annual Reports for 2004 (p. 54) and 2005 (p. 45).

20. Meyerton falls outside Emfuleni. Vereeniging, Vanderbijlpark and all the major black settlements of the Vaal Triangle north of the river fall within Emfuleni.

21. The narrow definition of unemployment includes only those of working age who are actively seeking work. The broad definition includes those who would like to work but have given up looking. The 'economically active population' is the number of people employed added to the number unemployed. The 'potential' work force includes all adults of working age who are able to work.

22. This observation accords well with the results of the Durban Health Study (see Chapter 3). The findings suggest that long-term exposure makes people more vulnerable to further exposure. People's bodies do not get used to pollution.

23. This figure combines industrial, mining and power station sources. See Scorgie (2004: Table 5.1).

24. Industrial combustion does not include particulate dust blown off industrial waste and coal-heaps. See Scorgie (2004: Tables 3.35 and 3.36).

25. By comparison, Patrick Bond notes that in Europe people pay twice as much as industry (2005: 8).

26. 'The rich list', *Sunday Times*, 6 August 2006.

27. groundWork and VEJA air quality workshop, 4 May 2006.

28. Community exchange strategy meeting, Bosco Centre, 26–28 July 2006. The meeting included participants from south Durban, Secunda and other pollution hot spots.

CHAPTER 3

1. This view was advanced by, among others, the Industrial Strategy Project, see Joffe et al. 1993.
2. The ANC won the first democratic elections handsomely but invited other parties to join it in the Government of National Unity, which included the former ruling National Party.
3. 'Who is the dominant class in South Africa?' *Mail and Guardian*, 28 July–3 August 2006.
4. Davis notes that Habitat was conservative in its definition of slums and so also in numbering the people who live in them.
5. See Trevor Manuel in the *Sunday Times*, 13 August 2006.
6. Brent Johnson, 'The idiot's guide to EIAs', *Mail and Guardian*, 18–24 August 2006.
7. The technical definition of a recession is two-quarters of contraction. This, however, is merely a bureaucratic convention given a supposed objectivity by the word 'technical'.
8. The distinctions are not always clear, particularly when it comes to choosing ministerial cars.
9. Published in mid-2009 under the presidency of Zuma.
10. IODSA 2002: 'King report on corporate governance for South Africa'.
11. These struggles are documented in greater detail in *The groundWork Report 2008*. Primary sources include the archive of EJNF and groundWork newsletters.
12. See *The groundWork Report 2008*, which draws on Butler (1997) as well as EJNF and groundWork newsletters.
13. *The groundWork Report 2002*, p. 36.
14. Tony Carnie, 'Suffer little children', *The Mercury*, 10 September 2000.
15. In 2011, the vulnerability of the new air quality system was starkly revealed. Having painstakingly developed one of the best local monitoring and regulatory systems, Durban City Health abruptly cut the staff complement by more than half. Pollution control, it seems, has once more been abandoned.
16. 'From the smoke stack', *groundWork Newsletter*, March 2003.
17. Ambient limits are set for different timeframes and the concentration of each pollutant is averaged out over that duration. Short-term limits are much higher than long-term limits. Thus, for sulphur dioxide, the annual limit is 19 parts per billion, the daily limit is 48 and the ten-minute limit is 191. The standards system thus allows pollution peaks up to the relevant short-term limit. The assumption is that higher levels of exposure can be tolerated for shorter time periods.
18. PM_{10} is generally visible as smoke or dust. $PM_{2.5}$ is not visible and is produced from higher temperature combustion such as in industrial processes, incinerators or motor engines.
19. At present, Benzene, toluene, ethyl benzene and xylene (BTEX) are monitored in Durban.
20. Speech by Marthinus van Schalkwyk, Minister of Environmental Affairs and Tourism, at the Eighth Conference of the International Network for Environmental Compliance and Enforcement, Cape Town, 7 April 2008.
21. In 2002, the Air Quality Bill had already gone through six drafts. One of these was leaked to civil society organisations by a source in the business sector.

22. Parliamentary Monitoring Group, Meeting report on discussion on incineration, 4 March 2008, at www.pmg.org.za.

CHAPTER 4

1. Old Mutual followed by Sanlam and Liberty Life.
2. Damian Brett, 'Trends and players in the mining industry', presentation to UNCTAD Expert Meeting on FDI in Natural Resources, 20–22 November 2006.
3. Moody is managing editor of the Mines and Communities website at www.minesand communities.org.
4. The US uses imperial measures: hence 'tons' rather than metric 'tonnes'.
5. About 378 million litres.
6. Terence Creamer, 'Acid drainage response team to seek "affordable" solutions', *Engineering News*, 19 August 2010. The issue has received extensive coverage in *Mining Weekly* over the last two years and, in 2010, increased coverage in the mainstream media.
7. Department of Water Affairs, 'Final Report of the Interdepartmental Committee Regarding Dolomitic Mine Water: Far West Rand', November 1960 (known as the Jordaan Report).
8. Interview by Victor Munnik, 7 April 2008.
9. See 'South Africa: Paying the price for mining', *Irin*, 15 February 2008, at www.irinnews. org/Report.aspx?ReportId=76780.
10. AngloPlatinum produces 40% of global platinum supply and controls 60% of platinum resources.
11. Interview with Zav Rustomjee by Victor Munnik in May 2006. See also documents on the Competition Tribunal's website: www.comptrib.co.za.
12. *Business Report*, 25 August 2004. Iscor's name changed first to Ispat Iscor, then to Mittal Steel and finally to ArcelorMittal.
13. Terence Creamer, 'Minister Davies to seek urgent meetings with Kumba, ArcelorMittal and Anglo', *Engineering News*, 27 February, 2010. At the time of writing, the matter was not resolved.
14. In chemistry, 'organic' refers to compounds containing carbon. These are found in living organisms, as well as the hydrocarbons that make up fossil fuels and are used as the building blocks of plastics.
15. These figures, and those below, are given in the 'Draft Scoping Report' for an environmental impact assessment on the closure of the slag-heap and the opening of a new one (right next to it) at the Vanderbijlpark site, prepared by consultants Strategic Environmental Focus.
16. Maggie Fox, 'Air pollution damages across generations – study', Reuters, 11 December 2002, at www.planetark.com.
17. Jan de Lange, 'BHP pulls rug under metal firms', www.Fin24.com, 2 May 2010.
18. Mozambique's own energy production is chiefly from the Cabora Bassa Dam, which adds 2 000 MW to South Africa's 40 000 MW installed capacity. The World Resource Institute puts Mozambique's CO_2 emissions at 1.2 mt/y in 2000. Its overall CO_2e emissions

are put at 15 mt/y – but this includes methane from cattle farting and is really a measure of how the industrialised world is trying to make the non-industrial world co-responsible for climate change.

19. Statement issued by Pieter van Dalen MP, Democratic Alliance, 21 April 2010.

20. Sector-specific sustainability reports are not evident on Billiton's website for later years.

21. groundWork, 'Comments on: Environment Conservation Act, 1989: Waste tyre regulations', 23 March 2007.

22. F. Carrasco, N. Bredin and M. Heitz, (2002) 'Atmospheric pollutants and trace gases: Gaseous contaminant emissions as affected by burning scrap tyres in cement manufacturing' in *Journal of Environmental Quality* 31: 1484–90.

23. DEAT media statement, 'Green Scorpions embark on countrywide "clean cement" campaign', 27 May 2008.

24. Following the Beijing spectacular, Olympic officials challenged London to beat it.

25. Abahlali baseMjondolo (AbM), the Western Cape Anti-Eviction League, the Landless People's Movement and the Anti-Privatisation Forum, amongst others, made the point. AbM, for example, saw the KwaZulu-Natal Slums Act as the means to 'clean' the city of shack settlements ahead of the World Cup (press release 21 June 2007). AbM subsequently won a Constitutional Court ruling against the Act.

26. Martin Creamer, 'Raw export rise flies in face of beneficiation calls', *Engineering News*, 8 December 2010.

CHAPTER 5

1. The seven sisters were: Standard of New Jersey (later called Exxon), Socony-Vacuum (Mobil), Standard of California (Chevron), Texaco, Royal Dutch Shell, BP and Total. The first three were all successors to John Rockefeller's Standard Oil, the original monopoly broken up under US anti-trust legislation.

2. David Teather, 'Open skies deal will undo curbs on CO_2, say greens', *The Guardian* (London), 24 March 2007.

3. These figures are on the low side. The most recent research estimates 2005 emissions at 7.9 Gt of carbon, which is just short of 29 Gt of carbon dioxide (see Levin and Pershing 2007).

4. Total CO_2 emissions, including industrial process and agricultural emissions, in 2004 were just over 30 Gt while total greenhouse gas emissions were 49 Gt of CO2e, according to the IPCC 2007 report.

5. This rationality is discussed in relation to South Africa in Chapter 9.

6. Steve Connor, 'Carbon emissions set to be the highest in history', Reuters, 22 November 2010.

7. 'Summary for policymakers of the synthesis report of the IPCC Fourth Assessment Report', p. 21.

8. The authors intentionally made optimistic assumptions about reduced deforestation and reductions in greenhouse gases other than carbon dioxide. If these assumptions failed, neither 450 nor 550 would be possible.

9. In its discussion of the 450 scenario, the WEO says annual reductions of 0.67 Gt would be needed after 2020 (446) when it puts energy emissions at 33 Gt. This is just over 2%. Elsewhere, it says the 450 target requires that greenhouse gas emissions 'peak in the next few years followed by annual reductions of "6% or more"' (411). This does not seem to inform its number crunching.

10. Meinshausen et al., published in *Nature* in April 2009, cited by George Monbiot, 'Not even wrong', *The Guardian* (London), 31 August 2009.

11. WEO 2008 converts this to 400 ppm CO_2e [411, note 4], which implies that other greenhouse gas emissions are substantially reduced as indicated by Hansen et al. (2008: 12). For a high probability of avoiding 2 °C, Meinshausen calculates that stabilisation should be at 350 CO_2e (cited in Anderson and Bowes 2008).

12. Allen et al., published in *Nature* in April 2009, cited by George Monbiot, 'Not even wrong', *The Guardian* (London), 31 August 2009.

13. IEA, Oil Market Reports: 10 July 2009 and 13 October 2010. This report revised 2007 figures upwards as actual data came in higher than previously anticipated.

14. Colin Campbell, *ASPO Newsletter 97*, January 2009.

15. Associated Press, 'Rising demand may outpace Saudi output capacity', www.gulfnews.com, 5 April 2006.

16. Greg Muttitt, 'Iraq's oil field bid rounds – Development or "stabilization"?' *Middle East Economic Survey*, Vol. LII. No. 19, 11 May 2009.

17. Low-quality oil is more dense (thicker) and has a higher sulphur content. Refineries must 'crack' high-density oils to make petrol and diesel, a process that is both energy-intensive and highly polluting. A high sulphur content produces high sulphur dioxide emissions at refineries or, if the sulphur is removed in compliance with environmental regulations, ever-growing mountains of waste sulphur. Despite efforts to find new uses for sulphur, the market is already glutted.

18. John McQuaid, 'The Gulf of Mexico oil spill: An accident waiting to happen', posted at *Yale Environment 360*, 10 May 2010.

19. David Strahan, 'Pipe dreams', *The Guardian* (London), 3 December 2008. Strahan is a trustee of the Oil Depletion Analysis Centre (ODAC).

20. The solid wastes settle out in the ponds.

21. *H₂Oil* Factsheet: 'Health and human impacts' at http://h2oildoc.com.

22. Quoted in David Luhnow and Peter Millard, 'How Chávez aims to weaken US', *Wall Street Journal*, 1 May 2007.

23. See *The groundWork Report 2006*.

24. WEO's figures for gas are given in billion cubic metres. They are converted here to enable comparison with oil.

25. Chris Nelder, 'Step on the gas', *Energy and Capital*, 24 July 2009. See also Arthur Berman, 'Lessons from the Barnett shale suggest caution in other shale plays', posted at *Aspo-USA*, 10 August 2009.

26. Sunita Dubey, 'Trojan horse: Hidden costs of coal-to-liquids in the USA', *groundWork Newsletter*, June 2007.

27. Vaclav Smil, 'Long-range energy forecasts are no more than fairy tales', *Letter to Nature*, Vol. 453, 8 May 2008.

28. See *Creamer's Mining Weekly*, 16–22 March 2007.

29. Most Internet comment is by investment touts who talk up the investment opportunity in uranium created by the disaster. See, for example, Energy and Capital website (www.energyandcapital.com). See Cameco's website (www.cameco.com) for its version.

30. Julio Godoy, 'Niger: French state-owned company "poisoning" poor', Inter-Press Service News Agency, 12 April 2010.

31. A Greenpeace review of nuclear construction shows that 75 US reactors 'were predicted to cost $45bn but the actual cost was closer to $145bn'. Indian costs have similarly run 300% over budget. See Terry Macalister, 'Nuclear power a "dangerous distraction" says Greenpeace', *The Guardian* (London), 3 May 2007.

32. www.precaution.org/lib/nuke_ghg_emissions.060224.pdf.

33. 'Chernobyl radiation killed nearly one million people: New book', *Environmental News Service*, 26 April 2010. The book draws on East European scholarship not previously published in the West: Alexey Yablokov, Vassily Nesterenko and Alexey Nesterenko, 2010. *Chernobyl: Consequences of the Catastrophe for People and the Environment*, New York Academy of Sciences.

34. See *The groundWork Report 2006*, p. 159ff.

35. World Bank Press Release: 'This can be a century of African growth and opportunity, World Bank president says following Africa visit', 20 August 2009.

36. Brady Yauch, 'More odious debts for the Democratic Republic of Congo if the World Bank gets its way', *Probe International*, Wednesday, 19 August 2009.

37. Terri Hathaway, 'Congo's Inga: Great power for whom?' International Rivers Network, 1 August 2006; and 'Scramble to dam the Congo keeps Africans in the dark', Pan-African Press Association, 23 April 2008. See also 'International Rivers' at www.international rivers.org/en/africa/grand-inga-dam.

38. Republic of Tanzania website: 'The economic survey 2005'.

39. Originally the Owen Falls Dam on the Nile just below Lake Victoria. British colonial engineers blasted out the lip of Lake Victoria, so turning the whole lake into a reservoir subject to the management regime at the dam.

40. International Rivers Network (IRN) press release, 26 April 2007.

41. Ann David and Russell Gold, 'US biofuel boom running on empty', *Wall Street Journal*, 27 August 2009.

42. The study in question is Farrel et al. 2006 in *Science*, 311: 506–508.

43. George Monbiot, 'The most destructive crop on earth is no solution to the energy crisis', *The Guardian* (London), 6 December 2005.

44. Statement by organisations and social movements of Brazil, Bolivia, Costa Rica, Colombia, Guatemala and the Dominican Republic, gathered at a forum on the expansion of the sugarcane industry in Latin America, 28 February 2007. Available at www.mstbrazil.org.

45. Edivan Pinto, Marluce Melo and Maria Luisa Mendonça, 'The myth of biofuels', posted at www.viacampesina.org and www.mstbrazil.org, 13 March 2007.

46. 'Response to UK Department for Transport consultation on the draft Renewable Transport Fuel Obligation', from: *Africa Biodiversity Network*, Kenya; *Melca Mahiber*, Ethiopia; *Envirocare*, Tanzania; *Climate and Development Initiatives*, Uganda; *Nature Tropicale*, Benin. May 2007.

47. See Hussein Bogere, Emmanuel Gyezaho, Mercy Nalugo and Zurah Nakabugo, 'MPs arrested', *The Monitor* (Kampala), 17 April 2007; International Rivers Network, 'Fast facts on Mabira Forest, Uganda', 17 April 2007; Fred Pearce, 'Biofuel plantations fuel strife in Uganda', *New Scientist*, 19 April 2007; Xan Rice, 'Uganda "averts tragedy" with reversal of decision to clear virgin forest for biofuel', *The Guardian* (London), 29 October 2007.

48. Lester R. Brown, 'Supermarkets and service stations now competing for grain', *Eco-Economy Update*, 13 July 2006, and 'Massive diversion of US Grain to fuel cars is raising world food prices', *Eco-Economy* update, 21 March 2007 at www.earth-policy.org.

49. John Ross, 'The plot against Mexican maize: Big biotech takes advantage of corn crisis to force farmers to buy GM seeds', *The Independent*, 22 February 2007.

50. Sam Burcher, 'FAO promotes organic agriculture', Institute of Science in Society Press Release, 10 September 2007, at www.i-sis.org.uk.

51. Grain, 2008. 'Seized: The 2008 land grab for food and financial security', Briefing October 2008, at www.grain.org/go/landgrab.

52. Small-scale biodigesters are widely used in rural China and India. There are also a very few large-scale biodigesters. The City of Stockholm, for example, runs its buses and municipal vehicles off the gas.

53. Kate Ravilious, 'Weather hots up under wind farms', www.newscientist.com, 4 November 2004.

54. Tadit Anderson, 'The Nature of the Jevons Paradox', posted at www.economics.arawakcity.org on 7 May 2009.

CHAPTER 6

1. This and all subsequent references to the NEPAD document are to paragraph numbers rather than pages.

2. GCIS IRPS Media Briefing, 6 May 2005.

3. *The groundWork Report 2005* gives a detailed account of the ecological and human assault at every step in the upstream production process.

4. MEND statement quoted by Reuters, 'Oil blow to Nigeria as rebels hit offshore facility', 20 June 2008.

5. The calculation of the 'landed price' contains a number of fictional costs that turn into guaranteed profits for the oil refineries.

6. The oil supermajors are, of course, making super profits from crude oil but their refineries in South Africa still have to pay the going rate.

7. Quoted by Kevin Davie, 'The clawback which wasn't', *Mail and Guardian*, 4–10 August 2006. The information on subsidies is mostly based on a number of articles by Davie in the *Mail and Guardian*: 'Govt's R6bn gift to Sasol', 23–29 September 2005; 'From Ogies

with love', 4–10 August 2006; 'No windfalls here', 18–24 August 2006. See also Hallowes 2005.

8. Deregulation is the word used in the policy. Reregulation would perhaps be more accurate since government would effectively empower a limited number of very powerful corporations. Government regulation in phase 3 would then mediate between these corporations.

9. Muzi Mkhize, 'Chief director: Hydrocarbons, is South Africa ready for market pricing/ deregulation?' Oil Summit, Midrand, 9 October 2008.

10. Ann Crotty, 'Fierce rivalries keep powerful oil players in check', *Business Report*, 2 November 2005.

11. See SAPIA Annual Report 2006, Appendices 1 and 2. Subsequent annual reports do not give the industry's financial results or value-added statement, purportedly because this requires information sharing and so facilitates price collusion.

12. OCGT plants gulped 346 million litres of diesel in 2008. This was reduced to 29 million as economic recession relieved Eskom's spinning margins.

13. This calculation assumes a strong rand at seven to the dollar and a weak rand at eleven to the dollar. The rand could of course go much lower.

14. Formerly Kellogs, Brown and Root.

15. Sasol presents the information in a bar chart. These numbers are the best reading I can make of it.

16. This is the last year that *The groundWork Report* was able to access figures for local site emissions as opposed to global emissions.

17. Sasol's 2005 Sustainable Development Report said that such information was available on request. When requested, corporate officials acknowledged that this was 'a bit misleading' as it wasn't.

18. Sasol is a signatory to the UN Global Compact initiated by General Secretary Kofi Annan and widely criticised by environmental justice organisations as corporate 'blue wash'.

19. Sasol's 2009 SDR says nothing about it and the Sasolburg SH&E report has not been updated since 2006.

20. E-mail, Sandra Redelinghuys to Bobby Peek, 5 November 2008.

21. R.J. Morris, 2003. 'Sulphur surplus in the making impacts refineries', The Sulphur Institute.

22. Organic chemicals are based on carbon, inorganic chemicals are not.

23. At www.plasfed.co.za.

24. Packaging Council of South Africa submission, 15 November 2007.

25. At www.stopcorporateabuse.org.

26. eThekwini Municipality 2004, 'Integrated Waste Management Plan'. eThekwini has subsequently introduced recycling with privatised curbside collection thus far confined to upmarket areas. It is motivated primarily by the need to reduce the municipal waste stream and so save 'air space' at the dumps. Waste managers have next to zero influence on the production system and hence on how much waste is produced in the first place.

CHAPTER 7

1. Louise Flanagan, 'Worst load shedding day so far', *Daily News*, 25 January 2008.
2. Interview with Wendy Annecke and Nthabiseng Mohlakoana by David Hallowes and Victor Munnik, 20 April 2007.
3. Minister Alec Erwin, 'Economic cluster: Higher growth, sustained growth, and shared growth', Parliamentary Media Briefing, 17 February 2005.
4. Siseko Njobeni, 'Minister's electricity revenue shock for cities', *Business Day*, 4 March 2009.
5. In 2011, government abruptly cancelled the REDs project so it appears that the distribution industry will not in fact be changed any time soon.
6. Figures are taken from Sustainable Energy Africa 2003 and DME's Energy Statistics 2006.
7. City of Cape Town, *State of Cape Town 2006*, p. 37.
8. 'Towards 2010: Going global leaves poor further behind', *Cape Argus*, 23 May 2007.
9. 'Down with ANC policies!' www.news24.com, 17 July 2007.
10. *NERSA Newsletter*, August 2006. The finding did not refer to the bolt in the rotor incidents which, according to NERSA, was subject to another government enquiry.
11. Areva press kit, 'SA government reception for the first group of engineers returning from training with Areva', 2 July 2003.
12. 'Western Cape electricity crisis: Declaration of meeting of 8 March 2006'.
13. Interview, 18 April 2007.
14. 'Western Cape recovery plan', p. 23.
15. Open letter to the minister and officials from DME, NERSA, Eskom, CEF and city electricity departments from Sustainable Energy Africa and twenty other organisations, 26 June 2006.
16. The US Environmental Protection Agency standard is 5 milligrams, so 5 million CFLs contain 25 kilograms of mercury. However, poor-quality CFLs contain more. Mercury is an excellent conductor of electricity and has found many applications but has also caused severe health problems for workers. See D'Itri and D'Itri 1977.
17. Micheal Sheridan, '"Green" lightbulbs poison workers', *The Sunday Times* (London), 3 May 2009.
18. The Cape Town Partnership's advocacy on behalf of CBD businesses is evident in an email from CEO Alex Boraine to the PMT, 9 June 2006.
19. Interview with David Nicol, 18 April 2007.
20. Reported in the *Cape Times*, 31 March 2006.
21. *Cape Business News* quoted in Von Ketelholdt (2006: 15).
22. These farms and cellars are a minority but represent the most profitable leading edge of the industry. Bulk wine producers have struggled to adapt to the new order and are now feeling the competition from imports.
23. Willemse, Power Point presentation: 'Economic position of agriculture in Western Cape', 2006.
24. Ewart and Du Toit (2005) report increasing use of mechanised harvesting, with each machine replacing 70 pickers and enabling quicker harvesting when sugar levels are

right. However, this advantage is offset because mechanical harvesting cannot discriminate for grape quality and also results in foreign matter in the harvest.

25. Interview with Carl Opperman, CEO Agri-Weskaap, 20 April 2007.
26. See FTFA website at www.trees.co.za. PricewaterhouseCooper's own website is silent on its accreditation.
27. City of Cape Town, 2006, 'Energy and Climate Change Strategy'.
28. Obama chose the supposedly 'good' war in Afghanistan over the 'bad' war in Iraq. The withdrawal from Iraq, however, involves a sleight of hand. Large US bases remain in remote areas along with the green zone defined by the US embassy precinct in Baghdad.

CHAPTER 8

1. Louise Flanagan, 'Eskom documents show business knew of the crisis', *The Star*, 11 February 2008.
2. James Myburgh, 'Eskom: The real cause of the crisis', *Politicsweb*, 1 February 2008.
3. Eskom's response is carried as an annex to NERSA's report, p. 44ff. The utility rather quaintly avoids understanding NERSA's criticism that national energy security had been subordinated to its 'business objectives' – that is, promoting BEE entrants, cutting short-term costs and boosting management bonuses.
4. Terence Creamer, 'Power prices have to begin reflecting true costs – Godsell', *Engineering News*, 6 February 2009.
5. See Terence Creamer, 'Electricity crisis response team hits turbulence', *Engineering News*, 3 November 2009, and 'Government at fault for inertia around power-crisis structures, Manuel admits', 8 December 2009.
6. The MYPD stipulates annual price increases over a three-year period, supposedly to bring certainty to both Eskom and its customers.
7. See McDaid 2010; Lynley Donnelly, 'Cloud over power plan', *Mail and Guardian*, 19 March 2010; Chris Yelland, 'National integrated resource plan for electricity, or conflict brewing?' *EE Publishers*, 22 April 2010.
8. Louise Flanagan, 'Eskom couldn't afford to buy back power – Maroga', *Business Report*, 31 March 2010.
9. BHP Billion statement quoted in 'BHP, Eskom review aluminium smelter power agreements', *Engineering News*, 5 April 2010.
10. This estimate was based on the assumed cost of 'unserved' energy and not on any direct evidence. See NERSA, 'Inquiry into the national electricity supply shortage and load shedding', 12 May 2008, p. 8.
11. Neva Makgetla, 'SA can't afford to neglect energy needs of mines', *Business Day*, 25 June 2008. Makgetla is now at the newly formed Department of Economic Development.
12. The source of this quote is quaintly appropriate: Ian McDonald, 'Hitting the sweet spot', *South Africa: The Good News*.

13. 'Where could emerging-market contagion spread next?' *The Economist*, 26 February 2009.

14. In the 1990s, the carry trade was sourced from Japan as it attempted to revive its stagnant economy with near zero interest rates. It is an integral feature of Ponzi capital and its replication on a larger scale is symptomatic of the global attempt to revive the bubble economy. The logic of doing so is that there is no other viable basis for growth – which is to say finally that there is no viable basis for growth.

15. 'Trade gap jumps to record on oil imports', Reuters, 30 November 2007.

16. Tumi Makgetla, 'Forget oil, look at food prices', *Mail and Guardian*, 22 October 2006; and 'Food price inflation under scrutiny', *Mail and Guardian*, 20 May 2007; Neva Makgetla, 'Hunger stalks poorer households as maize prices fuel inflation', *Business Day*, 6 June 2007; 'Farmers expected to plant more maize', *Business Report*, 25 September 2007.

17. A. d'Angelo, 'Politics stands to undermine effectiveness of World Summit', *Business Report*, 8 May 2002. Rio refers to the 1992 United Nations Conference on Environment and Development in Rio de Janeiro.

18. Address by the President of South Africa, Thabo Mbeki, at the Sixty-Second Session of the United Nations' General Assembly, New York, 25 September 2007.

19. See Omnia's Annual Report 2008. One tonne of nitrous oxide is equivalent to 310 tonnes of carbon dioxide, so such projects provide cheap credits and have been favoured in the carbon market. The return no doubt declined with the carbon price in 2008 but by then it was money for free.

20. 'Sasol clinches world first with emissions reducing project', *Engineering News*, 23 July 2007; and *Sasol Facts 2009*.

21. Robert Zoellick, President of the World Bank Group, speech to the Twelfth Ordinary Session of the Assembly of the African Union, Addis Ababa, Ethiopia, 9 February 2009.

22. World Bank (IBRD, IFC and MIGA), 'Country partnership strategy for the Republic of South Africa for the period 2008–2012', 12 December 2007, p. 4.

23. 'Alec Erwin, Minister of Public Enterprise: Generating electricity', 31 May 2006 at www.dpe.gov.za.

24. 25 June 2007.

25. Letter to the *Eastern Cape Herald*, 15 June 2007.

26. CDC Annual Reports, 2008/2009 and 2007/2008.

27. Terence Creamer, 'Powering down', *Engineering News*, 7–13 September 2007.

28. In 2009, Sere was 'delayed' to shave a splinter off the funding requirement. See *Engineering News*, 5 June 2009 and Eskom's *New Build News*, No. 7, March 2009.

29. Chanel Pringle, 'Electricity decision needed in the next year – Eskom', *Engineering News*, 7 September 2009.

30. Hilton Trollip, IDASA presentation on the IRP 2010, Durban, 22 October 2010; the DOE's draft IRP, published in October 2010, put forward several scenarios and recommended a 'revised balanced scenario'. This was again revised following a round of public consultation. The figures quoted here are from the 'policy adjusted IRP' contained in the final report approved by Cabinet in March 2011.

31. 'MTRM', p. 13. MTRM assumes even higher growth to 2016 than IRP.

32. Terence Creamer, 'Secunda to produce 800 MW of own power, sell 200 MW to Eskom', *Engineering News*, 21 October 2010.

33. Keith Campbell, 'SA squanders lead just as nuclear sector begins to grow globally', *Engineering News*, 26 March 2010.

34. Eskom CEO Jacob Maroga quoted by *Engineering News*, 27 July 2007. The comment echoes the notorious internal memo circulated by World Bank official Lawrence Summers, in which he argued that poor countries were under-polluted and 'the economic logic behind dumping a load of toxic waste in the lowest-wage country is impeccable . . .'.

35. DWA used to be DWAF before Zuma's reshuffle. Then, for a short while, it was DWEA (water and environment), then the departments did not amalgamate after all, although the ministries did, so now DWAF is DWA and DEAT is DEA (minus tourism).

36. For details of DWA planning, see 'Mokolo and Crocodile water augmentation project: Presentation to the Waterberg forum meeting', 23 November 2009. Broader plans are given in *Strategic Planning for Water Resources in South Africa, 2009*.

37. Transformation Resource Centre, 'Too many dams, too little water – Lesotho's rivers could become "waste water drains"', www.internationalrivers.org, 31 October 2000.

38. 'Emerging energy heavyweight', *Mining Weekly*, 11–17 August 2006.

39. 'Eskom says SA needs "at least" 40 new coal mines', SAPA, 11 August 2009.

40. Esmarie Swanepoel, 'Over 100 South African mines operating without water licences', *Engineering News*, 29 September 2009.

41. Terence Creamer, 'Transnet to spend R52 billion on rail unit in the next five years', *Engineering News*, 16 March 2010.

42. Sonal Patel, 'Powering the people: India's capacity expansion plans', *Power Magazine* at www.powermag.com, 1 May 2009.

43. Driven by the stimulus package, China's net imports for 2009 came in at over 100 mt, putting it in third place behind Europe and Japan, as compared with net exports of 34 mt in 2006. More moderate imports are expected over the next few years.

44. Northern country export credit agencies guarantee debt to secure contracts for their home industries – in this case for the boilers and turbines for Medupi and Kusile. They eliminate the risk to banks, effectively taking over unpaid debts, but not to the recipient country. They now hold a substantial proportion of Southern debt.

45. Calculated at R7.5 to the dollar.

46. This was the figure generally attributed to Eskom sources. Manuel put it at a marginally more modest R356 billion in the 2009 budget speech.

47. A power station with six boiler-generator sets.

48. The revised MYPD2 application (30 November 2009) compares spending based on the 45% and the 35% tariff applications. It gives total capital spend for the seven years from 2008/2009 to 2014/2015 as R755 billion in the 45% case (p. 64) and R619 billion in the 35% case (p. 65). Eskom's 2008 Annual Report gives total spending to 2007/2008 as 26 billion (p. 64).

49. Revised MYPD application p. 35ff.

50. Budget speech 2009.

51. Eskom, 'Presentation to the portfolio committee on public enterprises, funding model and World Bank loan', 4 May 2010.

52. Quoted by Terence Creamer, 'SA would "cope" should World Bank fail to grant Eskom loan', *Engineering News*, 1 April 2010.

53. Responding to requests for information from the Naspers Media24 group, Eskom said disclosure would be detrimental to Billiton's commercial interests. See, amongst others, Carli Lourens, 'Naspers sues Eskom to disclose BHP power discount', *Bloomberg*, 18 March 2010; SAPA, 'Hogan doesn't give Eskom deal details', *The Times* (Johannesburg), 19 March 2010.

54. Quoted by Tony Carnie, 'Sour taste over sweetheart deals', *The Mercury*, 18 March 2010. See also Melanie Gosling, 'Eskom's secret business deals come to light', *Business Report*, 10 March 2010.

55. Earthlife Africa, 'Sustainable energy briefing 18: Eskom costs and tariffs, 2009', submission to NERSA on Eskom's MYPD2 application, 30 November 2009.

56. Rob Rose, 'Chancellor House cloud over $3.75bn Eskom loan', *Business Times*, 11 April 2010.

57. Personal communication, Sarwat Hussain, Senior Communications Officer, Africa Region External Affairs, World Bank, 2 June 2009.

58. Eskom, 'Application for an interim price increase', April 2009, p. 7.

CHAPTER 9

1. Copenhagen Accord, UNFCCC, 18 December 2009.

2. La Via Campesina Press Release, 'Traders failed in Copenhagen: The future lies in people's hands', Copenhagen, 19 December 2009.

3. The critique of the LTMS is based on a paper written for the Sustainable Energy and Climate Change Project of Earthlife Africa, Johannesburg (Hallowes 2008a).

4. See Lohmann (2006: pp. 57ff and 84ff).

5. There is no real cap without it being global. Without the US and the big Southern countries, cap-and-trade has no basis – even if it is believed to work in principle.

6. Sinks such as forests, land, oceans and ice absorb carbon and so prevent it entering the atmosphere. CDM sink projects are focused on tree plantations, begging several questions including: Whose land will be taken to create the sink? What confidence can there be that the carbon will stay sunk? Is carbon emitted by industry in fact equivalent to carbon circulating through ecological systems?

7. Sasol reports 'good earnings' from reducing nitrous oxide emissions (Annual Review 2009: 61), but reports the full reduction in its performance data (Sustainable Development Report 2009: 56).

8. Denial was subsequently reasserted by the Republican 'tea party' and the election of several representatives to Congress, backed by corporate financing, who made good their promise to block climate legislation. That the proposed legislation, based on cap-and-trade, was itself regressive is another question.

9. Ma Kai, head of the National Development and Reform Commission, quoted by Jonathan Watts, *The Guardian* (London), 4 June 2007.

10. Pan Jiahua, Chen Ying and Li Chenxi, 'Balancing the carbon budget', posted at *WorldChanging*, 15 December 2009. The 'Carbon budget proposal' assumes global peak emissions in 2020 and a 50% reduction on 2005 emissions by 2050. This indicates that the budget of 2.33 tonnes CO_2 per person per year for the 150 years is overly generous. The authors propose that the budget be balanced through a combination of reductions and financial transfers.

11. George Monbiot, 'Copenhagen negotiators bicker and filibuster while the biosphere burns', *The Guardian* (London), 19 December 2009.

12. As is so often the case, the title REDDs belies the real content. For a short critique, see the World Rainforest Movement, 'From REDD to HEDD', November 2008.

13. As well as 'displacing' coal, the landfill gas projects convert methane (a more powerful GHG) into carbon dioxide, resulting in an overall decline in CO_2e. Lohmann (2008) argues that the equivalence of other GHGs to CO_2 (e.g. 1 tonne methane = 22 tonnes CO_2) is itself arbitrary, but making this equivalence was necessary for a carbon market to function.

14. George Monbiot, 'So far the Bali deal is worse than Kyoto', *The Guardian* (London), 17 December 2007.

15. The boundary between production and finance is now so blurred that this is not strictly a choice.

16. Undated document titled 'Landfill gas to electricity project', prepared by L. Strachan (eThekwini Municipality), B. Couth (Consultant) and R. Chronowski (Prototype Carbon Fund, World Bank).

17. Lohmann (2006: 219ff.) documents a series of carbon-offset land-grabs in Latin America, Africa and Asia. For the Mount Elgon case, see also Stephan Faris, 'The other side of carbon trading', *Fortune Magazine*, 29 August 2007.

18. Coal subsidies are enumerated in Johnston et al. 2010.

19. The US statement at Copenhagen, reiterated in its letter signing on to the Copenhagen Accord, said it would reduce emissions by 17%. This, however, is against the baseline year of 2005 as against the 1990 baseline used by all other Annex I countries.

20. Letter to World Bank president Robert Zoellick signed by nine Southern country executive directors (undated).

21. Memorandum to the Government of India on the UNFCCC's Fifteenth Conference of the Parties at Copenhagen, signed by the National Alliance of People's Movements and eighteen other organisations, 24 November 2009.

22. The State Department leads US climate negotiations but the Treasury Department has been intimately involved from the start.

23. Walden Bello, 'The anti-climate summit', posted at *truthout*, 15 July 2008.

24. Walden Bello, 'The environmental movement in the Global South: The pivotal agent in the fight against global warming', *Focus on the Global South*, 12 October 2007.

25. Note to Bolivian Press Release, 'Bolivia rejects US blame game on Copenhagen and calls for the people to decide', 18 January 2010.

26. Tina Gerhardt, 'Bolivia's people's conference calls for system change, not climate change', posted at *Grist*, 25 April 2010.
27. Letter to UNFCCC Executive Secretary from Alf Wills on behalf of the South African Focal Point, 29 January 2010.
28. The LTMS (2007) produced three documents known as the Scenarios Document (SD) that gives the conclusions to the study, the Technical Report (TR), which discusses methods and gives detailed findings, and a Process Report (PR). Discussion of the LTMS is based on a more detailed paper prepared for Sustainable Energy and Climate Change Project of Earthlife Africa Johannesburg (Hallowes 2008).
29. LTMS produces a graph without giving the actual figures for these dates, but this approximation was subsequently confirmed by Department of Environmental Affairs figures presented to NEDLAC.
30. Department of Environmental Affairs, 'South Africa's desired greenhouse gas mitigation outcomes – to define or not to define'. Presentation to the 3rd meeting of NEDLAC National Climate Change Response Green Paper Joint Task Team, 25 March 2011.
31. See Table SPM.5 in AR4 Working Group III report on mitigation, p. 23.
32. The effect of climate policy is specifically excluded from the scenarios although policies addressing other environmental concerns, primarily pollution, are included.
33. For a critique on the inflation index, see Dick Forslund, 'Official inflation reports and "the government of the wealthy"', *Amandla*, March/April 2010, Issue No. 13. The method of calculating inflation is actually called 'plutocratic indexing' – hence 'government of the wealthy'.
34. IRP puts renewables capacity at 14.5% in 2030 but production is lower from variable renewables than it is from constant base load.
35. Eskom got around to developing a climate strategy in 2005.
36. This means that the electric energy produced by the new power stations will amount to 40% of the primary energy embodied in the coal.
37. Eskom's interest in Underground Coal Gasification originates with poor planning for its Majuba plant. Majuba was designed as a pithead power station but a fault in the coal seam made the proposed mine unviable. Coal is now trucked in by road at considerable economic and environmental cost. Gasification would enable Eskom to use the original coal resource to fuel the plant. Underground Coal Gasification involves controlled burning of the coal in situ in a low-oxygen environment – much the same technique as is used to produce charcoal. It thus replaces the entire mining operation and is being considered for other areas where coal is difficult or expensive to extract including deep deposits on the Waterberg. Long-term environmental costs, including the possibility of uncontrolled underground fires, are uncertain.
38. Quoted in Earthlife Africa's submission to NERSA on Eskom's MYPD2 application, 30 November 2009.
39. Carbon savings are represented on a graph (p. 70) that cannot be precisely read.

CHAPTER 10

1. Catherine Brahic, 'Organic farming could feed the world', *New Scientist*, 12 July 2007.
2. Institute of Science in Society (ISIS) Press Release, 10 September 2007. The report is titled 'Organic agriculture and food security'. It does not represent the dominant view within FAO but rather a challenge to that view.
3. *China Daily*, 14 June 2006.
4. Bagasse is organic waste from sugar and is commonly used to power sugar mills. Its alternative use is as a compost, indicating the choice of power or soil fertility.
5. SDCEA, 'Comments on the draft central spatial development plan', 20 October 2009.
6. Amit Sengupta, 'The train stops at Nandigram', *Hardnews*, April 2007 at www.hardnews media.com.
7. Press release issued by Richard Spoor, attorney for the Mohlohlo Community, 'Anglo Platinum mining community members acquitted', 31 July 2009.
8. Maandagshoek Development Committee, Press Statement, 9 September 2007.
9. Personal communication.
10. Email network communication, 6 August 2007.
11. Email network communication, 15 July 2007.
12. Inflation on those goods that the poor spend most of their money on is higher than official inflation.

Select bibliography

ActionAid. 2008. 'Precious metal: The impact of Anglo Platinum on poor communities in Limpopo, South Africa' (Mark Curtis). Johannesburg: ActionAid.

Alexander, N. 2009. 'Let us return to the source: In quest of a humanism of the twenty-first century'. Sipho Maseko Memorial Lecture, 8 October 2009, University of the Western Cape, Cape Town.

Allison, I., N. Bindoff, R. Bindschadler, P. Cox, N. de Noblet, M. England, J. Francis, N. Gruber, A. Haywood, D. Karoly, G. Kaser, C. le Quéré, T. Lenton, M. Mann, B. McNeil, A. Pitman, S. Rahmstorf, E. Rignot, H. Schellnhuber, S. Schneider, S. Sherwood, R. Somerville, K. Steffen, E. Steig, M. Visbeck and A. Weaver. 2009. *The Copenhagen Diagnosis, 2009: Updating the World on the Latest Climate Science*. Sydney: University of New South Wales Climate Change Research Centre (CCRC).

Alston, D. 1993. 'Environment and development: An issue of justice'. In *Hidden Faces: Environment, Development, Justice: South Africa and the Global Context*, ed. D. Hallowes. Scottsville: Earthlife Africa.

Altvater, E. 2006. 'The social formation of capitalism, fossil energy, and oil-imperialism'. Colloquium on the Economy, Society and Nature, Centre for Civil Society, University of KwaZulu-Natal, Durban.

Anderson, K. and A. Bows. 2008. 'Reframing the climate change challenge in light of post-2000 emission trends'. Philosophical Transactions of the Royal Society. doi:10.1098/rsta.2008.0138, published online at http://rsta.royalsocietypublishing.org/content/366/1882/3863.full.pdf+html.

Annecke, W. 2006. 'The ethics and economics of including women in the ESI'. In *ESI Africa* 3.

Arrighi, G. 2005. 'Hegemony unravelling–1'. In *New Left Review* 32, March–April 2005.

———. 1994. *The Long Twentieth Century: Money, Power and the Origins of Our Times*. London: Verso.

Baird, E. 2003. *World Petroleum Congress Handbook 2004*. Fifth Dewhurst Lecture given to the Seventeenth World Petroleum Congress on 3 September 2002, Rio de Janeiro, Brazil.

Bénit, C. and P. Gervais-Lambony. 2005. 'The poor and the shop window: Globalisation, a local political instrument in the South African city?' In *Transformation* 57.

Bertelsen, E. 1998. 'Ads and amnesia: Black advertising in the new South Africa'. In *Negotiating the Past: The Making of Memory in South Africa*, eds S. Nuttall and C. Coetzee. Cape Town: Oxford University Press.

BHP Billiton. 2006. *Aluminium Customer Sector Group (CSG) Sustainable Development Report 2005–2006*. BHP Billiton.

Bond, P. 2005. 'What's wrong with our energy system?' In *Trouble in the Air*, eds P. Bond and R. Dada. Durban, Centre for Civil Society, University of KwaZulu-Natal.

Brenner, R. 2003. 'Towards the precipice'. *London Review of Books*, Vol. 25, No. 3.

Butler, M. 1997. 'Lessons from Thor Chemicals'. In *The Bottom Line: Industry and the Environment in South Africa*, eds L. Bethlehem and M. Goldblatt. Cape Town: University of Cape Town Press and International Development Research Centre.

Campbell, C. 2008. *Newsletter 90*, June 2008. Association for the Study of Peak Oil (ASPO), Available at www.peakoil.ie.

———. 2007. *Newsletter 75*, March.

Canadian Centre for Policy Alternatives (CCPA). 2006. 'Fuelling fortress America: A report on the Athabasca tar sands and US demand for Canada's energy'. H. McCullum for CCPA.

Castells, M. 2000a. *The Information Age: Economy, Society and Culture, Volume I: The Rise of the Network Society*. Second edition. Oxford: Blackwell.

———. 2000b. *End of Millennium. The Information Age: Economy, Society and Culture, Volume III*. Second edition. Oxford: Blackwell.

Chipkin, I. 2003. 'The South African nation'. In *Transformation* 51.

Christian Aid. 1999. *Who Owes Who? Climate Change, Debt, Equity and Survival*. London: Christian Aid.

Cock, J. 2001. 'Gun violence and masculinity in South Africa'. In *Changing Men in South Africa*, ed. R. Morrell. Pietermaritzburg: University of Natal Press and London: Zed Books.

Cock, J. and V. Munnik. 2006. *Throwing Stones at a Giant: An Account of the Struggle of the Steel Valley Community against Pollution from the Vanderbijlpark Steel Works*. Durban: Centre for Civil Society, University of KwaZulu-Natal.

Danmarks Naturfredningsforening (DN) and South Durban Community Environmental Alliance (SDCEA). 2003. *Comparison of Refineries in Denmark and South Durban in an Environmental and Societal Context: A 2002 Snapshot*. Durban: SDCEA and DN.

Davis, D. 2002. *When Smoke Ran Like Water: Tales of Environmental Deception and the Battle against Pollution*. New York: Basic Books.

Davis, M. 2004. 'Planet of slums'. In *New Left Review* 26, March/April.

———. 2002. *Late Victorian Holocausts: El Nino Famines and the Making of the Third World*. London: Verso.

De los Reyes, R. 2006. *Impacts of Coal Plants on Communities.* World Wildlife Fund.

Department of Environmental Affairs and Tourism (DEAT). 2007. Long-Term Mitigation Scenarios (LTMS). *Strategic Options for South Africa* (the Scenarios Document or SD); *Technical Report* (TR), ed. H. Winkler, Energy Research Centre Report for the DEAT.

———. 2006. *South Africa Environment Outlook: A Report on the State of the Environment.*

———. 2004. *A National Climate Change Response Strategy for South Africa.*

Department of Minerals and Energy (DME). 2007. *Energy Security Master Plan: Liquid Fuels.*

———. 2006. *Digest of South African Energy Statistics.*

———. 2005. *Digest of South African Energy Statistics.*

———. 2004. *Status of Radio-Active Waste Management in South Africa.*

———. 2002. *Energy Outlook.* DME, Eskom, Energy Research Institute, University of Cape Town.

———. 2001. *Electricity Distribution Industry Blueprint Report.*

Department of Trade and Industry (DTI). 2010. *Industrial Policy Action Plan* (IPAP2).

———. 2007. *Implementation of Government's National Industrial Policy Framework: Industrial Policy Action Plan.* September.

———. 2002. *Accelerating Growth and Development: The Contribution of an Integrated Manufacturing Strategy* (IMS).

Desai, A. and R. Pithouse. 2004. *'But We Were Thousands': Dispossession, Resistance, Repossession and Repression in Mandela Park.* Durban: Centre for Civil Society, University of Natal.

D'Itri, P and F. D'Itri. 1977. *Mercury Contamination: A Human Tragedy.* New York: John Wiley and Sons.

Dobreva, R. 2006. *Trade and Poverty in South Africa: The Link between Trade and Poverty: The Case of Polymers.* Cape Town: University of Cape Town, Southern Africa Labour and Development Research Unit (SALDRU).

Down, C.G. and J. Stocks. 1977. *Environmental Impact of Mining.* London: Applied Science Publishers.

Doyle, J. 2002. *Riding the Dragon: Royal Dutch Shell and the Fossil Fire.* Boston: Environmental Health Fund.

Dugard, J. 2010. 'Power to the people? A rights-based analysis of South Africa's electricity services'. In *Electric Capitalism: Recolonising Africa on the Power Grid*, ed. D. McDonald. London: Earthscan and Pretoria: HSRC Press.

Eberhard, A. 2005. *From State to Market and Back Again: South Africa's Power Sector Reforms.* Cape Town: Graduate School of Business, University of Cape Town.

Eberhard A. and C. Van Horen. 1995. *Poverty and Power: Energy and the South African State.* Cape Town: UCT Press and East Haven: Pluto Press.

Emfuleni Local Municipality. 2005. *Integrated Development Plan 2005/2006.*

Environmental Rights Action (ERA) and Oilwatch. 2000. *Pipe Dream: The West African Gas Pipeline Project and the Environment.* Environmental Rights Action/Friends of the Earth Nigeria and Oilwatch.

EnviroServ. *Annual Report 2007.*

———. *Annual Report 2006.*

Eskom. *Annual Reports 2009, 2008, 2007.*

Ewart, J. and A. du Toit. 2005. 'New fault lines in the countryside: Restructuring in the Western Cape wine industry'. In *Beyond the Apartheid Workplace: Studies in Transition,* eds E. Webster and K. von Holdt. Pietermaritzburg: University of KwaZulu-Natal Press.

Extractive Industries Review. 2003. *Striking a Better Balance. Volume I: The World Bank Group and Extractive Industries. Volume II: Stakeholder Inputs: Converging Issues and Diverging Views on the World Bank Group's Involvement in Extractive Industries.* EIR.

Farrell, A. and A. Brandt. 2006. *Risks of the Oil Transition.* Environmental Research Letters. Berkeley, California: University of California, Energy and Resources Group.

Ferguson, J. 2005. 'Seeing like an oil company'. In *American Anthropologist,* Vol. 107, No. 3.

Fig, D. 2005. *Uranium Road: Questioning South Africa's Nuclear Direction.* Johannesburg: Jacana.

Fiil-Flynn, M. and P. Naidoo. 2004. 'Nothing for mahala: The forced installation of prepaid water metres in Stretford, Extension 4, Orange Farm, Johannesburg'. Centre for Civil Society Research Report 16. Public Citizen, Anti-Privatisation Forum, Coalition Against Water Privatisation. Durban: Centre for Civil Society, University of Natal.

Fine, B. 2008a. 'Engaging the MEC'. Paper for School of Development Studies Workshop on the Minerals and Energy Complex, University of KwaZulu-Natal, June.

———. 2008b. 'The minerals-energy complex is dead: Long live the MEC?' Paper for Amandla Colloquium, April.

Fine, B. and Z. Rustomjee. 1996. *The Political Economy of South Africa: From Minerals-Energy Complex to Industrialisation.* Johannesburg: Witwatersrand University Press.

Gary, I. and T. Karl. 2003. *Bottom of the Barrel: Africa's Oil Boom and the Poor.* Catholic Relief Services.

Gentle, L. 2009. 'Escom to Eskom: From racial Keynesian capitalism to neo-liberalism'. In *Electric Capitalism: Recolonising Africa on the Power Grid,* ed. D. McDonald. London: Earthscan and Pretoria: HSRC Press.

German Advisory Council on Global Change (WBGU). 2009. *Solving the Climate Dilemma: The Carbon Budget Approach.* Berlin: WBGU.

Graham, S. 2007. 'War and the city'. In *New Left Review* 44, March/April.

Greenberg, S. 2010. *Status Report on Land and Agricultural Policy in South Africa.* Research Report 40, Institute for Poverty, Land and Agrarian Studies (Plaas). Cape Town: University of the Western Cape.

———. 2006. *The State, Privatisation and the Public Sector in South Africa.* SAPSN and AIDC.

Greenpeace. 2006. *Plastic Debris in the World's Oceans.* Written by M. Allsopp, A. Walters, D. Santillo and P. Johnston.

The groundWork Reports

———. 2008. *Wasting the Nation: Making Trash of People and Places,* (D. Hallowes and V. Munnik).

———. 2007. *Peak Poison: The Elite Energy Crisis and Environmental Injustice,* (D. Hallowes and V. Munnik).

———. 2006. *Poisoned Spaces: Manufacturing Wealth, Producing Poverty,* (D. Hallowes and V. Munnik).

———. 2005. *Whose Energy Future? Big Oil against People in Africa,* (D. Hallowes and M. Butler).

———. 2004. *The Balance of Rights: Constitutional Promises and Struggles for Environmental Justice,* (D. Hallowes and M. Butler).

———. 2003. *Forging the Future: Industrial Strategy and the Making of Environmental Injustice,* (D. Hallowes and M. Butler).

———. 2002. *Corporate Accountability in South Africa: The Petrochemical Industry and Air Pollution,* (M. Butler and D. Hallowes).

groundWork. 2003. *Air Quality Report 2003.*

Hallowes, D. 2009. *The World Bank and Eskom: Banking on Climate Destruction.* groundWork.

———. 2008a. *A Critical Appraisal of the LTMS.* Johannesburg: Earthlife Africa.

———. 2008b. *Capital Meltdown, Kyoto and Civil Society.* Johannesburg: Earthlife Africa.

———. 2005. 'Sustainable energy: Towards a civil society review of South African energy policy'. Discussion Paper for Earthlife Africa, Johannesburg.

———. 1993. *Hidden Faces: Environment, Development, Justice: South Africa and the Global Context.* Johannesburg: Earthlife Africa.

Hansen, J. 2006. 'The threat to the planet'. In *New York Review of Books,* 13 July.

Hansen, J., M. Satoa, P. Kharechaa, D. Beerling, R. Bernerd, V. Masson-Delmottee, M. Paganid, M. Raymof, D. Royerg and J. Zachosh. 2008. *Target Atmospheric CO_2: Where Should Humanity Aim?* Submitted at arXiv.org, 7 April 2008 and revised 18 June 2008 (ref: arXiv:0804.1126v2).

Harley, J. 2006. 'Report on cement kilns', prepared for groundWork.

Harvey, D. 2008. *The Right to the City.* Available at http://davidharvey.org/media/righttothe city.pdf.

———. 2005. *The New Imperialism.* London: Oxford University Press.

Heinberg, R. 2007. 'Bridging peak oil and climate change activism'. Paper for International Forum on Globalisation Seminar, London, 23–27 February.

———. 2005. *The Party's Over: Oil, War and the Fate of Industrial Societies.* Second edition. Gabriola Island, Canada: New Society Publishers.

———. 2004. *Powerdown: Options and Actions for a Post-Carbon World.* Gabriola Island, Canada: New Society Publishers.

Hendler, P., J. Holliday, S. Ratcliffe and J. Wakeford. 2007. 'Current global challenges and alternative futures for South Africa: The inter-connection between oil depletion, climate change and global financial imbalances'. ASPO South Africa.

Hildyard, N. 2008. *A (Crumbling) Wall of Money: Financial Bricolage, Derivatives and Power.* Sturminster Newton, Dorset: The Corner House.

Hirsch, A. 2005. *Season of Hope: Economic Reform under Mandela and Mbeki.* Pietermaritzburg: University of KwaZulu-Natal Press.

Hirsch, R. 2007. 'Peaking of world oil production: Recent forecasts'. In *World Oil Magazine*, April.

Hirsch, R., R. Bezdek and R. Wendling. 2005. 'Peaking of world oil production: Impacts, mitigation and risk management'. Report prepared for the National Energy Technology Laboratory of the US Department of Energy.

Hobbs, P., S. Oelofse and J. Rascher. 2008. 'Management of environmental impacts from coal mining in the upper Olifants River catchment as a function of age and scale'. In *International Journal of Water Resources Development* 24: 3.

Holm, D., D. Banks, J. Schäffler, R. Worthington and Y. Afrane Okese. 2008. 'Renewable energy briefing paper: Potential of renewable energy to contribute to national electricity emergency response and sustainable development'. Prepared for Trade and Industry Policy Studies (TIPS).

Hopkins, R. 2006. 'Energy descent pathways: Evaluating potential responses to peak oil'. MSc. dissertation for the University of Plymouth.

———. Undated. 'Powerdown and permaculture: At the cusp of transition'. In *Permaculture Magazine* 50.

Human Rights Watch. 2005. *Rivers and Blood: Guns, Oil and Power in Nigeria's Rivers State.* HRW Briefing Paper.

———. 2002. *The Niger Delta: No Democratic Dividend.* HRW Briefing Paper.

Intergovernmental Panel on Climate Change (IPCC). 2007a. *Climate Change 2007: The Physical Science Basis.* Summary for Policymakers, Contribution of Working Group I to the Fourth Assessment Report of the Intergovernmental Panel on Climate Change.

———. 2007b. *Climate Change 2007: Climate Change Impacts, Adaptation and Vulnerability.* Summary for Policymakers, Contribution of Working Group II to the Fourth Assessment Report of the Intergovernmental Panel on Climate Change.

———. 2000. *Special Report on Emissions Scenarios* (SRES).

International Energy Agency (IEA). *World Energy Outlook 2010*, Executive Summary.

———. *World Energy Outlook 2008.*

———. *World Energy Outlook 2006.*

Joffe, A., D. Kaplan, R. Kaplinsky and D. Lewis. 1993. 'Meeting the global challenge: A framework for industrial revival in South Africa'. In *South Africa and the World Economy in the 1990s*, eds P. Baker et al. Pretoria: IDASA and The Aspen Institute.

Johnston, L., L. Hamilton, M. Kresowik, T. Sanzillo and D. Schlissel. 2010. 'Phasing out federal subsidies for coal'. Report by Synapse Energy Economics.

Joubert, L. 2006. *Scorched: South Africa's Changing Climate.* Johannesburg: Witwatersrand University Press.

Keegan, T. 1986. *Rural Transformations in Industrializing South Africa.* Johannesburg: Ravan Press.

Kemmer, F. 1971. 'Pollution control in the steel industry'. In *Industrial Pollution Control Handbook*, ed. H.F. Lund. Columbus, Ohio: McGraw-Hill.

Khor, M. 2010. 'Complex implications for the Cancun climate conference'. In *Economic and Political Weekly*, Vol. XIV, No. 52.

Kovel, J. 2002. *Enemy of Nature.* London: Zed Books.

Kretzmann, S. and I. Nooruddin. 2005. *Drilling into Debt: An Investigation into the Relationship between Debt and Oil.* Oilchange International, Jubilee USA Network, Institute for Public Policy Research, Milieu Defensie, Amazon Watch.

Kropotkin, P. 1913. *The Conquest of Bread.* London: Chapman and Hall.

Kull, D. 2006. 'Connections between recent water level drops in Lake Victoria, dam operations and drought', available at International Rivers Network website.

Lahiff, E. 2003. 'The politics of land reform in southern Africa'. Sustainable Livelihoods in Southern Africa Programme.

Leggett, J. 2005. *Half Gone: Oil, Gas, Hot Air and the Global Energy Crisis.* London: Portobello Books.

Leibbrandt, M., I. Woolard, A. Finn and J. Argent. 2010. 'Trends in South African income distribution and poverty since the fall of apartheid'. OECD Social, Employment and Migration Working Papers No. 101.

Leonard, A. 2008. *The Story of Stuff.* Video and referenced and annotated script, at www.storyofstuff.com.

Levin, K. and J Pershing. 2007. *Climate Science 2006: Major New Discoveries.* World Resource Institute.

Lohmann, L. 2008. 'Toward a different debate in environmental accounting: The cases of carbon and cost-benefit'. In *Accounting, Organizations and Society* (forthcoming) and available at www.thecornerhouse.org.uk.

———. 2006. 'Carbon trading: A critical conservation on climate change, privatisation and power'. In *Development Dialogue* 48.

Lucas, C., A. Jones and C. Hines. 2006. *Fuelling a Food Crisis: The Impact of Peak Oil on Food Security.* The Greens/European Free Alliance in the European Parliament.

Lucas, C., V. Shiva and C. Hines. 2005. 'Making poverty inevitable: The consequences of the UK government's damaging approach to global trade'. At www.carolinelucas mep.org.uk.

MacKenzie, D. 2007. 'The political economy of carbon trading'. In *London Review of Books*, Vol. 29, No. 7, 5 April.

Marais, H. 2001. *South Africa Limits to Change: The Political Economy of Transition*. London: Zed Books and Cape Town: UCT Press.

Maré, G. 2003. 'The state of the state'. In *State of the Nation: South Africa 2003–2004*, eds J. Daniel, A. Habib and R. Southall. Pretoria: HSRC Press.

McCarthy T. and K. Pretorius. Undated. 'Coal mining on the highveld and its implications for future water quality in the Vaal River system'. Paper for South African Environmental Observation Network (SAEON).

McCluney, R. 2005. 'Renewable energy limits'. In *The Final Energy Crisis*, eds A. McKillop and S. Newman. East Haven: Pluto.

McDaid, L. 2010. *Power to the People: Raising the Voice of Civil Society in Electricity Planning – Integrated Resources Plan 2010 Inputs and Departmental Responses*. Cape Town: Institute for Security Studies, September.

McDonald, D. and L. Smith. 2004. 'Privatising Cape Town: From apartheid to neo-liberalism in the mother city'. In *Urban Studies*, Vol. 41, No. 8, July.

McKibben, B. 2007. 'Warning on warming'. In *New York Review of Books*, 15 March.

McKillop, A. 2006. 'Peak oil to peak gas is a short ride'. In *Energy Bulletin*, 12 December.

McRae, I. 2006. *The Test of Leadership: Fifty Years in the Electricity Supply Industry in South Africa*. Nooitgedacht, Gauteng. EE Publishers.

Metsimaholo Municipality. 2003. *Integrated Development Plan 2003*.

Ministry of Finance. 1995. *Growth, Employment and Redistribution: A Macroeconomic Strategy*.

Mobbs, P. 2005. *Energy beyond oil*. Leicester: Matador Publishing.

Mohlakoana, N. and W. Annecke. 2008. 'Finally breaking the barriers: South African case study on LPG use by low-income urban households'. Istanbul Pre-Conference Workshop on Clean Cooking Fuels, 16–17 June.

Monbiot, G. 2006. *Heat: How to Stop the Planet Burning*. London: Penguin/Allen Lane.

Moody, R. 2007. *Rocks and Hard Places: The Globalization of Mining*. Cape Town: David Philip and London: Zed Books.

Moore, C. 2003. 'Trashed: Across the Pacific Ocean, plastics, plastics, everywhere'. In *Natural History*, Vol. 112, No. 9, November.

Morgan, F. (dir.). 2006. *The Power of Community: How Cuba Survived Peak Oil*. Video: The Community Solution.

Munn, A. 2004. 'Emission reductions at Engen Refinery in south Durban'. Paper delivered to the Eighth World Congress on Environmental Health held at International Convention Centre in Durban, 22–27 February.

Murray, R. 2002. *Zero Waste*. London: Greenpeace Environmental Trust.

Naidoo, R., N. Gqaleni, S. Batterman and T. Robins. 2006. *South Durban Health Study*. Centre for Occupational Health, University of KwaZulu-Natal; Department of Environmental Health Sciences, University of Michigan; Department of Environmental Health Sciences, Durban Institute of Technology.

National Nuclear Regulator (NNR). 2007. *The Wonderfontein Catchment Area Public Report, Results and Corrective Action*, TR-NTNS-07-0001.

NEPAD. October 2001. Available from www.nepad.org.

National Energy Regulator of South Africa (NERSA). 2008. *Inquiry into the National Electricity Supply Shortage and Load Shedding*. Report by the Energy Regulator. 12 May.

Nolan, P. and J. Zhang. 2010. 'Global competition after the financial crisis'. In *New Left Review* 64, July/August.

O'Hagan, A. 2007. 'The things we throw away'. In *London Review of Books*, 24 May 2007.

Okonta, I. 2006. *Behind the Mask: Explaining the Emergence of the MEND Militia in Nigeria's Oil-Bearing Niger Delta*. Economies of Violence Working Paper No. 11, Institute of International Studies, University of California, Berkeley.

———. 2005. 'Asari Dokubo: Insurgent or self-serving opportunist'. In *This Day*, 8 January 2005, and posted by Black Looks at http://okrasoup.typepad.com/black_looks/.

Okonta, I. and O. Douglas. 2003. *Where Vultures Feast: Shell, Human Rights and Oil in the Niger Delta*. London: Verso.

Pallister, D., S. Stewart and I. Lepper. 1987. *South Africa Inc. The Oppenheimer Empire*. London: Corgi.

Pelupessy, W. 2000. 'Can Restructuring of Industry Create Jobs in South African Townships?' Development Research Institute, Tilburg University.

Pichtel, J. 2005: *Waste Management Practices: Municipal, Hazardous and Industrial*. New York: CRC Press.

Pimentel, D. and T. Patzek. 2005. 'Ethanol production using corn, switchgrass and wood; biodiesel production using soybean and sunflower'. In *Natural Resources Research*, Vol. 14, No. 1, March.

Pimentel, D., T. Patzek and G. Cecil. 2007. 'Ethanol production: Energy, economic and environmental losses'. In *Reviews of Environmental Contamination and Toxicology* 189. New York: Springer.

Pithouse, R. 2006. *'Our Struggle is Thought, on the Ground, Running': The University of Abahlali baseMjondolo*. Durban: Centre for Civil Society, University of KwaZulu-Natal.

Polanyi, K. 2001. *The Great Transformation: The Political and Economic Origins of our Time*. Second edition. Boston: Beacon Press.

Pomeranz, K. 2009. 'The great Himalayan watershed'. In *New Left Review* 58, July/August.

Pone, J.D.N. et al. 2007. 'The spontaneous combustion of coal and its by-products in the Witbank and Sasolburg coalfields of South Africa'. In *International Journal of Coal Geology* 72.

PPC. *Annual Report 2007*.

———. *Annual Report 2006*.

The Presidency. 2008. *Towards a Fifteen Year Review*. South African Government.

———. 2003. *Towards a Ten Year Review*.

Raupach, M., G. Marland, P. Ciais, C. le Quéré, J. Canadell, G. Klepper and C. Field. 2007. *Global and Regional Drivers of Accelerating CO₂ Emissions*. Proceedings of the National Academy of Sciences, available at www.pnas.org.

Retort. 2005. *Afflicted Powers: Capital and Spectacle in a New Age of War*. Written by I. Boal, T. Clark, J. Matthews and M. Watts. London: Verso.

Roberts, S. and Z. Rustomjee. 2009. 'Industrial policy under democracy: Apartheid's grown up infant industries? Iscor and Sasol'. In *Transformation* 71.

Robinson, K.E. and G.C. Toland. 1979. 'Case histories of different seepage problems for nine tailings dams'. In *Mine Drainage: Proceedings of the First International Mine Drainage Symposium*, eds G.O. Argall and C.O. Brawner. San Francisco: Miller Freeman.

Roesner, T. et al. 2001. 'A preliminary assessment of pollution contained in the unsaturated and saturated zone beneath reclaimed gold-mine residue deposits'. Water Research Commission Report 797/1/01.

Rogers, H. 2005a. *Gone Tomorrow: The Hidden Life of Garbage*. New York: The New Press.

———. 2005b. 'A brief history of plastic'. In *The Brooklyn Rail*, May.

Rosset, P. 2000. 'Cuba: A successful case study of sustainable agriculture'. In *Hungry for Profit: The Agribusiness Threat to Farmers, Food and the Environment*, eds F. Magdoff, J. Bellamy Foster and F. Buttel. New York: Monthly Review Press. Also at the Cuba Organic Support Group site www.cosg.org.uk.

Ruiters, G. 2005. 'Knowing your place: Urban services and new modes of governability in South African cities'. Institute for Social and Economic Research, Rhodes University; paper for Centre for Civil Society Seminar, University of KwaZulu-Natal, 6 October.

Runge, C. F., and B. Senauer. 2007. 'How biofuels could starve the poor'. In *Foreign Affairs*, May/June.

Sachs, W. 2005. 'Equity in the greenhouse: How just is the Kyoto Protocol?' In *Reading the Kyoto Protocol: Ethical Aspects of the Convention on Climate Change*, ed. E. Vermeersch. Delft: Eburon Academic Publishers.

Sachs, W. and T. Santorius. 2007. *Fair Future: Resource Conflicts, Security and Global Justice*. London: Zed Books.

Sachs, W., H. Acselrad, F. Akhter, A. Amon, T. Egziabher, H. French, P. Haavisto, P. Hawken, H. Henderson, A. Khosla, S. Larrain, R. Loske, A. Roddick, V. Taylor, C. von Weiszaecker and S. Zabelin. 2002. *The Jo'burg Memo: Fairness in a Fragile World*. Second edition. Memorandum for the World Summit on Sustainable Development. Heinrich Böll Foundation, World Summit Papers.

Samson, M. 2008. 'Reclaiming livelihoods: The role of waste reclaimers in municipal waste management systems'. Report for groundWork.

———. 2004. *Organizing in the Informal Economy: A Case Study of the Municipal Waste Management Industry in South Africa*. International Labour Organisation.

Sasol. *Annual Reports 2009, 2007 and 2006* (AR 2009; AR 2007; AR 2006).

————. *Sustainable Development Reports 2009, 2007 and 2006* (SDR 2009; SDR 2007; SDR 2006).

————. *Sasolburg Safety, Health and Environment (SH&E) Briefs 2006 and 2005* (SH&E 2006; SH&E 2005).

————. 2005. *Sustainable Development Report 2002–2004.*

————. 2004. *Annual Review 2004.*

Scorgie, Y. 2004. 'Air quality situation assessment for the Vaal Triangle region'. Report for the Legal Resource Centre.

Seekings, J. 2007. *Poverty and Inequality after Apartheid.* Cape Town: University of Cape Town, Centre for Social Science Research.

Simms, A. 2001. 'Ecological debt: Balancing the environmental budget and compensating developing countries'. Pamphlet for the World Summit on Sustainable Development. International Institute for Environment and Development (IIED).

Simms, A., J. Oram and P. Kjell. 2004. *The Price of Power: Poverty, Climate Change, the Coming Energy Crisis and the Renewable Revolution.* London: New Economics Foundation.

Slabbert, T. 2004. 'An investigation into the state of affairs and sustainability of the Emfuleni economy'. Doctoral thesis, Faculty of Economic and Management Sciences, University of Pretoria.

Solomon, S., G-K. Plattner, R. Knuttic and P. Friedlingstein. 2009. *Irreversible Climate Change Due to Carbon Dioxide Emissions.* Proceedings of the National Academy of Sciences available at www.pnas.org.

South African Government (SAG). 2009. *South Africa Year Book 2008/09.*

————. 2006. 'Background document: A catalyst for Accelerated and Shared Growth – South Africa (ASGISA)'. Media Briefing by Deputy President Phumzile Mlambo-Ngcuka, 6 February.

————. 2004. *South Africa Year Book 2003/04.*

South African Labour Bulletin (SALB). 2002. 'The DTI's integrated manufacturing strategy: Is it all just packaging?' In *SALB*, Vol. 26, No. 3, June.

Statistics South Africa (Stats SA). 2008. *Income and Expenditure of Households 2005/2006: Analysis of Results.* Statistics South Africa.

Steffen, W., A. Sanderson, P. Tyson, J. Jager, P. Matson, B. Moore, F. Oldfield, K. Richardson, H. Schellnhuber, B. Turner and R. Wasson. 2004. *Global Change and the Earth System: A Planet under Pressure.* The IGBP Series. New York: Springer Press.

Stern, N. 2006. 'The economics of climate change' (*The Stern Report*). Review for the UK Treasury.

Stockman, L. and G. Muttitt. 2005. *Pumping Poverty: Britain's Department for International Development and the Oil Industry.* London: Platform Research, Friends of the Earth and Plan B.

Stover, R., K. Evans and K. Pickett. 1996. *Report of the Berkeley Plastics Task Force.* Berkeley: The Ecology Centre.

Supiot, A. 2006. 'Law and labour: A world market of norms?' In *New Left Review* 39, May/June.

Sustainable Energy Africa (SEA). 2003. *City of Cape Town State of Energy Report.* Available at www.sustainable.org.za.

Swilling, M. 2007. *Cape Town 2025: A City of Sustainable Neighbourhoods.* Cape Town: Sustainability Institute.

———. 2006. *Sustainability and Infrastructure Planning in South Africa: A Cape Town Case Study.* Cape Town: Environment and Urbanisation.

Taylor, T. 2007. *The Political Economy of Power.* Johannesburg: Earthlife Africa.

The Economics of Ecosystems and Biodiversity (TEEB). 2009. 'Factsheet', prepared for Fifth Intergovernmental Conference: Biodiversity in Europe, 22–24 September 2009, Liege, Belgium.

———. 2008. *An Interim Report.* European Communities.

Thompson, J. and H. Anthony. 2005. *The Health Effects of Waste Incinerators.* Fourth Report of the British Society for Ecological Medicine.

Tissington, K. 2009. 'The deficiency of reality in the Joe Slovo judgment'. Johannesburg: University of the Witwatersrand, Centre for Applied Legal Studies.

Turner, T. and L. Brownhill. 2004. 'Why women are at war with Chevron: Nigerian subsistence struggles against the international oil industry'. In *Journal of Asian and African Studies*, Vol. 39, No. 1/2.

Turton, A.R. 2008. *Water and Mine Closure in South Africa: Development that is Sustainable?* Available at www.sidint.org/development.

Tyndall Centre. 2005. *Decarbonising the UK: Energy for a Climate Conscious Future.* Tyndall Centre for Climate Change Research at www.tyndall.ac.uk.

United Nations Environmental Programme (UNEP). 2002. *Global Environmental Outlook 3.* UNEP and London: Earthscan.

Vallette, J. and S. Kretzmann. 2004. *The Energy Tug of War: The Winners and Losers of World Bank Fossil Fuel Finance.* Washington DC: Institute for Policy Studies, Sustainable Energy and Economy Network.

Von Holdt, K. 2003. *Transition from Below: Forging Trade Unionism and Workplace Change in South Africa.* Pietermaritzburg: University of KwaZulu-Natal Press.

Von Ketelhodt, A. 2006. 'The long term impact of electricity shortage on small and medium sized businesses in Cape Town'. MBA thesis, University of Pretoria.

Walsh, S. (dir). 2009. *H₂Oil.* Loaded Pictures. Web site: www.h2Oildoc.com.

Watts, M. 2009. 'Tipping point: Slipping into darkness'. Economies of Violence Working Paper No. 23. Berkeley: University of California, Institute of International Studies.

———. 2008. 'Blood oil: The anatomy of a petro-insurgency in the Niger Delta, Nigeria', Economies of Violence Working Paper No. 22. Berkeley: University of California, Institute of International Studies.

Webster, E. and K. von Holdt. 2005. 'Work restructuring and the crisis of social reproduction: A southern perspective'. In *Beyond the Apartheid Workplace: Studies in Transition*, eds E. Webster and K. von Holdt. Pietermaritzburg: University of KwaZulu-Natal Press.

Whyte, Anne V. (ed.). 1995. *Building a New South Africa: Volume 4: Environment, Reconstruction and Development: A Report from the International Mission on Environmental Policy*. International Development Research Centre.

Wiley, D., C. Root and S. Peek. 2002. 'Contesting the urban industrial environment in South Durban in a period of democratisation and globalisation'. In *(D)urban Vortex: A South African City in Transition*, eds W. Freund and V. Padayachee. Pietermaritzburg: University of Natal Press.

Winkler, H. 2007. 'Energy policies for sustainable development in South Africa'. In *Energy for Sustainable Development*, Vol. XI, No. 1, March.

Wolmarans, J. 1984. 'Ontwatering van die dolomietgebied aan die Verre Wesrand (Dewatering of the dolomite area of the Far West Rand)'. Unpublished DSc. (Geology) thesis, Potchefstroom University.

World Bank. 2009. *Draft Safeguards Diagnostic Review for South Africa Eskom Investment Support Project*, 3 November.

World Business Council for Sustainable Development (WBCSD). 2010. *Vision 2050: The New Agenda for Business*. WBCSD.

World Wildlife Fund (WWF). 2006. *Living Planet Report 2006*.

Worthington, R. 2009. 'Cheap at half the cost: Coal and electricity in South Africa'. In *Electric Capitalism*, ed. D. McDonald. London: Earthscan.

Xaba, T. 2001. 'Masculinity and its malcontents'. In *Changing Men in South Africa*, ed. R. Morrell. Pietermaritzburg: University of Natal Press and New York: Zed Books.

Yergin, D. 1991. *The Prize: The Epic Quest for Oil, Money and Power*. New York: Free Press.

Zibechi, R. 2007. *The Militarization of the World's Urban Peripheries*. Americas Policy Programme, 9 February 2007, at www.americaspolicy.org.

Zittel, W. and J. Schindler. 2007. 'Crude oil: The supply outlook'. Report for Energy Watch Group (EWG).

Zizek, S. 2002. *Welcome to the Desert of the Real*. London: Verso.

Index